Path Integral Quantization

Path Integral Quantization

Mark S Swanson
Emeritus Professor of Physics, University of Connecticut, Connecticut, USA

IOP Publishing, Bristol, UK

ISBN 978-0-7503-3547-8 (ebook)
ISBN 978-0-7503-3545-4 (print)
ISBN 978-0-7503-3548-5 (myPrint)
ISBN 978-0-7503-3546-1 (mobi)

DOI 10.1088/978-0-7503-3547-8

Version: 20200801

IOP ebooks

British Library Cataloguing-in-Publication Data: A catalogue record for this book is available from the British Library.

Published by IOP Publishing, wholly owned by The Institute of Physics, London

IOP Publishing, Temple Circus, Temple Way, Bristol, BS1 6HG, UK

US Office: IOP Publishing, Inc., 190 North Independence Mall West, Suite 601, Philadelphia, PA 19106, USA

Contents

Preface

Quantum mechanics and its application to the behavior of atomic and subatomic matter has dominated the last century of experimental and theoretical physics. As part of these efforts, the path integral method for analyzing quantum processes was first formulated and applied by Richard Feynman, who used it to develop a perturbative representation of quantum electrodynamics (QED). His work enabled a convenient and powerful space–time visualization of particle processes in terms of Feynman diagrams. Since its initial appearance, the path integral approach to quantum systems has become a staple of modern theoretical physics, particularly in the area of quantum field theory, where it provides a powerful tool to analyze the dynamics of both condensed matter and the subatomic world.

The purpose of this monograph is to present the reader with a reasonably brief introduction to functional methods and path integral techniques for analyzing quantized systems. It has been structured to provide the reader with a concise collection of basic concepts which are also complete enough to present many of the fundamental applications that path integrals have. Its intended audience is both the student and practitioner who desires a basic and concise introductory overview of path integral and functional methods. In that regard, one of the central themes of this monograph is the relationship of the path integral to other methods for analyzing quantum systems, in particular the operator and state formulation of quantized systems. It is the author's hope that emphasizing this relationship will help clarify both the nature of the path integral as well as demonstrating its utility. After some preliminary developments, the book is broken into two general areas: the path integral as applied to quantum mechanics, and the path integral as applied to quantum field theory. The first half of the monograph assumes that the reader has a background in the basic mathematics and physics of classical and quantum mechanics at the level of the many excellent introductory texts in those areas. The second half of the monograph assumes the reader has an understanding of the mathematics and physics of *classical* relativistic fields. Such a background is available from Landau and Lifshitz, the author's IOP monograph, *Classical Field Theory and the Stress–Energy Tensor*, or in the introductory sections of the many other excellent treatises on quantum field theory. The goal of brevity precludes all but a cursory review of these two general areas in order to clarify notation. However, the mathematical tools and physical concepts specific to formulating and analyzing the path integral are developed throughout this monograph. In addition, the canonical quantization of fields using basic operator techniques is presented at the beginning of the second half of this monograph. It is hoped that doing so will connect many of the results and techniques of the first half of the monograph to the particle content of quantum field theory.

There are many excellent and detailed books that apply path integral techniques to a myriad of problems in quantum mechanics, quantum field and particle theory, and condensed matter physics, beginning with the original monograph on the subject by Feynman and Hibbs and emended by Styer. The book by Schulman has long been a valued reference on the subject. More recently, the encyclopedic monograph by Kleinert contains a wealth of path integral applications primarily in the area of quantum mechanics and statistical mechanics. Most introductory texts on quantum field theory, such as those of Peskin and Schroeder or Greiner and Reinhardt, introduce path integral quantization and functional techniques at varying levels of detail, while the monograph by Shankar applies path integral techniques to problems in condensed matter physics. In that regard, it is the ambitious intent of this brief monograph to present the reader with a very concise but sufficiently thorough introduction to the concepts of path integrals in both quantum mechanics and quantum field theory that will enable the reader to study the many and diverse areas of path integral applications available in other sources. The book ends each chapter with a list of some of the books, texts, and articles that contain further details and numerous applications that were not included in this monograph. It is the author's sincere hope that the reader will be sufficiently prepared by this text to use these recommended books and articles for further study of the myriad details and applications that path integrals possess.

A set of notational conventions for theoretical physics has developed, and this monograph attempts to observe them whenever it would not lead to confusion. For example, this monograph uses of $\boldsymbol{x}$ and $\boldsymbol{p}$ to represent the canonically conjugate position and momentum vectors in an arbitrary dimension, although $\boldsymbol{q}$ and $\boldsymbol{p}$ are often used in other texts. Likewise, $\mathrm{d}x$ is the volume element in one dimension, while $\mathrm{d}\boldsymbol{x}$ is the volume element in an arbitrary dimension and $\mathrm{d}^n\boldsymbol{x}$ is the volume element in n dimensions. The summation convention for repeated indices characterizes notation in relativistically invariant theories, while the summation sign $\sum$ is typically used in non-relativistic theories. Where there is possible doubt or ambiguity, the summation sign is explicitly displayed. In the case of definite integrals the limits should be understood to be $(-\infty, \infty)$ *if they are not explicitly displayed.* The exception to this convention are integrals over Grassmann variables, which are all indefinite. These conventions and the exceptions to them will be discussed when it is relevant to the continuity or clarity of the presentation to do so. There are several non-standard symbols used in the text, including '$\Longrightarrow$' for 'implies that', '$\Longleftrightarrow$' for 'are equivalent to each other', '$\equiv$' for 'is defined as', '$\approx$' for 'approximately equal to', '$\rightarrow$' for 'becomes', and '$\cdots$' for 'all the product terms in between.'

Because this monograph is primarily pedagogical in nature and designed to be concise, many of the original research papers on path integrals and their applications are not listed. In that regard, some concepts presented have become so widely used that they are known simply by the name of their originator, and this form of attribution is followed whenever possible. The author sincerely apologizes to the giants of physics, both living and dead, whose brilliant insights and hard work laid the foundations of this monograph, but whose well-deserved recognition may have been omitted. The author also extends his heartfelt thanks to IOP for the opportunity to write this book.

Further reading

The concept of the path integral was first presented in R P Feynman 1948 Space–time approach to non-relativistic quantum mechanics *Rev. Mod. Phys.* **20** 367.

There have been many texts and monographs that focus on the derivation and the nature of quantum mechanical path integrals. In order of original publication date those mentioned in this preface are R P Feynman, A R Hibbs and D Styer 2010 *Quantum Mechanics and Path Integrals* (New York: Dover); L S Schulman 2005 *Techniques and Applications of Path Integration* (New York: Dover); and H Kleinert 2004 *Path Integrals in Quantum Mechanics, Statistics, Polymer Physics, and Financial Markets*, (Singapore: World Scientific). Derivations and applications of path integrals in the context of quantum field theory are found in M Peskin and D Schroeder 1995 *An Introduction to Quantum Field Theory* (Reading, MA: Addison-Wesley); W Greiner and J Reinhardt 1996 *Field Quantization* (Berlin: Springer); and R Shankar 2017 *Quantum Field Theory and Condensed Matter: An Introduction* (Cambridge: Cambridge University Press).

Author biography

Mark S Swanson

Mark S Swanson received his PhD in physics from the University of Missouri at Columbia in 1976 under the supervision of Justin Huang. After postdoctoral appointments at the University of Alberta and the University of Connecticut he held a faculty appointment at the University of Connecticut at Stamford from 1983 to 2014, during which time he spent six years in university administrative roles. He is the author of 25 research articles and three monographs, with an emphasis on field theory and path integral techniques. He is currently Emeritus Professor of Physics at the University of Connecticut and lives in Monroe, Connecticut.

IOP Publishing

Path Integral Quantization

Mark S Swanson

Chapter 1

Preliminaries

This chapter presents some mathematical tools and physical concepts commonly used in the development and application of the path integral, but which may not be familiar to the reader. These include functionals, functional derivatives, operator identities, Grassmann variables, Jacobians, Gaussian integrals, and Poisson resummation. Whenever possible and relevant, these concepts are placed in the context of classical and quantum mechanics.

1.1 Functionals and classical mechanics

The concept of the *functional* plays a central role both in the Lagrangian formulation of classical mechanics and in the formulation of the path integral. A functional is a means to map a collection of functions and their derivatives into a real or complex number. The calculus of variations was developed to understand the behavior of functionals and to determine those functions for which the associated functional attains an extremum value.

Given a function $f(x)$ of the independent variable x, the functional $F[f(x)]$ denotes a method to obtain a real or complex number from this function. Although the function depends on x, the functional does not. The simplest functional consists of an integral of the function $f(x)$ and its derivatives. An important physical example is the Lagrangian formulation of classical mechanics for the motion of a point particle with mass m, where the functions are the real valued time-dependent position vectors of the particle, denoted $\boldsymbol{x}(t)$. The parameter t corresponds to time, and the components of the position vector are denoted $x_i(t)$, where i indexes the coordinates for the particular choice of coordinate system. The Lagrangian approach to classical mechanics constructs the *action functional* S from the position vector, $\boldsymbol{x}(t)$, and its first derivative, denoted $\dot{\boldsymbol{x}}(t) = \mathrm{d}\boldsymbol{x}(t)/\mathrm{d}t$, so that $S = S[\boldsymbol{x}(t), \dot{\boldsymbol{x}}(t)]$. The action functional for classical mechanics is given by

doi:10.1088/978-0-7503-3547-8ch1

$$S[\boldsymbol{x}(t), \dot{\boldsymbol{x}}(t)] = \int_{t_i}^{t_f} \mathrm{d}t\ \mathcal{L}(\boldsymbol{x}(t), \dot{\boldsymbol{x}}(t), t), \tag{1.1}$$

where t_i and t_f are the initial and final times for the motion of the particle under consideration and $\mathcal{L}$ is a *real valued* quantity known as the *Lagrangian density*, or more commonly just the *Lagrangian*. For the sake of notational brevity the position vector functions will often be written $\boldsymbol{x}$, with the time-dependence suppressed, so that $\boldsymbol{x}$ is shorthand for $\boldsymbol{x}(t)$. In the most general case $\mathcal{L}$ may have an explicit time dependence. However, in typical elementary applications the Lagrangian takes the general form $\mathcal{L}(\boldsymbol{x}, \dot{\boldsymbol{x}}) = K(\boldsymbol{x}, \dot{\boldsymbol{x}}) - V(\boldsymbol{x}, \dot{\boldsymbol{x}})$, where K is the kinetic energy and V is the potential energy of the particle in the system being analyzed, so that $\partial\mathcal{L}/\partial t = 0$. The *calculus of variations* provides a differential equation which the *trajectories* $\boldsymbol{x}(t)$ must solve in order to make the action functional (1.1) an extremum. This equation generalizes Newton's equation of motion, $\boldsymbol{F}(\boldsymbol{x}, t) = m\ddot{\boldsymbol{x}}(t)$, where $\boldsymbol{F}(\boldsymbol{x}, t)$ is the force and $\ddot{\boldsymbol{x}}$ is the second derivative.

The first step in this process is to introduce the *functional differential* of the action, denoted δS. This is done analogously to the infinitesimal differential of a function of several variables, $f(x, y)$, which elementary calculus defines as

$$\begin{aligned} \mathrm{d}f(x, y) &= f(x + \mathrm{d}x, y + \mathrm{d}y) - f(x, y) \\ &= \frac{\partial f(x, y)}{\partial x}\mathrm{d}x + \frac{\partial f(x, y)}{\partial y}\mathrm{d}y. \end{aligned} \tag{1.2}$$

In defining the functional differential of equation (1.1), it is important to note that both $\boldsymbol{x}$ and $\dot{\boldsymbol{x}}$ are treated as *independent* degrees of freedom, similarly to x and y in equation (1.2). For simplicity, the case of a single particle moving in one dimension will be considered, so that the vector $\boldsymbol{x}$ is replaced by the function x. A definition for functionals identical in structure to equation (1.2) is then given by

$$\begin{aligned} \delta S[x, \dot{x}] &= S[x + \delta x, \dot{x} + \delta\dot{x}] - S[x, \dot{x}] \\ &= \int_{t_i}^{t_f} \mathrm{d}t \left\{ \frac{\delta S[x, \dot{x}]}{\delta x(t)}\delta x(t) + \frac{\delta S[x, \dot{x}]}{\delta\dot{x}(t)}\delta\dot{x}(t) \right\}, \end{aligned} \tag{1.3}$$

where $\delta x(t)$ is an *infinitesimal variation* of the function $x(t)$ such that $(\delta x(t))^2 \approx 0$. It follows that $\delta\dot{x}(t) = \mathrm{d}(\delta x(t))/\mathrm{d}t$, so that $\delta x(t)\,\delta\dot{x}(t) = \frac{1}{2}\mathrm{d}(\delta x(t)^2)/\mathrm{d}t \approx 0$. It is important to note that the *functional derivatives* appearing in (1.3), denoted $\delta S[x, \dot{x}]/\delta x(t)$ and $\delta S[x, \dot{x}]/\delta\dot{x}(t)$, have an explicit dependence on t. By applying the first line of (1.3) to the action functional (1.1), it follows that, to first order in $\delta x(t)$ and $\delta\dot{x}(t)$,

$$\begin{aligned} \delta S[x, \dot{x}] &= \int_{t_i}^{t_f} \mathrm{d}t\ \{\mathcal{L}(x(t) + \delta x(t), \dot{x}(t) + \delta\dot{x}(t)) - \mathcal{L}(x(t), \dot{x}(t))\} \\ &= \int_{t_i}^{t_f} \mathrm{d}t \left\{ \frac{\partial\mathcal{L}(x(t), \dot{x}(t))}{\partial x(t)}\delta x(t) + \frac{\partial\mathcal{L}(x(t), \dot{x}(t))}{\partial\dot{x}(t)}\delta\dot{x}(t) \right\}. \end{aligned} \tag{1.4}$$

By comparison to equation (1.3), the functional derivatives of the action (1.1) are therefore given by

$$\frac{\delta S[x,\dot{x}]}{\delta x(t)}=\frac{\partial \mathcal{L}(x(t),\dot{x}(t))}{\partial x(t)}\quad\text{and}\quad\frac{\delta S[x,\dot{x}]}{\delta \dot{x}(t)}=\frac{\partial \mathcal{L}(x(t),\dot{x}(t))}{\partial \dot{x}(t)},\tag{1.5}$$

where once again $x(t)$ and $\dot{x}(t)$ are treated as independent variables.

The particle trajectories of classical mechanics are denoted $x_c(t)$, and in the Lagrangian or functional formulation of classical mechanics they correspond to an extremum of the classical action. This is easily stated using the functional differential by demanding that $\delta S[x, \dot{x}] = 0$ when evaluated using x_c and a fluctuation that vanishes at the end points, so that $\delta x(t_i) = \delta x(t_f) = 0$. Using this property shows that integrating by parts in the second term in the integral of equation (1.4) yields

$$\begin{aligned}\int_{t_i}^{t_f} \mathrm{d}t\, \frac{\partial \mathcal{L}(x(t),\dot{x}(t))}{\partial \dot{x}(t)}\,\delta\dot{x}(t) &= \int_{t_i}^{t_f} \mathrm{d}t\, \frac{\partial \mathcal{L}(x(t),\dot{x}(t))}{\partial \dot{x}(t)}\frac{\mathrm{d}}{\mathrm{d}t}\delta x(t)\\ &= -\int_{t_i}^{t_f} \mathrm{d}t\,\delta x(t)\frac{\mathrm{d}}{\mathrm{d}t}\frac{\partial \mathcal{L}(x(t),\dot{x}(t))}{\partial \dot{x}(t)}\end{aligned}\tag{1.6}$$

Combining equation (1.6) with the first term in the integral of equation (1.4) shows that $\delta S[x, \dot{x}] = 0$ if $x_c(t)$ satisfies the Euler–Lagrange equation,

$$\left.\left(\frac{\partial \mathcal{L}(x(t),\dot{x}(t))}{\partial x(t)}-\frac{\mathrm{d}}{\mathrm{d}t}\frac{\partial \mathcal{L}(x(t),\dot{x}(t))}{\partial \dot{x}(t)}\right)\right|_{x=x_c}=0.\tag{1.7}$$

In effect, equation (1.7) is the generalization of Newton's law of motion. In addition to the Euler–Lagrange equation (1.7), the functional approach allows the definition of the momentum p *canonically conjugate* to x as $p = \delta S[x, \dot{x}]/\delta\dot{x} = \partial\mathcal{L}(x, \dot{x}, t)/\partial\dot{x}$. It is assumed that this expression can be inverted to find $\dot{x}$ in terms of p and x, so that $\dot{x} = \dot{x}(p, x)$. Using this expression for the canonically conjugate momentum, the generalization of mechanical energy, referred to as the *Hamiltonian H*, is given by a Legendre transformation,

$$\begin{aligned}&H(p,x)=p\,\dot{x}(p,x)-\mathcal{L}(x,\dot{x}(p,x),t),\\ &\Longrightarrow\frac{\mathrm{d}H}{\mathrm{d}t}=\dot{p}\dot{x}+p\ddot{x}-\frac{\partial\mathcal{L}}{\partial x}\dot{x}-\frac{\partial\mathcal{L}}{\partial\dot{x}}\ddot{x}-\frac{\partial\mathcal{L}}{\partial t}=\left(\dot{p}-\frac{\partial\mathcal{L}}{\partial x}\right)\dot{x}-\frac{\partial\mathcal{L}}{\partial t}.\end{aligned}\tag{1.8}$$

If $\mathcal{L}$ has no explicit time-dependence, so that $\partial\mathcal{L}/\partial t = 0$, then it is straightforward to show that $\mathrm{d}H/\mathrm{d}t = 0$ when p is evaluated for a classical trajectory, since for that case the Euler–Lagrange equation gives

$$\dot{p}(x_c)=\left.\frac{\mathrm{d}}{\mathrm{d}t}\frac{\partial\mathcal{L}}{\partial\dot{x}}\right|_{x=x_c}=\left.\frac{\partial\mathcal{L}}{\partial x}\right|_{x=x_c},\tag{1.9}$$

so that equation (1.8) shows that $\dot{H} = 0$ along the classical trajectory.

Using equation (1.8) the Lagrangian may be rewritten in its *phase space version*,

$$\mathcal{L}(p, x) = p\,\dot{x}(p, x) - H(p, x), \tag{1.10}$$

and used in the definition (1.1) to obtain a second version of the action functional, denoted $S[p, x]$. Demanding that $\delta S[p, x] = S[p + \delta p, x + \delta x] - S[p, x] = 0$ and enforcing the independence of δx and δp yields *Hamilton's equations of motion*,

$$\dot{x} = \frac{\partial H(p, x)}{\partial p}, \quad \dot{p} = -\frac{\partial H(p, x)}{\partial x}. \tag{1.11}$$

Both forms of the Lagrangian, equations (1.1) and (1.10), are important in the path integral. The classical trajectory, denoted $x_c(t)$, found from either version will depend on the end points x_i and x_f as well as the time interval of motion, $T = t_f - t_i$. Once the classical trajectory is known the energy along the trajectory, denoted E_c, can be found from the Hamiltonian. Even if E_c is constant along the trajectory, its specific value will depend on the boundary conditions and time T of the classical trajectory, so that $E_c = E_c(x_i, x_f, T)$. Similarly, it is assumed that the momentum along the trajectory, denoted p_c, can expressed terms of x_c and E_c. Version (1.10) *for the case of a conserved Hamiltonian* gives a useful expression for the action evaluated using the classical trajectory, denoted $S_c(x_f, x_i, T)$. Setting $t_i = 0$, it is given by

$$\begin{aligned} S_c(x_f, x_i, T) &= \int_0^T \mathrm{d}t\; p_c \dot{x}_c - \int_0^T \mathrm{d}t\; H(p_c, x_c) \\ &= \int_{x_i}^{x_f} \mathrm{d}x\; p_c(x) - \int_0^T \mathrm{d}t\; H(p_c, x_c), \end{aligned} \tag{1.12}$$

where $p_c(x)$ is the momentum of the particle at the position x. The form (1.12) gives the following useful results,

$$\frac{\partial S_c}{\partial x_f} = p_c(x_f) = p_f, \quad \frac{\partial S_c}{\partial x_i} = -p_c(x_i) = -p_i, \quad \frac{\partial S_c}{\partial T} = -E_c(x_i, x_f, T). \tag{1.13}$$

Another useful classical definition is the *Poisson bracket*. If $F(\boldsymbol{p}, \boldsymbol{x})$ and $G(\boldsymbol{p}, \boldsymbol{x})$ are two functions of n canonical variables, then their Poisson bracket is defined as

$$\{F, G\} = \sum_{i=1}^{n} \left(\frac{\partial F}{\partial x_i} \frac{\partial G}{\partial p_i} - \frac{\partial G}{\partial x_i} \frac{\partial F}{\partial p_i} \right). \tag{1.14}$$

The Poisson bracket satisfies $\{F, G\} = -\{G, F\}$ and $\dot{F} = \{H, F\}$, with the latter following from Hamilton's equations (1.11).

In order to generalize the functional derivative of equation (1.3) to an arbitrary functional in a manner consistent with equation (1.5), Volterra made use of the Dirac delta, $\delta(t - t')$, which has the following property. If t' is contained in the interval (t_i, t_f), then integrating the Dirac delta with a function $x(t)$ yields

$$\int_{t_i}^{t_f} \mathrm{d}t\; x(t)\, \delta(t - t') = x(t'), \tag{1.15}$$

and is zero if t' is not in the interval $(t_i,\ t_f)$. The Dirac delta can be represented as the derivative of the discontinuous Heaviside step function, which is defined as

$$\theta(t - t') = \begin{cases} 1 & \text{if} \quad t - t' > 0 \\ 0 & \text{if} \quad t - t' < 0 \end{cases}. \tag{1.16}$$

It is assumed that the step function converges in the mean to $\theta(0) = \frac{1}{2}$. Using the step function's properties and integrating by parts shows that

$$\lim_{\epsilon \to 0} \int_{t'-\epsilon}^{t'+\epsilon} \mathrm{d}t\, f(t) \frac{\mathrm{d}}{\mathrm{d}t} \theta(t - t') = f(t') \Longrightarrow \frac{\mathrm{d}}{\mathrm{d}t} \theta(t - t') = \delta(t - t'). \tag{1.17}$$

This relationship allows the derivation of several important properties of the Dirac delta, which include $\delta(-t) = \delta(t)$ as well as

$$\delta(a(t - t')) = \frac{1}{|a|} \delta(t - t'), \tag{1.18}$$

$$\delta(f(t)) = \sum_j \left(\left| \frac{\mathrm{d}f(t_j)}{\mathrm{d}t_j} \right| \right)^{-1} \delta(t - t_j), \tag{1.19}$$

where the sum in equation (1.19) is over all the roots t_j of $f(t)$ such that $f(t_j) = 0$.

The Dirac delta can be given representations adapted to the class of *test functions* against which it is integrated. This can be understood in the context of Fourier series and transforms. For a function $f(t)$ that is piecewise continuous over the interval $(-T,\ T)$, the Fourier series representation of $f(t)$ is given by

$$f(t) = \sum_{n=-\infty}^{\infty} f_n \exp\left(\frac{i\pi n t}{T} \right), \tag{1.20}$$

where the coefficients are given by

$$f_n = \frac{1}{T} \int_{-T}^{T} \mathrm{d}t\, f(t) \exp\left(-\frac{i\pi n t}{T} \right). \tag{1.21}$$

Choosing $f(t) = \delta(t)$ in equation (1.21) immediately gives $f_n = 1/T$, and the Fourier series for the delta function that is periodic over the interval $(-T,\ T)$ is given by

$$\delta(t) = \frac{1}{T} \sum_{n=-\infty}^{\infty} \exp\left(\frac{i\pi n t}{T} \right). \tag{1.22}$$

Similarly, for a function $f(t)$ that has no periodicity, the Fourier integral representation of $f(t)$ is given by

$$f(t) = \int_{-\infty}^{\infty} \frac{\mathrm{d}\omega}{2\pi} \tilde{f}(\omega)\, e^{i\omega t}, \tag{1.23}$$

where the Fourier transform $\tilde{f}(\omega)$ is given by

$$\tilde{f}(\omega) = \int_{-\infty}^{\infty} \mathrm{d}t\, f(t) e^{-i\omega t}. \tag{1.24}$$

Choosing $f(t) = \delta(t)$ in equation (1.24) immediately gives $\tilde{\delta}(\omega) = 1$, so that the Fourier integral representation of the Dirac delta for an infinite range is given by

$$\delta(t) = \int_{-\infty}^{\infty} \frac{\mathrm{d}\omega}{2\pi} e^{i\omega t}. \tag{1.25}$$

Both equations (1.25) and (1.22) demonstrate that $\delta(0)$ is a divergent quantity. With care, the Dirac delta can be visualized as zero everywhere except at the points where its argument vanishes, and there it is singular. In that regard, the integral of $t\,\delta(t)$ yields zero, and so $t\,\delta(t)$ is often discarded as zero. However, if $t\,\delta(t)$ is integrated against a test function that is singular at $t = 0$, it will not vanish, so care must be taken in claiming $t\,\delta(t) = 0$. The Dirac delta can be extended to n dimensions by using the product,

$$\delta^n(\boldsymbol{x}) = \delta(x_1)\cdots\delta(x_n) = \int \frac{\mathrm{d}^n p}{(2\pi)^n} \exp(i\boldsymbol{p} \cdot \boldsymbol{x}). \tag{1.26}$$

The Heaviside step function used to define the Dirac delta can be given the Fourier representation

$$\theta(t) = \lim_{\eta \to 0^+} \int_{-\infty}^{\infty} \frac{\mathrm{d}\omega}{2\pi i} \frac{1}{\omega - i\eta} e^{i\omega t}. \tag{1.27}$$

Cauchy's theorem shows that equation (1.27) vanishes for $t < 0$ since the contour encloses the lower half of the complex ω plane where there is no pole. Results (1.27) and (1.25) satisfy definition (1.17).

Using the Dirac delta, Volterra defined the functional derivative in a manner similar to the usual derivative of a function. The functional derivative of the functional $F[J]$ with respect to the function $J(t')$ is defined as

$$\frac{\delta F[J]}{\delta J(t')} = \lim_{\epsilon \to 0} \frac{1}{\epsilon}(F[J(t) + \epsilon\, \delta(t - t')] - F[J(t)]) \implies \frac{\delta J(t)}{\delta J(t')} = \delta(t - t'). \tag{1.28}$$

Definition (1.28) reproduces the previous result (1.5) when applied to the action (1.1) by treating x and $\dot{x}$ as independent functions. It is also easily generalized to higher order functional derivatives. Because it obeys the Leibniz or product rule, equation (1.28) yields results similar in form to the derivatives of functions, but with the added feature that the argument of the function appears in the final result. For the interval of integration $(-\infty, \infty)$, applying equation (1.28) gives two useful results,

$$\frac{\delta}{\delta J(t)} \int_{-\infty}^{\infty} \mathrm{d}t'\, J^n(t') = nJ^{n-1}(t), \tag{1.29}$$

$$\frac{\delta}{\delta J(t)}\exp\left(\int_{-\infty}^{\infty} \mathrm{d}t'\; K(t')J(t')\right) = K(t)\exp\left(\int_{-\infty}^{\infty} \mathrm{d}t'\; K(t')J(t')\right). \tag{1.30}$$

The definition (1.28) is consistent with the *functional chain rule*, given by

$$\frac{\delta F[J]}{\delta K(t)} = \int_{-\infty}^{\infty} \mathrm{d}t' \frac{\delta F[J]}{\delta J(t')}\frac{\delta J(t')}{\delta K(t)}. \tag{1.31}$$

It is often useful to represent a functional in terms of a functional Taylor series, which takes a form quite similar to the Taylor series for functions. For the case that $F[J]$ is well behaved in the neighborhood of $J(t) = 0$, it can be expanded around $J(t) = 0$ and the functional Taylor series takes the form

$$F[J] = \sum_{n=0}^{\infty} \frac{1}{n!}\int_{-\infty}^{\infty} \mathrm{d}t_1 \cdots \int_{-\infty}^{\infty} \mathrm{d}t_n\; J(t_1)\cdots J(t_n) \frac{\delta^n F[J]}{\delta J(t_1)\cdots \delta J(t_n)}\bigg|_{J=0}. \tag{1.32}$$

The validity of equation (1.32) is readily demonstrated at each order by using equation (1.28). It is also possible to use equation (1.32) to expand a functional around a function rather than zero.

1.2 The operator formulation of quantum mechanics

It is assumed the reader is familiar with basic quantum mechanics. This section reviews the operator formalism to clarify notation and remind the reader of some relevant aspects of single particle quantum mechanics. In the Dirac formulation of quantum mechanics the particle's position x and its canonically conjugate momentum p become operators, denoted $\hat{X}$ and $\hat{P}$ respectively. These operators act on a *Hilbert space* that contains all the possible *states* of the particle. The eigenstates of $\hat{X}$ and $\hat{P}$ are denoted $|\, x \,\rangle$ and $|\, p \,\rangle$, and they satisfy the eigenvalue equations $\hat{X}|\, x \,\rangle = x|\, x \,\rangle$ and $\hat{P}|\, p \,\rangle = p|\, p \,\rangle$, where x and p are the *eigenvalues* of the position and momentum operators. In the Copenhagen interpretation of quantum mechanics, $|\, x \,\rangle$ and $|\, p \,\rangle$ represent possible *position and momentum states* for the particle, and the eigenvalues represent the outcome of measuring the particle's position or momentum when it is in that state. For the moment these eigenvalues are assumed to exist with an infinite continuum of possible values.

These states possess a dual space, denoted $|\, x \,\rangle^\dagger = \langle\, x \,|$ and $|\, p \,\rangle^\dagger = \langle\, p \,|$, and the action of the dual space is realized by the inner product, denoted $\langle\, x \,|\, x' \,\rangle$ and $\langle\, p \,|\, p' \,\rangle$. The *adjoint operators* $\hat{X}^\dagger$ and $\hat{P}^\dagger$ act on these members of the dual space, so that $\langle\, x \,|(\hat{X}|\, x' \,\rangle) = (\langle\, x \,|\hat{X}^\dagger)|\, x' \,\rangle$. In order to have real eigenvalues, it is required that $\hat{X}^\dagger = \hat{X}$ and $\hat{P}^\dagger = \hat{P}$, and this immediately gives $(x - x')\langle\, x \,|\, x' \,\rangle = 0$ and $(p - p')\langle\, p \,|\, p' \,\rangle = 0$. It follows that the inner products are given by $\langle\, x \,|\, x' \,\rangle = \delta(x - x')$ and $\langle\, p \,|\, p' \,\rangle = \delta(p - p')$. These states are therefore *orthogonal* in the sense that the Dirac delta of equation (1.25) vanishes everywhere except where its argument vanishes, where its singular nature demonstrates that these states have an infinite norm. The delta function normalization is consistent with *unit projection operators*,

$$\hat{\mathbf{1}}_x = \int_{-\infty}^{\infty} \mathrm{d}x \mid x \rangle\langle x \mid, \quad \hat{\mathbf{1}}_p = \int_{-\infty}^{\infty} \mathrm{d}p \mid p \rangle\langle p \mid. \tag{1.33}$$

Result (1.33) represents the *completeness* of both the position and momentum states, so that the Hilbert space of all possible particle states is *spanned* by either set of states. That these are unit projection operators follows from

$$\hat{\mathbf{1}}_x| x \rangle = \int_{-\infty}^{\infty} \mathrm{d}x'| x' \rangle\langle x' \mid x \rangle = \int_{-\infty}^{\infty} \mathrm{d}x' \mid x' \rangle\delta(x' - x) = \mid x \rangle, \tag{1.34}$$

where the value of x is arbitrary. A similar result for the momentum projection operator follows for the momentum states.

The key assumption of quantum mechanics is that the two operators $\hat{X}$ and $\hat{P}$ do not *commute*. Instead, they obey the *Heisenberg algebra*, given by

$$[\hat{X}, \hat{P}] \equiv \hat{X}\hat{P} - \hat{P}\hat{X} = i\hbar. \tag{1.35}$$

Relation (1.35) can be combined with the eigenstates $| x \rangle$ and $| p \rangle$ to place the operators $\hat{X}$ and $\hat{P}$ into a specific *representation*. For example,

$$i\hbar\, \delta(x - x') = \langle x |i\hbar| x' \rangle = \langle x |[\hat{X}, \hat{P}]| x' \rangle = (x - x')\langle x |\hat{P}| x' \rangle, \tag{1.36}$$

and this shows that $\langle x |\hat{P}| x' \rangle = -i\hbar\frac{\partial}{\partial x}\delta(x - x')$, the usual representation of momentum in wave mechanics. Using this result, it can be shown that

$$\langle x \mid p \rangle = \frac{1}{\sqrt{2\pi\hbar}} e^{ipx/\hbar}, \tag{1.37}$$

where the Dirac delta function normalization of equation (1.25) has been employed so that the unit projection operator requirement (1.33) holds. This follows from

$$\langle x' |\hat{\mathbf{1}}_p| x \rangle = \int_{-\infty}^{\infty} \frac{\mathrm{d}p}{2\pi\hbar} \exp\left\{\frac{i}{\hbar}p(x' - x)\right\} = \delta(x' - x). \tag{1.38}$$

Results (1.36) and (1.37) are detailed in any standard quantum mechanics text.

The replacements $p \to \hat{P}$ and $x \to \hat{X}$ in the definition (1.8) give the operator version of the Hamiltonian. In its most general form it is denoted $\hat{H}(t)$, where the possibility of explicit time-dependence will be considered. In the *Schrödinger picture* the time-dependent state of the system is denoted $| \psi, t \rangle$, and it is related to the wave function $\psi(x, t)$ of wave mechanics by $\psi(x, t) = \langle x \mid \psi, t \rangle$. Its time-dependence is determined by the Schrödinger equation, which states that

$$\hat{H}(t)| \psi, t \rangle = i\hbar\frac{\partial}{\partial t}| \psi, t \rangle. \tag{1.39}$$

The formal solution to equation (1.39) begins by specifying the initial state of the system at the initial time t_i as $| \psi, t_i \rangle$. The state at any later time t is given by the action of a unitary operator, $|\,\psi,\, t\,\rangle = \hat{U}(t,\, t_i)|\,\psi,\, t_i\,\rangle$, where $\hat{U}(t,\, t_i)$ is referred to as the *evolution operator* and must satisfy the initial condition $\hat{U}(t_i,\, t_i) = 1$. The evolution operator gives $|\,\psi,\, t\,\rangle = \hat{U}(t,\, t')|\,\psi,\, t'\,\rangle = \hat{U}(t,\, t')\hat{U}(t',\, t_i)|\,\psi,\, t_i\,\rangle$. This gives

$$\hat{U}(t,\, t')\hat{U}(t',\, t_i) = \hat{U}(t,\, t_i) \implies \hat{U}(t_i,\, t)\hat{U}(t,\, t_i) = \hat{U}(t_i,\, t_i) = 1, \tag{1.40}$$

revealing that $\hat{U}(t_i,\, t) = \hat{U}^{-1}(t,\, t_i)$. Using equation (1.39) shows that the Schrödinger equation is satisfied only if

$$i\hbar\frac{\partial}{\partial t}\hat{U}(t,\, t_i) = \hat{H}(t)\; \hat{U}(t,\, t_i). \tag{1.41}$$

If the Hamiltonian has no explicit time-dependence, $\partial\hat{H}/\partial t = 0$, then equation (1.41) is solved by $\hat{U}(t,\, t_i) = \exp(-i\hat{H}(t - t_i)/\hbar)$. For such a case, the evolution operator can be viewed as $\hat{U}(t,\, t_i) = \hat{U}(t)\hat{U}^{-1}(t_i)$, where $\hat{U}(t) = \exp(-i\hat{H}t/\hbar)$. If the Hamiltonian is explicitly time-dependent, $\hat{H} = \hat{H}(t)$, then equation (1.41) can be solved by integration followed by iteration. The first step is to note that integrating both sides of equation (1.41) from t_i to t and using the initial condition gives

$$\int_{t_i}^{t} \mathrm{d}t'\, \frac{\partial}{\partial t'}\hat{U}(t',\, t) = \hat{U}(t,\, t_i) - 1 = -\frac{i}{\hbar}\int_{t_i}^{t} \mathrm{d}t'\, \hat{H}(t')\hat{U}(t',\, t_i). \tag{1.42}$$

Equation (1.42) can now be iterated by repeatedly inserting the result for $\hat{U}(t,\, t_i)$ into the right-hand side. The result is

$$\hat{U}(t,\, t_i) = 1 - \frac{i}{\hbar}\int_{t_i}^{t} \mathrm{d}t_1\, \hat{H}(t_1) + \left(-\frac{i}{\hbar}\right)^2 \int_{t_i}^{t} \mathrm{d}t_1\, \hat{H}(t_1)\int_{t_i}^{t_1} \mathrm{d}t_2\, \hat{H}(t_2) + \cdots. \tag{1.43}$$

Result (1.43) can be written much more compactly by introducing the operation known as *time-ordering*, which uses the previously defined Heaviside step function (1.16). The time-ordering operator $\mathcal{T}$ acts on products of time-dependent operators and orders them from latest to earliest. For the product of two operators it is given by $\mathcal{T}\{\hat{H}(t_1)\hat{H}(t_2)\} = \theta(t_1 - t_2)\hat{H}(t_1)\hat{H}(t_2) + \theta(t_2 - t_1)\hat{H}(t_2)\hat{H}(t_1)$, so that

$$\int_{t_i}^{t} \mathrm{d}t_1\, \hat{H}(t_1)\int_{t_i}^{t_1} \mathrm{d}t_2\, \hat{H}(t_2) = \frac{1}{2!}\int_{t_i}^{t} \mathrm{d}t_1 \int_{t_i}^{t} \mathrm{d}t_2\, \mathcal{T}\big\{\hat{H}(t_1)\hat{H}(t_2)\big\}. \tag{1.44}$$

Result (1.44) is demonstrated by inserting the time-ordered product, using the definition of the step function to change the limits on the integrals, and then relabeling the dummy variables of integration. The generalization to arbitrary products is straightforward, and for products of n operators the result is $n!$ possible orderings with $n - 1$ step functions for each ordering. The upshot of time-ordering is that the integrations in equation (1.43) become identical, so that the nth order term becomes

$$\int_{t_i}^{t} \mathrm{d}t_1 \int_{t_i}^{t_1} \mathrm{d}t_2 \cdots \int_{t_i}^{t_{n-1}} \mathrm{d}t_n \, \hat{H}(t_1)\hat{H}(t_2)\cdots\hat{H}(t_n) \\ = \frac{1}{n!} \int_{t_i}^{t} \mathrm{d}t_1 \int_{t_i}^{t} \mathrm{d}t_2 \cdots \int_{t_i}^{t} \mathrm{d}t_n \, \mathcal{T}\left\{\hat{H}(t_1)\hat{H}(t_2)\cdots\hat{H}(t_n)\right\}. \tag{1.45}$$

Because the time-ordering operator acts only on the Hamiltonians, the infinite series for the evolution operator can be summed to obtain

$$\hat{U}(t, t_i) = \mathcal{T}\left\{\exp\left(-\frac{i}{\hbar}\int_{t_i}^{t} \mathrm{d}\tau \, \hat{H}(\tau)\right)\right\}, \tag{1.46}$$

where $e^x = \sum_{n=1}^{\infty} x^n/n!$ was used. This verifies that $\hat{U}$ is a unitary operator, since $\hat{U}^\dagger(t, t_i) = \hat{U}(t_i, t) = \hat{U}^{-1}(t, t_i)$. Result (1.46) is central to deriving the path integral representation of quantum transition elements. For the case that $\partial\hat{H}/\partial t = 0$, equation (1.46) reduces to

$$\frac{\partial \hat{H}}{\partial t} = 0 \implies \hat{U}(t, t_i) = \exp\{-i\hat{H}(t - t_i)/\hbar\}, \tag{1.47}$$

which is consistent with $\mathcal{T}\{\hat{H}^n\} = \hat{H}^n$.

The evolution operator plays an important role in evaluating time-ordered products of observables in the *Heisenberg picture*, where the states are static and the observables evolve in time. If the Heisenberg states and the Schrödinger picture states coincide at t_i, then the Schrödinger picture observable $\hat{O}_S(t)$ is related to the Heisenberg picture observable $\hat{O}_H(t)$ according to

$$\hat{O}_H(t) = \hat{U}^{-1}(t, t_i)\, \hat{O}_S(t)\, \hat{U}(t, t_i) = \hat{U}(t_i, t)\, \hat{O}_S(t)\, \hat{U}(t, t_i). \tag{1.48}$$

Combining the time derivative of equation (1.48) with equation (1.41) and its complex conjugate gives

$$\frac{\mathrm{d}}{\mathrm{d}t}\hat{O}_H(t) = \frac{i}{\hbar}[\hat{H}(t), \hat{O}_H(t)] + \hat{U}(t_i, t)\, \frac{\partial \hat{O}_S(t)}{\partial t}\, \hat{U}(t, t_i). \tag{1.49}$$

If $\hat{O}_S$ has no explicit time dependence the second term vanishes.

Path integrals are commonly used to evaluate time-ordered products of Heisenberg picture operators in the general case of a time-dependent Hamiltonian. Using the properties of time-ordering, result (1.40), and definition (1.48), it follows that

$$\begin{aligned} &\hat{U}(t_f, t_i)\, \mathcal{T}\left\{\hat{O}_H(t_1)\, \hat{O}_H(t_2)\right\} \\ &= \hat{U}(t_f, t_i)\left\{\theta(t_1 - t_2)\, \hat{O}_H(t_1)\, \hat{O}_H(t_2) + \theta(t_2 - t_1)\, \hat{O}_H(t_2)\, \hat{O}_H(t_1)\right\} \\ &= \theta(t_1 - t_2)\, \hat{U}(t_f, t_1)\, \hat{O}_S(t_1)\, \hat{U}(t_1, t_2)\, \hat{O}_S(t_2)\, \hat{U}(t_2, t_i) \\ &+ \theta(t_2 - t_1)\, \hat{U}(t_f, t_2)\, \hat{O}_S(t_2)\, \hat{U}(t_2, t_1)\, \hat{O}_S(t_1)\, \hat{U}(t_1, t_i), \end{aligned} \tag{1.50}$$

where it is assumed that t_1 and t_2 are contained in the interval (t_i, t_f). As a result, the last step in equation (1.50) follows from $\hat{U}(t_f, t_i) = \hat{U}(t_f, t_2)\hat{U}(t_2, t_1)\hat{U}(t_1, t_i) = \hat{U}(t_f, t_1)\hat{U}(t_1, t_2)\hat{U}(t_2, t_i)$. More complicated time-ordered products follow similarly.

1.3 Operator identities

The exponential of the Hamiltonian plays a key role in quantum mechanics since it determines the time evolution of the states and operators. One means of evaluating the action of the evolution operator on states is to break it into a product of exponentials that are more easily evaluated. This process is complicated since the individual factors typically do not commute as a result of the fundamental commutator (1.35). As a result, the process of factorization in quantum mechanics is almost always characterized by powers of $\hbar$ that occur as a result of this commutation relation, and these are commonly referred to as *quantum effects*.

For the case of two arbitrary operators, $\hat{A}$ and $\hat{B}$, their commutator $[\hat{A}, \hat{B}]$ will typically not vanish. A useful relation for analyzing an exponential involving these operators is obtained by considering the operator $\hat{F}(\tau)$, defined as

$$\hat{F}(\tau) = e^{\tau\hat{A}}\hat{B}e^{-\tau\hat{A}} = \sum_{n=0}^{\infty}\frac{1}{n!}\hat{F}_n\tau^n, \tag{1.51}$$

where the operators $\hat{F}_n$ in the Taylor series representation of $\hat{F}(\tau)$ will be found so that the two operators coincide. Differentiating the first expression for $\hat{F}(\tau)$ gives

$$\frac{d}{d\tau}\hat{F}(\tau) = \hat{A}\, e^{\tau\hat{A}}\hat{B}e^{-\tau\hat{A}} - e^{\tau\hat{A}}\hat{B}e^{-\tau\hat{A}}\hat{A} = [\hat{A}, \hat{F}] = \sum_{n=0}^{\infty}\frac{1}{n!}[\hat{A}, \hat{F}_n]\tau^n. \tag{1.52}$$

Differentiating the second expression for $\hat{F}(\tau)$ gives

$$\frac{d}{d\tau}\hat{F}(\tau) = \sum_{n=1}^{\infty}\frac{1}{(n-1)!}\hat{F}_n\tau^{n-1} = \sum_{n=0}^{\infty}\frac{1}{n!}\hat{F}_{n+1}\tau^n. \tag{1.53}$$

Equating the coefficients appearing in the two Taylor series expressions (1.52) and (1.53) gives the recurrence relation $\hat{F}_{n+1} = [\hat{A}, \hat{F}_n]$. Setting $\tau = 0$ in equation (1.51) immediately gives $\hat{F}_0 = \hat{B}$, and all the higher order coefficients therefore follow from the recurrence relation, e.g. $\hat{F}_1 = [\hat{A}, \hat{F}_0] = [\hat{A}, \hat{B}]$ and so on. The final result is

$$e^{\tau\hat{A}}\hat{B}e^{-\tau\hat{A}} = \frac{1}{0!}\hat{B} + \frac{\tau}{1!}[\hat{A}, \hat{B}] + \frac{\tau^2}{2!}[\hat{A}, [\hat{A}, \hat{B}]] + \cdots, \tag{1.54}$$

so that the expression is determined by a sequence of *nested commutators*.

The process of factorizing two exponentiated operators is governed by the Baker–Campbell–Hausdorff (BCH) theorem. The Zassenhaus variant of the BCH theorem states that if $\hat{A}$ and $\hat{B}$ are two operators and τ is a parameter, then

$$e^{\tau(\hat{A}+\hat{B})} = e^{\tau\hat{A}}e^{\tau\hat{B}}e^{-\frac{1}{2}\tau^2[\hat{A},\,\hat{B}]}e^{\frac{1}{6}\tau^3(2[\hat{B},\,[\hat{A},\,\hat{B}]]+[\hat{A},\,[\hat{A},\,\hat{B}]])}\cdots, \tag{1.55}$$

where the ellipsis refers to an infinite series of exponential factors whose arguments are nested commutators with increasing powers of τ. The BCH theorem and the method for generating the higher order terms follows from equation (1.54), but for brevity the details will be omitted. The interested reader should consult the bibliography.

It is sometimes the case that the higher order factors in equation (1.55) reduce to unity because all the higher order nested commutators vanish. An example is the choice of the Hamiltonian for a linear potential, so that $\hat{H} = \hat{A} + \hat{B}$, where $\hat{A} = -f\hat{X}$ and $\hat{B} = \hat{P}^2/2m$. Using equation (1.35) it follows that $[\hat{A},\,\hat{B}] = -i\hbar f\hat{P}/m$. From this it follows that $[\hat{B},\,[\hat{A},\,\hat{B}]] = 0$ and $[\hat{A},\,[\hat{A},\,\hat{B}]] = -\hbar^2 f^2/m$. Because the double commutators are not operators, all higher order nested commutators will vanish. For this simple Hamiltonian, it follows that

$$e^{\tau\left(\hat{P}^2/2m-f\hat{X}\right)} = e^{-\tau f\hat{X}}e^{\tau\hat{P}^2/2m}e^{i\hbar\tau^2 f\hat{P}/2m}e^{-\hbar^2 f^2\tau^3/6m}. \tag{1.56}$$

The utility of doing this is that the actions of the individual exponentials on either $|\,x\,\rangle$ or $|\,p\,\rangle$ are well defined, so that using equation (1.56) gives

$$\langle\, x\,|e^{\tau\left(\hat{P}^2/2m-f\hat{X}\right)}|\,p\,\rangle = e^{\tau(p^2/2m-fx+i\hbar\tau fp/2m-\hbar^2\tau^2 f^2/6m)}\langle\, x\,|\,p\,\rangle. \tag{1.57}$$

Setting $\tau = -it/\hbar$ immediately gives the matrix element of the evolution operator, or the *quantum transition amplitude*, between the initial momentum state $|\,p\,\rangle$ and the final position state $|\,x\,\rangle$ for the case of the linear potential in one dimension.

Very few Hamiltonians are amenable to analysis in this manner. However, for the case that τ is infinitesimal, the BCH expansion (1.55) is dominated by the first two exponential factors. This occurs since the other exponential factors have arguments that are suppressed by a factor of τ^2 or higher, thereby reducing to factors of unity in the limit $\tau \to 0$. The *Trotter product formula* states that if $\hat{A}$ and $\hat{B}$ are two non-singular operators, then

$$e^{\tau(\hat{A}+\hat{B})} = \lim_{N\to\infty}\left(e^{(\tau/N)\hat{A}}e^{(\tau/N)\hat{B}}\right)^N. \tag{1.58}$$

The demonstration of equation (1.58) follows from the BCH theorem (1.55). Defining $\epsilon = \tau/N$, the BCH theorem gives

$$e^{\tau(\hat{A}+\hat{B})} = \lim_{N\to\infty}\left(e^{\epsilon(\hat{A}+\hat{B})}\right)^N = \lim_{N\to\infty}\left(e^{\epsilon\hat{A}}e^{\epsilon\hat{B}}e^{-\frac{1}{2}\epsilon^2[\hat{A},\,\hat{B}]}\cdots\right)^N. \tag{1.59}$$

For the case that the operators and their commutators are not singular, the exponential factors in the BCH expansion involving commutators all possess arguments suppressed by ever greater powers of ϵ compared to the first two terms in the parentheses. As a result, they reduce to factors of unity in the large N limit since $N\epsilon^n = \tau^n/N^{(n-1)} \to 0$ for $n > 1$. It is important to note that the Trotter product formula fails if the commutators are singular, since then they may offset the $1/N^{(n-1)} \to 0$ limit in the neighborhood of the singularity.

There are variants of the Trotter formula that are useful in developing path integrals. In particular, the exponential of the sum of two operators at different times, $\hat{A}(t_1) + \hat{A}(t_2)$, can be factorized *if time-ordering is applied.* time-ordering of the exponentiated operators is given by

$$\mathcal{T}\left\{e^{\tau(\hat{A}(t_1)+\hat{A}(t_2))}\right\} = \theta(t_2 - t_1)e^{\tau\hat{A}(t_2)}e^{\tau\hat{A}(t_1)} + \theta(t_1 - t_2)e^{\tau\hat{A}(t_1)}e^{\tau\hat{A}(t_2)}. \tag{1.60}$$

This is because time-ordering the commutators in the BCH theorem yields zero,

$$\mathcal{T}\left\{[\hat{A}(t_1), \hat{A}(t_2)]\hat{B}(t_3)\right\} = \mathcal{T}\left\{\hat{A}(t_1)\hat{A}(t_2)\hat{B}(t_3) - \hat{A}(t_2)\hat{A}(t_1)\hat{B}(t_3)\right\} = 0, \tag{1.61}$$

a result that holds for all times t_1, t_2, and t_3. Time-ordering the exponential of a sum of time-dependent operators allows the commutators to be ignored, so that it becomes the time-ordered product of individual exponentials, as in equation (1.60).

1.4 Grassmann variables

An important component of modern theoretical physics is the use of Grassmann variables. Their presence in path integrals allows the *Pauli exclusion principle* associated with fermions to be modeled, and they are central to formulating supersymmetry and superstrings. Because they have important but unusual properties, their definition and manipulation will be introduced here and used for a Grassmann variable version of classical mechanics.

It is a standard assumption that two real or complex variables, a and b, are commutative, so that $[a, b] = ab - ba = 0$. These will now be referred to as c-number variables to reflect their commutativity. Unlike c-numbers, two Grassmann variables, denoted η and ζ, have the property that they *anticommute*, so that $\eta\zeta = -\zeta\eta$. This is written $\{\eta, \zeta\} = \eta\zeta + \zeta\eta = 0$, where $\{\eta, \zeta\}$ is referred to as the *anticommutator*. The definition of a Grassmann number has the immediate consequence that η must anticommute with itself, so that $\eta^2 = -\eta^2 \Longrightarrow \eta^2 = 0$. A Grassmann variable is therefore *nilpotent*. It is worth noting that the number zero both commutes and anticommutes with itself and all other standard numbers, and so zero can be thought of as both a Grassmann number and a c-number. If a is a c-number variable and η a Grassmann variable, then $[a, \eta] = 0$. There are some standard mathematical operations that cannot be applied to Grassmann numbers. For example, $\sqrt{\eta^2} = \sqrt{0}$ appears to give $\eta = \sqrt{0}$, and this whimsical result indicates that square root operations on Grassmann variables are not easily defined.

The generalization to a collection of n Grassmann variables $\eta_1, \ldots, \eta_n$ is straightforward, so that $\{\eta_i, \eta_j\} = 0$ and $\eta_i^2 = 0$ for all η_i. It follows that the product of any two distinct Grassmann variables will commute with any other Grassmann variable since $\eta_1\eta_2\eta_3 = -\eta_1\eta_3\eta_2 = \eta_3\eta_1\eta_2 \Longrightarrow [\eta_1\eta_2, \eta_3] = 0$. This reveals that products of even numbers of Grassmann variables behave like a c-number variable. Similarly, products of odd numbers of Grassmann variables retain their Grassmann property and anticommute with other Grassmann variables. For the case of n Grassmann variables there are 2^n distinct nonzero products, where $\eta_i^0 = 1$ is counted once. This is demonstrated by induction, where adding a new Grassmann variable to the collection simply doubles the number of possible combinations by adding the new variable to all the previous combinations.

A function of a single Grassmann variable $f(\eta)$ is derived from its *c*-number counterpart $f(x)$. This is accomplished by simply making the replacement $f(x) \to f(\eta)$ in the case of a function of one variable. The result of doing this can be evaluated by using the Taylor series representation of the original function and simply making the replacement $x \to \eta$ to obtain a *superfunction*. For the case of a single Grassmann variable this means that all terms η^2 or higher will vanish. As an example, the *c*-number exponential function e^x possesses an infinite series representation $e^x = \sum_{n=0}^{\infty} x^n/n!$. Substituting $x \to \eta$ gives the superfunction $e^\eta = 1 + \eta$. It is important to note that superfunctions are typically a mixture of a *c*-number and a Grassmann variable, which DeWitt referred to as the *body* and *soul* of the superfunction. For the case of a single Grassmann variable $f(\eta)$, it follows that $f(\eta) = a + b\eta$, where both a and b are *c*-numbers. Functions of multiple *c*-number variables become superfunctions of multiple Grassmann variables, with even products joining the body and odd products joining the soul. In mathematics these are known as *graded algebras*. The exponential function of two Grassmann variables reveals this underlying algebra. For two Grassmann variables satisfying $\{\eta_1, \eta_2\} = 0$, their commutator is $[\eta_1, \eta_2] = \eta_1\eta_2 - \eta_2\eta_1 = 2\,\eta_1\eta_2$. This gives

$$\begin{aligned} e^{\eta_1+\eta_2} &= 1 + \eta_1 + \eta_2 = (1 + \eta_1)(1 + \eta_2)(1 - \frac{1}{2}[\eta_1, \eta_2]) \\ &= e^{\eta_1}e^{\eta_2}e^{-\frac{1}{2}[\eta_1, \eta_2]}. \end{aligned} \tag{1.62}$$

Result (1.62) is the BCH theorem (1.55) for two Grassmann variables.

The definition of superfunctions allows the definition of Grassmann differentiation. The differential of a superfunction of n Grassmann variables, $\mathrm{d}f(\eta_1, \dots, \eta_n)$, is defined identically to the *c*-number version (1.2), so that

$$\mathrm{d}f(\eta_1, \dots, \eta_n) = \sum_{i=1}^{n} \mathrm{d}\eta_i \frac{\partial}{\partial\eta_i} f(\eta_1, \dots, \eta_n). \tag{1.63}$$

Setting $f(\eta_1, \dots, \eta_n) = \eta_1$ in equation (1.63) yields the expected result $\frac{\partial}{\partial\eta_1}\eta_1 = 1$. It is possible to define a Grassmann partial derivative that acts from the right, as opposed to the more traditional version (1.63). However, in this text the derivative will always be assumed to act from the left. Choosing $f(\eta_1, \eta_2) = \eta_1\eta_2$ in equation (1.63) gives

$$\mathrm{d}(\eta_1\eta_2) = \mathrm{d}\eta_1 \frac{\partial}{\partial\eta_1}(\eta_1\eta_2) + \mathrm{d}\eta_2 \frac{\partial}{\partial\eta_2}(-\eta_2\eta_1) = \mathrm{d}\eta_1\, \eta_2 - \mathrm{d}\eta_2\, \eta_1. \tag{1.64}$$

In order to maintain the Leibniz property of differentiation, $\mathrm{d}(\eta_1\eta_2) = \mathrm{d}\eta_1\, \eta_2 + \eta_1\, \mathrm{d}\eta_2$, it is necessary that $\mathrm{d}\eta_2\, \eta_1 = -\eta_1\, \mathrm{d}\eta_2$. As a result, $\mathrm{d}\eta_j$ is also a Grassmann variable. The Grassmann algebra requires that the partial derivative with respect to a Grassmann variable anticommutes with the other Grassmann variables and their derivative operators so that, for an arbitrary superfunction $f(\eta_1, \dots, \eta_n)$,

$$\left\{\frac{\partial}{\partial\eta_j}, \frac{\partial}{\partial\eta_k}\right\} f(\eta_1, \dots, \eta_n) = 0, \tag{1.65}$$

$$\left\{\frac{\partial}{\partial\eta_j}, \eta_k\right\} f(\eta_1, \dots, \eta_n) = \delta_{jk} f(\eta_1, \dots, \eta_n), \tag{1.66}$$

where δ_{jk} is the *Kronecker delta*, defined by

$$\delta_{jk} = \begin{cases} 1 & \text{if} \quad j = k, \\ 0 & \text{if} \quad j \neq k. \end{cases} \tag{1.67}$$

Choosing $f(\eta_1, \dots, \eta_n) = \eta_k$ shows that equation (1.66) is consistent with $\eta_k^2 = 0$.

Using the definition (1.63) with a function $f(\eta)$ of a single Grassmann variable allows Grassmann integration to be defined. A critical property of integration in physical applications is the ability to use integration by parts and discard the surface terms, as in equation (1.6). Doing so requires that the integral of an exact differential vanishes. The vanishing of the integral of $\mathrm{d}(fg) = \mathrm{d}f\ g + f\ \mathrm{d}g$ simplifies integration by parts, since then $\mathrm{d}f\ g = -f\ \mathrm{d}g$ upon integration. For the case of c-number functions this is accomplished by using integrands that vanish at the limits of the integral. For Grassmann variables this requirement must be implemented in another way. Since a general function of a single Grassmann variable has the form $f(\eta) = a + b\eta$, it follows that $\mathrm{d}f = b\ \mathrm{d}\eta$. The vanishing of this integral requires

$$\int \mathrm{d}f = b \int \mathrm{d}\eta = 0 \implies \int \mathrm{d}\eta = 0. \tag{1.68}$$

Because both $\mathrm{d}\eta$ and zero have Grassmann properties, result (1.68) is consistent. The second Grassmann integral of importance is defined by demanding that the integral of the Grassmann exponential $e^\eta = 1 + \eta$ yields unity. Combining this with equation (1.68) gives

$$\int \mathrm{d}\eta\ e^\eta = \int \mathrm{d}\eta + \int \mathrm{d}\eta\ \eta = 1 \implies \int \mathrm{d}\eta\ \eta = 1. \tag{1.69}$$

Because $\mathrm{d}\eta\ \eta$ commutes with all other Grassmann variables, it is a c-number, and so equation (1.69) is consistent with the Grassmann algebra. Result (1.69) has the valuable property that it is invariant under translation of η, $\eta \to \eta' = \eta + \xi$, where ξ is a *constant* Grassmann variable satisfying $\mathrm{d}\xi = 0$. This follows from equations (1.68) and (1.69),

$$\int \mathrm{d}\eta'\ \eta' = \int \mathrm{d}(\eta + \xi)\ (\eta + \xi) = \int \mathrm{d}\eta\ \eta - \xi \int \mathrm{d}\eta = 1. \tag{1.70}$$

These results can be extended to an arbitrary collection of real Grassmann variables. For example,

$$\int d\eta_1\, d\eta_2\, (\eta_1\eta_2) = -\int d\eta_2\Big(\int d\eta_1\, \eta_1\Big)\eta_2 = -\int d\eta_2\, \eta_2 = -1. \tag{1.71}$$

It should be noted that the integrals (1.68), (1.69), and (1.71) are identical to applying Grassmann derivatives to the integrand, so that there is an equivalence $\int d\eta_j \Longleftrightarrow \partial/\partial\eta_j$. Both Grassmann derivatives *and* Grassmann integrals are dependent on the order in which they are performed.

There is an important property of the differential $d\eta$ which is revealed by evaluating the integral of $e^{a\eta}$, where a is a c-number. The evaluation begins by noting that $e^{a\eta} = 1 + a\eta$. This expansion is combined with results (1.68) and (1.69) to find $\int d\eta\, e^{a\eta} = a$. This integral can also be evaluated by scaling the η variable to $\eta = \xi/a$, where ξ is *also* a Grassmann variable. For a c-number differential this would yield $d\eta = d\xi/a$, reducing the integral to $\int d\eta\, e^{a\eta} = a^{-1}\int d\xi\, e^{\xi} = 1/a$. However, this is inconsistent with the Grassmann result a just obtained by using the Taylor expansion. The error lies in the treatment of the differential under the change of variables. The correct rule for Grassmann differentials is

$$\eta \to a\,\xi \Longrightarrow d\eta \to \frac{1}{a}\, d\xi. \tag{1.72}$$

This behavior preserves equation (1.69), since it gives $d\eta\, \eta = d\xi\, \xi$. All Grassmann variables related by equation (1.72) will then satisfy both equations (1.68) and (1.69). Changing variables $\eta = \xi/a$ then gives $\int d\eta\, e^{a\eta} = a\int d\xi\, e^{\xi} = a$, which is the correct Grassmann result. Maintaining the equivalence $\int d\eta \Longleftrightarrow \partial/\partial\eta$ for the change of variable requires

$$\eta \to a\,\xi \Longrightarrow \frac{\partial}{\partial\eta} \to \frac{1}{a}\frac{\partial}{\partial\xi}. \tag{1.73}$$

Rules (1.72) and (1.73) are consistent with the definition of the Grassmann differential (1.63). This can be seen by choosing $f(\eta) = \eta = a\xi$, which gives

$$d\eta = df(\eta) = d\eta\, \frac{\partial}{\partial\eta} f(\eta) = \frac{1}{a^2}\, d\xi\, \frac{\partial}{\partial\xi} f(a\xi) = \frac{1}{a} d\xi. \tag{1.74}$$

This is consistent with the rule (1.72) that was used in evaluating equation (1.74).

So far, the Grassmann variables and the rules for their manipulation have assumed that they are real. By definition, complex conjugation has no effect on real Grassmann variables, so that $\eta^* = \eta$ and $(i\eta)^* = -i\eta$. Using two real Grassmann variables, η_R and η_I, a complex Grassmann variable is defined as $\zeta = (\eta_R + i\eta_I)/\sqrt{2}$, so that the complex conjugate is given by $\zeta^* = (\eta_R - i\eta_I)/\sqrt{2}$. The factor of $1/\sqrt{2}$ is for normalization. From the nilpotency and anticommutativity of η_R and η_I, these definitions satisfy both $\zeta^2 = \zeta^{*2} = 0$ and the necessary property $\zeta^{**} = \zeta$. The product

$\zeta^*\zeta$ is found to be $\zeta^*\zeta = i\eta_R\eta_I$, while $\zeta\zeta^* = -i\eta_R\eta_I$. This immediately yields $\zeta\zeta^* + \zeta^*\zeta = \{\zeta, \zeta^*\} = 0$, so that ζ and ζ^* are distinct Grassmann variables. This is consistent with the fact that two real Grassmann degrees of freedom are used to define them. The extension of these results to a collection of complex Grassmann variables yields two copies of the Grassmann algebra, $\{\zeta_i, \zeta_j\} = \{\zeta_i, \zeta_j^*\} = \{\zeta_i^*, \zeta_j^*\} = 0$.

The additional assumption that $(\zeta^*\zeta)^* = \zeta^*\zeta$ is often made with complex Grassmann variables. This assumption avoids the result that $(\zeta^*\zeta)^* = \zeta\zeta^* = -\zeta^*\zeta$, which would make $\zeta^*\zeta$ behave as if it were a pure imaginary number. However, this assumption requires that $(\zeta^*\zeta)^* = (i\eta_R\eta_I)^* = -i(\eta_R\eta_I)^* = \zeta^*\zeta = i\eta_R\eta_I$, and this is satisfied only if the two real Grassmann variables satisfy

$$(\eta_R\eta_I)^* = -\eta_R\eta_I \tag{1.75}$$

The product of the two real Grassmann variables that make up ζ must therefore behave as a pure imaginary c-number, similar to the product of two pure imaginary c-numbers behaving as a real c-number. Using the real nature of each of these variables, this can be written as $(\eta_R\eta_I)^* = -\eta_R\eta_I = \eta_I\eta_R = \eta_I^*\eta_R^*$. The rule for complex conjugation of Grassmann variable products is then identical in form to the Hermitian conjugation of operator products, where $(\hat{A}\hat{B})^\dagger = \hat{B}^\dagger\hat{A}^\dagger$. Using this result with complex Grassmann variables gives

$$\begin{aligned}
2(\zeta_1\zeta_2)^* &= ((\eta_{1R} + i\eta_{1I})(\eta_{2R} + i\eta_{2I}))^* \\
&= (\eta_{1R}\eta_{2R} + i\eta_{1R}\eta_{2I} + i\eta_{1I}\eta_{2R} - \eta_{1I}\eta_{2I})^* \\
&= \eta_{2R}\eta_{1R} - i\eta_{2I}\eta_{1R} - i\eta_{2R}\eta_{1I} - \eta_{2I}\eta_{1I} \\
&= (\eta_{2R} - i\eta_{2I})(\eta_{1R} - i\eta_{1I}) = 2(\zeta_2^*\zeta_1^*),
\end{aligned} \tag{1.76}$$

so that the same rule holds for complex Grassmann variable products.

This result yields an unusual rule for the complex conjugation of differentials. For a real Grassmann variable η, result (1.69) requires that $(\mathrm{d}\eta\ \eta)^* = \mathrm{d}\eta\ \eta$ in order that the integral is a real number. However, the newly established rule for the complex conjugation of products gives $(\mathrm{d}\eta\ \eta)^* = \eta\ (\mathrm{d}\eta)^*$. Comparison of the two forms yields the requirement that $(\mathrm{d}\eta)^* = -\mathrm{d}\eta$ in order that $(\mathrm{d}\eta\ \eta)^* = \mathrm{d}\eta\ \eta$. A similar result then follows for complex conjugation of the differential of a complex Grassmann variable, $(\mathrm{d}\zeta)^* = (\mathrm{d}\eta_R + i\ \mathrm{d}\eta_I)^*/\sqrt{2} = -(\mathrm{d}\eta_R - i\mathrm{d}\eta_I)/\sqrt{2} = -\mathrm{d}\zeta^*$. Finally, complex conjugation of the derivative must obey $(\partial/\partial\eta)^* = -\partial/\partial\eta$. This is required in order that the differential of a real Grassmann function obeys

$$(\mathrm{d}f)^* = \left(\mathrm{d}\eta\ \frac{\partial f}{\partial\eta}\right)^* = \left(\frac{\partial f}{\partial\eta}\right)^*(\mathrm{d}\eta)^* = \frac{\partial f}{\partial\eta}\ \mathrm{d}\eta = -\mathrm{d}\eta\ \frac{\partial f}{\partial\eta} = -\mathrm{d}f. \tag{1.77}$$

Similarly, for complex variables $(\partial/\partial\zeta)^* = -\partial/\partial\zeta^*$.

There are differences from real Grassmann variables when integrals over complex Grassmann variables are evaluated. Using equations (1.68) and (1.69) shows that

$$\int d\zeta\, \zeta = \frac{1}{2}\int (d\eta_R + i\, d\eta_I)(\eta_R + i\,\eta_I) \\ = \frac{1}{2}\int d\eta_R\, \eta_R - \frac{1}{2}\int d\eta_I\, \eta_I = 0, \tag{1.78}$$

while

$$\int d\zeta\, \zeta^* = \frac{1}{2}\int (d\eta_R + i\, d\eta_I)(\eta_R - i\,\eta_I) \\ = \frac{1}{2}\int d\eta_R\, \eta_R + \frac{1}{2}\int d\eta_I\, \eta_I = 1. \tag{1.79}$$

Taking the complex conjugate of both sides of equation (1.79) and using $(d\zeta\, \zeta^*)^* = \zeta\, (d\zeta)^* = -\zeta\, d\zeta^* = d\zeta^*\, \zeta$ shows that the integral of $d\zeta^*\, \zeta$ is also one. For the case of complex Grassmann variables the equivalence between integration and differentiation is given by $\int d\zeta = (\partial/\partial\eta_R + i\, \partial/\partial\eta_I)/\sqrt{2} \Longleftrightarrow \partial/\partial\zeta^*$. This follows by using $\eta_R = (\zeta + \zeta^*)/\sqrt{2}$ and $\eta_I = i(\zeta^* - \zeta)/\sqrt{2}$ in the chain rule,

$$\frac{\partial}{\partial\zeta^*} = \frac{\partial\eta_R}{\partial\zeta^*}\frac{\partial}{\partial\eta_R} + \frac{\partial\eta_I}{\partial\zeta^*}\frac{\partial}{\partial\eta_I} = \frac{1}{\sqrt{2}}\left(\frac{\partial}{\partial\eta_R} + i\frac{\partial}{\partial\eta_I}\right) \Longleftrightarrow \int d\zeta. \tag{1.80}$$

A similar result shows $\int d\zeta^* \Longleftrightarrow \partial/\partial\zeta$. Once again, integration and differentiation give identical results. For example, expanding the exponential and using the complex Grassmann integrals gives

$$\int d\zeta^*\, d\zeta\, e^{a\zeta^*\zeta} = \frac{\partial}{\partial\zeta}\frac{\partial}{\partial\zeta^*}\, e^{a\zeta^*\zeta} = a. \tag{1.81}$$

The same result is obtained by scaling $\zeta \to \xi/\sqrt{a}$ and $\zeta^* \to \xi^*/\sqrt{a}$, and then using equation (1.72), which gives $d\zeta = d\xi\, \sqrt{a}$ and $d\zeta^* = d\xi^*\, \sqrt{a}$.

A very useful tool in dealing with Grassmann variables is the *Levi-Civita symbol*, denoted $\varepsilon^{abc\cdots}$. The Levi-Civita symbol can have an arbitrary number N of superscripts, but once N is specified each superscript belongs to the set of integers 1 through N. The Levi-Civita symbol is then defined as

$$\varepsilon^{abc\cdots} = \begin{cases} 1 & \text{if } abc\cdots \text{is an evenpermutation of 1 through } N, \\ -1 & \text{if } abc\cdots \text{is an odd permutation of 1 through } N, \\ 0 & \text{if } abc\cdots \text{has any repeated index.} \end{cases} \tag{1.82}$$

For example, ε^{ab} has the values $\varepsilon^{12} = -\varepsilon^{21} = 1$ and $\varepsilon^{11} = \varepsilon^{22} = 0$. Depending on the dimension of the Levi-Civita symbol, it will satisfy various identities, but one of the more important ones is given by

$$\sum_{a_1, a_2, \ldots = 1}^{n} \varepsilon^{a_1 \ldots a_n}\varepsilon^{a_1 \ldots a_n} = n!. \tag{1.83}$$

One of its important uses with Grassmann variables comes from ordering an arbitrary product of *all* the members of a collection of n Grassmann variables. It can be shown that

$$\eta_{a_1}\cdots\eta_{a_n} = \varepsilon^{a_1\cdots a_n}\eta_1\cdots\eta_n. \tag{1.84}$$

Result (1.84) is easy to see for small values of n. For $n = 2$ it follows that $\eta_1\eta_2 = \varepsilon^{12}\eta_1\eta_2$, $\eta_2\eta_1 = \varepsilon^{21}\eta_1\eta_2$, $\eta_1\eta_1 = \varepsilon^{11}\eta_1\eta_2$, and $\eta_2\eta_2 = \varepsilon^{22}\eta_1\eta_2$. Higher values of n follow similarly. Result (1.84) applies to any quantity with Grassmann properties, such as the partial derivatives associated with a collection of n Grassmann variables,

$$\frac{\partial^n}{\partial\eta_{a_1}\cdots\partial\eta_{a_n}} = \varepsilon^{a_1\cdots a_n}\frac{\partial^n}{\partial\eta_1\cdots\partial\eta_n}. \tag{1.85}$$

Result (1.85) is useful in evaluating integrals by using the correspondence $\int \mathrm{d}\zeta^* \Longleftrightarrow \partial/\partial\zeta$ for complex Grassmann variables.

The Dirac delta is quite simple for a single real Grassmann variable. It is given by the Grassmann equivalent of equation (1.25),

$$\delta(\eta - \eta') = \int \mathrm{d}\xi\ e^{\xi(\eta-\eta')} = \int \mathrm{d}\xi\ e^{\xi\eta}e^{-\xi\eta'} = \eta - \eta', \tag{1.86}$$

where ξ is a Grassmann variable. This is easily demonstrated by using the function $f(\eta) = a + b\eta$ and the results (1.68) and (1.69), so that

$$\int \mathrm{d}\eta\ \delta(\eta - \eta')f(\eta) = \int \mathrm{d}\eta\ (a\eta + a\eta' - b\eta'\eta) = a + b\,\eta' = f(\eta'). \tag{1.87}$$

Result (1.87) is valid only if the Dirac delta appears to the left of the function. It is straightforward to see that $\delta(\eta - \eta')f(\eta) = -f(\eta)\delta(\eta - \eta') = f(\eta)\delta(\eta' - \eta)$. In that regard, generalizing equation (1.86) to many complex Grassmann variables is beset by ordering issues and will not be pursued here.

Grassmann variables have been developed sufficiently to present Grassmann classical mechanics. For the case of a single real Grassmann degree of freedom, η, the Grassmann coordinate becomes time-dependent and is required to satisfy $\{\eta(t), \eta(t')\} = 0$ for all times. It is important to note that the Grassmann nature of η gives $\{\dot\eta, \eta\} = \dot\eta\eta + \eta\dot\eta = \mathrm{d}(\eta^2)/\mathrm{d}t = 0$, so that $\dot\eta$ is also a Grassmann variable. This prevents $\eta\dot\eta$ from being expressed as a total time derivative and also yields $\dot\eta^2 = 0$. In general, the action for Grassmann variables is assumed to take the usual integral form,

$$S[\eta, \dot\eta] = \int_{t_i}^{t_f} \mathrm{d}t\ \mathcal{L}(\eta, \dot\eta). \tag{1.88}$$

The equation of motion for the Grassmann system is then identified by the variational principle, which states that

$$\begin{aligned}\delta S[\eta, \dot{\eta}] &= S[\eta + \delta\eta, \dot{\eta} + \delta\dot{\eta}] - S[\eta, \dot{\eta}] \\ &= \int_{t_i}^{t_f} \mathrm{d}t \, \{\mathcal{L}(\eta + \delta\eta, \dot{\eta} + \delta\dot{\eta}) - \mathcal{L}(\eta, \dot{\eta})\} = 0.\end{aligned} \tag{1.89}$$

Expressing the variation of the action is complicated by the Grassmann algebra and the ordering chosen for the variables η and $\dot{\eta}$. In addition, the variations of the variables are also Grassmann variables, so that $\delta\eta$ and $\delta\dot{\eta}$ anticommute with all the other Grassmann variables. The Grassmann nature of the variables prevents expressing the variation in terms of the usual Taylor series expansion of the Lagrangian, since the choice of ordering will not necessarily correctly reproduce the variation. The functional derivative of the product of $\eta(t_1)\,\eta(t_2)$ demonstrates this problem. Since $\delta/\delta\eta(t)$ anticommutes with $\eta(t_1)$ and $\eta(t_2)$ it follows that

$$\begin{aligned}\frac{\delta}{\delta\eta(t)}(\eta(t_1)\,\eta(t_2)) &= \delta(t_1 - t)\,\eta(t_2) - \delta(t_2 - t)\,\eta(t_1) \\ &= -\frac{\delta}{\delta\eta(t)}(\eta(t_2)\,\eta(t_1)).\end{aligned} \tag{1.90}$$

To deal with this, the c-number ordering of (1.10) will be used as the convention with Grassmann variables. Introducing ρ as the momentum canonically conjugate to η, writing the Lagrangian as $\mathcal{L} = \dot{\eta}\rho - H(\rho, \eta)$ corresponds to

$$\rho = \frac{\partial\mathcal{L}}{\partial\dot{\eta}}. \tag{1.91}$$

The ability to use equation (1.91), the same definition as the c-number case, would disappear if the order of ρ and $\dot{\eta}$ were reversed. This ambiguity requires an ordering convention, and so all Lagrangians will be ordered with the time derivatives to the left to be consistent with equation (1.91). In analyzing equation (1.89) it is therefore understood that all terms $O(\delta\eta^2)$ and higher are dropped. The remaining variations are moved to the right while observing the Grassmann algebra, and the term containing $\delta\dot{\eta}$ is then integrated by parts. The remaining integrand must vanish, and this provides the equation of motion. The resulting equation of motion may or *may not be* consistent with the usual Euler–Lagrange equation (1.7).

An example clarifies this process. For a single Grassmann degree of freedom, the only quadratic terms that can be used involving $\dot{\eta}$ are $\eta\dot{\eta}$ and $\dot{\eta}\eta$. In the case of a c-number action, these terms would be a total time derivative and contribute nothing to the action when integrated. Noting that complex conjugation gives $(\eta\dot{\eta})^* = \dot{\eta}^*\eta^* = \dot{\eta}\eta = -\eta\dot{\eta}$, the only real c-number quadratic Lagrangian constructed from a single real Grassmann coordinate is given by

$$\mathcal{L}(\eta, \dot{\eta}) = -\frac{1}{2}i\dot{\eta}\eta, \tag{1.92}$$

where the factor of $\frac{1}{2}$ is for later convenience. Next, the action is varied similarly to the c-number case (1.3), using the infinitesimal Grassmann variable $\delta\eta$ that satisfies

$\delta\eta(t_f) = \delta\eta(t_i) = 0$. Ordering the variations to the right while observing the Grassmann algebra yields

$$\delta S[\eta, \dot{\eta}] = \int_{t_i}^{t_f} \mathrm{d}t \left(\frac{1}{2} i\eta\, \delta\dot{\eta} - \frac{1}{2} i\dot{\eta}\, \delta\eta\right) = 0. \tag{1.93}$$

After an integration by parts on the first term, the integrand becomes $-i\dot{\eta}\,\delta\eta$, which gives the equation of motion $\dot{\eta} = 0$. This simple equation is solved by a constant Grassmann variable. However, the important feature of this exercise is that the Grassmann equation of motion is a *first order* differential equation, as opposed to the second order differential equation given by the typical c-number action. The Lagrangian and equation (1.91) gives $\rho = -\frac{1}{2} i\eta$. The Grassmann Hamiltonian is found to be $H = \dot{\eta}\rho - \mathcal{L} = \dot{\eta}\rho - \dot{\eta}\rho = 0$. The Hamiltonian vanishes and is therefore trivially conserved.

A more interesting example involves two real Grassmann coordinates, η_1 and η_2. An extension of equation (1.92) is given by the Grassmann Lagrangian

$$\mathcal{L} = -\frac{1}{2} i\dot{\eta}_1\eta_1 - \frac{1}{2} i\dot{\eta}_2\eta_2 - \frac{1}{2} i\omega(\eta_1\eta_2 - \eta_2\eta_1), \tag{1.94}$$

where ω is a real c-number constant with the physical units of inverse time. Since $(\eta_1\eta_2)^* = \eta_2\eta_1$, the Lagrangian gives $\mathcal{L}^* = \mathcal{L}$. It is also important to note that the Lagrangian (1.94) is manifestly *antisymmetric* under the interchange $1 \leftrightarrow 2$ since $\mathcal{L} \to -\mathcal{L}$. Varying both η_1 and η_2 in the same manner as equation (1.89) and then integrating by parts yields two independent equations of motion from equation (1.94),

$$\delta S = \int_{t_i}^{t_f} \mathrm{d}t\ \{-i(\dot{\eta}_1 - \omega\eta_2)\delta\eta_1 - i(\dot{\eta}_2 + \omega\eta_1)\delta\eta_2\}, \tag{1.95}$$

so that the equations of motion are $\dot{\eta}_1 - \omega\eta_2 = 0$ and $\dot{\eta}_2 + \omega\eta_1 = 0$. Differentiating the first equation and using the second gives $\ddot{\eta}_1 = -\omega^2\, \eta_1$, which is the equation for a harmonic oscillator with frequency ω. The use of Grassmann variables has factorized the harmonic oscillator equation into two first order differential equations. A simple pair of solutions to the equations of motion of equation (1.95) is given by

$$\eta_1 = \alpha \cos(\omega t) + \beta \sin(\omega t), \qquad \eta_2 = -\alpha \sin(\omega t) + \beta \cos(\omega t), \tag{1.96}$$

where α and β are arbitrary constant real Grassmann variables. Using $\rho_1 = -\frac{1}{2} i\eta_1$ and $\rho_2 = -\frac{1}{2} i\eta_2$ gives the Hamiltonian $H = \rho_1\dot{\eta}_1 + \rho_2\dot{\eta}_2 - \mathcal{L} = \frac{1}{2} i\omega(\eta_1\eta_2 - \eta_2\eta_1)$, which becomes the constant $H = \frac{1}{2} i\omega(\alpha\beta - \beta\alpha)$ when evaluated using equation (1.96).

The oscillator Lagrangian (1.94) can expressed in terms of the complex Grassmann variable $\zeta = (\eta_1 + i\eta_2)/\sqrt{2}$. The complex Lagrangian is given by

$$\mathcal{L} = -\frac{1}{2}i\dot{\zeta}\zeta^{*} - \frac{1}{2}i\dot{\zeta}^{*}\zeta - \frac{1}{2}\omega(\zeta^{*}\zeta - \zeta\zeta^{*}), \tag{1.97}$$

as direct substitution and the Grassmann algebra verifies. The rules for the complex conjugation of Grassmann products give $(\dot{\zeta}\zeta^{*})^{*} = \zeta\dot{\zeta}^{*} = -\dot{\zeta}^{*}\zeta$, insuring $\mathcal{L}^{*} = \mathcal{L}$. The equations of motion for ζ and ζ^{*} are found identically to equation (1.95), yielding $-i\dot{\zeta} + \omega\zeta = 0$ and $i\dot{\zeta}^{*} + \omega\zeta^{*} = 0$. These are solved by $\zeta = \zeta_0 e^{-i\omega t}$ and $\zeta^{*} = \zeta_0^{*} e^{i\omega t}$, where ζ_0 is an arbitrary constant complex Grassmann variable. The conjugate momenta are $\rho = -\frac{1}{2}i\zeta^{*}$ and $\rho^{*} = -\frac{1}{2}i\zeta$, which give the Hamiltonian of equation (1.97) as

$$H = \dot{\zeta}\rho + \dot{\zeta}^{*}\rho^{*} - \mathcal{L} = \frac{1}{2}\omega(\zeta^{*}\zeta - \zeta\zeta^{*}) = \frac{1}{2}\omega(\zeta_0^{*}\zeta_0 - \zeta_0\zeta_0^{*}), \tag{1.98}$$

where the last step shows that the Hamiltonian is constant for the classical solution. The complex solutions are related to equation (1.96) since $\zeta_0 = (\eta_1(0) + i\eta_2(0))/\sqrt{2} = (\alpha + i\beta)/\sqrt{2}$. Using this form and its complex conjugate in the Hamiltonian gives $H = \frac{1}{2}\omega(\zeta_0^{*}\zeta_0 - \zeta_0\zeta_0^{*}) = \frac{1}{2}i\omega(\alpha\beta - \beta\alpha)$, which is the same result as before. Quantum aspects of Grassmann systems are developed in chapter 2.

1.5 Gaussian integrals

Gaussian integrations often occur in the evaluation of path integrals. In the case of a single real c-number variable x, they take the form

$$\begin{aligned} I(\alpha, J) &= \int_{-\infty}^{\infty} \mathrm{d}x\, e^{-\alpha x^2 + Jx} = e^{J^2/4\alpha} \int_{-\infty}^{\infty} \mathrm{d}x\, e^{-\alpha(x - J/2\alpha)^2} \\ &= e^{J^2/4\alpha} \int_{-\infty}^{\infty} \mathrm{d}y\, e^{-\alpha y^2} = e^{J^2/4\alpha} \sqrt{\frac{\pi}{\alpha}}, \end{aligned} \tag{1.99}$$

where α is a real c-number such that $\alpha > 0$. Adding the term Jx allows integrands of the form $x^n e^{-\alpha x^2 + Jx}$ to be evaluated easily, with their integrals simply given by $\partial^n I/\partial J^n$. The step in equation (1.99) where x undergoes the translation $x \to x - J/2$ is often referred to as *completing the square*. In evaluating path integrals it is often the case that both α and J are complex numbers. As a result, translating the real variable x by a complex number J would appear to change the limits of the integral. However, for the case that $J = i\beta$ is pure imaginary and the real part of α is positive, it follows from $\sin(-\beta x) = -\sin(\beta x)$ that

$$\int_{-\infty}^{\infty} \mathrm{d}x\, e^{-\alpha x^2} e^{i\beta x} = \int_{-\infty}^{\infty} \mathrm{d}x\, \cos(\beta x)\, e^{-\alpha x^2} = e^{-\beta^2/4\alpha} \sqrt{\frac{\pi}{\alpha}}. \tag{1.100}$$

This is the same result as completing the square. In fact, the form obtained by completing the square is still valid even if J and α are complex numbers as long as the real part of α is not negative. For example, if $\alpha = i\beta$ is pure imaginary, so that β is real, complex variable analysis can be used to prove *Fresnel's theorem*,

$$I(i\beta, 0) = \int_{-\infty}^{\infty} \mathrm{d}x \exp(-i\beta x^2) = \sqrt{\frac{\pi}{i\beta}}. \tag{1.101}$$

which shows that equation (1.99) is still valid for that case. Result (1.99) is singular at $\alpha = 0$ for all cases. It is useful to note that in the pure imaginary case (1.101) there is a discontinuity in I, since equation (1.99) shows that $\lim_{\beta\to 0}(I(i\beta, 0) - I(-i\beta, 0)) = -i\sqrt{2\pi/\beta}$.

Analyzing the multivariable generalization of equation (1.99) requires the reader to be familiar with basic matrix techniques. In particular, the determinant of an $n \times n$ matrix $\mathbf{A}$ is defined in terms of the Levi-Civita symbol (1.82) with n indices,

$$\begin{aligned} \det \mathbf{A} &= \sum_{a_1,\ldots,a_n=1}^{n} \varepsilon^{a_1\ldots a_n} A_{1a_1} \cdots A_{na_n} \\ &= \frac{1}{n!} \sum_{a_1,b_1,\ldots,a_n,b_n=1}^{n} \varepsilon^{a_1\ldots a_n} \varepsilon^{b_1\ldots b_n} A_{a_1b_1} \cdots A_{a_nb_n}, \end{aligned} \tag{1.102}$$

where A_{jk} is the element of the matrix $\mathbf{A}$ at the jth row and kth column. The determinant has the properties that $\det(\mathbf{A} \cdot \mathbf{B}) = \det \mathbf{A} \det \mathbf{B}$, $\det(\mathbf{A}^{-1}) = 1/\det \mathbf{A}$, and $\det \mathbf{I} = 1$, where $\mathbf{I}$ is the $n \times n$ unit matrix whose elements are $I_{jk} = \delta_{jk}$. If $\mathbf{A}^T$ is the transpose of $\mathbf{A}$, so that its elements are $A^T_{jk} = A_{kj}$, then $\det \mathbf{A}^T = \det \mathbf{A}$.

An important aspect of analyzing the multivariable Gaussian integral is the *Jacobian* that occurs with a change of integration variables. Assuming that the initial variables are Cartesian c-numbers, the n-dimensional volume element is given by $\mathrm{d}^n\boldsymbol{x} = \mathrm{d}x_1 \cdots \mathrm{d}x_n$. A change to new variables y_i is defined by a set of n functions, so that $x_i = x_i(y_1, \ldots, y_n)$. The n-dimensional volume element in terms of the new variables is given by a result that occurs constantly in the analysis of path integrals,

$$\mathrm{d}^n\boldsymbol{x} = \mathrm{d}^n\mathbf{y}\, J(y_1, \ldots, y_n), \qquad J = \det\left(\frac{\partial \boldsymbol{x}}{\partial \boldsymbol{y}}\right). \tag{1.103}$$

The factor J is known as the *Jacobian* of the variable change and is the determinant of the $n \times n$ matrix $\partial\boldsymbol{x}/\partial\boldsymbol{y}$, whose jk element is given by $\partial x_j/\partial y_k$. For example, the change from two-dimensional Cartesian coordinates to polar coordinates, $x_1 = r\cos\theta$ and $x_2 = r\sin\theta$, with $y_1 = r$ and $y_2 = \theta$, gives $\partial x_1/\partial y_1 = \cos\theta$, $\partial x_2/\partial y_1 = \sin\theta$, $\partial x_1/\partial y_2 = -r\sin\theta$, and $\partial x_2/\partial y_2 = r\cos\theta$. This gives the Jacobian

$$J = \det\begin{pmatrix} \cos\theta & -r\sin\theta \\ \sin\theta & r\cos\theta \end{pmatrix} = r(\cos^2\theta + \sin^2\theta) = r, \tag{1.104}$$

so that the area element in polar coordinates is given by the familiar elementary result, $\mathrm{d}^2\boldsymbol{x} = J\, dr\, \mathrm{d}\theta = r\, dr\, \mathrm{d}\theta$. If the Jacobian is a negative real or imaginary number, then the minus sign is dropped so that $J^* = J$.

The extension of equation (1.99) to n real c-number variables x_j is given by

$$I_n = \int_{-\infty}^{\infty} \prod_{i=1}^{n} \mathrm{d}x_i \exp\left(-\alpha \sum_{j,k=1}^{n} x_j M_{jk} x_k + \sum_{k=1}^{n} J_k x_k\right). \tag{1.105}$$

The coefficients M_{jk} can be treated as the elements of a real c-number $n \times n$ matrix, denoted $\mathbf{M}$. The x_k and J_k can be treated as the components of n-dimensional vectors $\boldsymbol{x}$ and $\boldsymbol{J}$ or, equivalently, $n \times 1$ matrices, so that $\boldsymbol{x}^T$ is a $1 \times n$ matrix. Finally, α is a complex number with a real part that is positive. In matrix notation the exponential in equation (1.105) can be written $\exp(-\alpha\, \boldsymbol{x}^T \cdot \mathbf{M} \cdot \boldsymbol{x} + \boldsymbol{J}^T \cdot \boldsymbol{x})$, where $\boldsymbol{x}^T \cdot \mathbf{M}$ has the components $(\boldsymbol{x}^T \cdot \mathbf{M})_k = \sum_{j=1}^{n} x_j M_{jk}$. The square matrix $\mathbf{M}$ can be expressed as the sum of a symmetric and an antisymmetric matrix, $\mathbf{M} = \frac{1}{2}(\mathbf{M} + \mathbf{M}^T) + \frac{1}{2}(\mathbf{M} - \mathbf{M}^T) \equiv \mathbf{S} + \mathbf{A}$, which obey $\mathbf{S}^T = \mathbf{S}$ and $\mathbf{A}^T = -\mathbf{A}$. Since the x_j commute, the antisymmetric part of $\mathbf{M}$ does not contribute to equation (1.105). This follows from the fact that $\boldsymbol{x}^T \cdot \mathbf{A} \cdot \boldsymbol{x}$ is a 1×1 matrix and must therefore equal its transpose. This gives

$$\boldsymbol{x}^T \cdot \mathbf{A} \cdot \boldsymbol{x} = (\boldsymbol{x}^T \cdot \mathbf{A} \cdot \boldsymbol{x})^T = \boldsymbol{x}^T \cdot \mathbf{A}^T \cdot \boldsymbol{x} = -\,\boldsymbol{x}^T \cdot \mathbf{A} \cdot \boldsymbol{x}. \tag{1.106}$$

Therefore, the antisymmetric part contributes nothing, and the matrix $\mathbf{M}$ can be treated as symmetric, satisfying $\mathbf{M}^T = \mathbf{M}$. It will be *assumed* that $M_{jk} = M_{jk}^*$. These two conditions mean the matrix $\mathbf{M}$ is Hermitian since it satisfies $\mathbf{M}^\dagger = \mathbf{M}$. The inverse of $\mathbf{M}$, denoted $\mathbf{M}^{-1}$, exists if $\det \mathbf{M} \neq 0$, and will also be symmetric, $(\mathbf{M}^{-1})^T = \mathbf{M}^{-1}$.

The Hermitian matrix $\mathbf{M}$ has n real eigenvalues, denoted $\lambda^{(j)}$, $j = 1, \ldots, n$, and each eigenvalue is associated with an eigenvector $\boldsymbol{e}^{(j)}$. When paired, these solve the *eigenvalue problem* $\mathbf{M} \cdot \boldsymbol{e}^{(j)} = \lambda^{(j)} \boldsymbol{e}^{(j)}$. The n eigenvalues $\lambda^{(j)}$ are found by solving the nth order polynomial generated by $\det(\mathbf{M} - \lambda^{(j)}\mathbf{I}) = 0$. Once the eigenvalues are known, the eigenvectors $\boldsymbol{e}^{(j)}$ are found by returning the $\lambda^{(j)}$ to the eigenvalue problem and solving for the $\boldsymbol{e}^{(j)}$. For a Hermitian matrix it can be shown that the eigenvectors associated with each eigenvalue are orthogonal, so that $\boldsymbol{e}^{(j)T} \cdot \boldsymbol{e}^{(k)} = 0$ if $j \neq k$. As a result, the eigenvectors can be chosen to satisfy the orthonormalization condition $\boldsymbol{e}^{(j)T} \cdot \boldsymbol{e}^{(k)} = \delta_{jk}$. Using the orthonormal eigenvectors, an $n \times n$ orthogonal matrix $\mathbf{U}$ can be constructed whose elements are $U_{jk} = e_k^{(j)}$. Because the eigenvectors are orthonormal, this matrix is orthogonal since its transpose $\mathbf{U}^T$ satisfies $\mathbf{U} \cdot \mathbf{U}^T = \mathbf{U}^T \cdot \mathbf{U} = \mathbf{I}$, so that $\mathbf{U}^T = \mathbf{U}^{-1}$. By virtue of using the eigenvectors, it has the property that $\mathbf{U} \cdot \mathbf{M} \cdot \mathbf{U}^T \equiv \mathbf{D}$ is a diagonal matrix whose diagonal elements are the eigenvalues of the matrix $\mathbf{M}$, $D_{jk} = \lambda^{(j)}\delta_{jk}$. The diagonal matrix $\mathbf{D}$ has the property that $\det \mathbf{D} = \prod_{j=1}^{n} \lambda^{(j)}$. Orthogonality immediately gives $(\det \mathbf{U})^2 = \det \mathbf{U} \det \mathbf{U}^T = \det(\mathbf{U} \cdot \mathbf{U}^T) = \det \mathbf{I} = 1$, so that $\det \mathbf{U} = \pm 1$. The negative case can be excluded since it corresponds to an inversion. In turn, this result gives

$$\det \mathbf{M} = \det(\mathbf{U}^T \cdot \mathbf{U} \cdot \mathbf{M}) = \det(\mathbf{U} \cdot \mathbf{M} \cdot \mathbf{U}^T) = \det \mathbf{D} = \prod_{j=1}^{n} \lambda^{(j)}, \tag{1.107}$$

so that the determinant of a Hermitian matrix is the product of its n eigenvalues.

Evaluating the generalized Gaussian integral (1.105) begins by completing the square, as in equation (1.99), and then using the translational invariance of the integrals. In matrix notation, this is accomplished by translating the variables of integration according to $\boldsymbol{x} \to \boldsymbol{x} - \frac{1}{2}\alpha^{-1}\mathbf{M}^{-1} \cdot \boldsymbol{J}$, so that $\boldsymbol{x}^T \to \boldsymbol{x}^T - \frac{1}{2}\alpha^{-1}\boldsymbol{J}^T \cdot \mathbf{M}^{-1}$. This simplifies equation (1.105) to

$$I_n = \exp\left(\frac{1}{4\alpha}\boldsymbol{J}^T \cdot \mathbf{M}^{-1} \cdot \boldsymbol{J}\right) \int \mathrm{d}^n\boldsymbol{x} \exp\left(-\alpha\, \boldsymbol{x}^T \cdot \mathbf{M} \cdot \boldsymbol{x}\right). \tag{1.108}$$

The remaining integral in equation (1.108) is evaluated by using the matrix $\mathbf{U}$ to define the new coordinates y_i by $\boldsymbol{x} = \mathbf{U} \cdot \boldsymbol{y}$. For this transformation the Jacobian (1.103) is given by $J = \det \mathbf{U} = 1$, while $\boldsymbol{x}^T \cdot \mathbf{M} \cdot \boldsymbol{x} = \boldsymbol{y}^T \cdot \mathbf{U}^T \cdot \mathbf{M} \cdot \mathbf{U} \cdot \boldsymbol{y} = \boldsymbol{y}^T \cdot \mathbf{D} \cdot \boldsymbol{y}$. Because $\mathbf{D}$ has the elements $D_{jk} = \lambda^{(j)}\delta_{jk}$, this gives $\boldsymbol{y}^T \cdot \mathbf{D} \cdot \boldsymbol{y} = \sum_{j,\,k=1}^{n} y_j \lambda^{(j)} \delta_{jk} y_k = \sum_{j=1}^{n} \lambda^{(j)} y_j^2$. Using these coordinates causes the remaining integral in equation (1.108) to become the product of n copies of equation (1.99), with the jth integral in the product contributing

$$P_j = \int_{-\infty}^{\infty} \mathrm{d}y_j \exp\left(-\alpha\lambda^{(j)} y_j^2\right) = \sqrt{\frac{\pi}{\alpha\lambda^{(j)}}}. \tag{1.109}$$

Result (1.109) assumes that all the eigenvalues of $\mathbf{M}$ are such that the real part of $\alpha\lambda^{(j)}$ is positive. If there are eigenvalues of $\mathbf{M}$ for which this condition does not hold, then the integral I_n is *undefined.* If α is a positive real number, this failure occurs if any of the eigenvalues are zero or negative. Inserting the product of the n factors P_j into equation (1.108) and using equation (1.107) yields the final result,

$$I_n = e^{\boldsymbol{J}^T \cdot \mathbf{M}^{-1} \cdot \boldsymbol{J}/4\alpha} \prod_{j=1}^{n} \sqrt{\frac{\pi}{\alpha\lambda^{(j)}}} = e^{\boldsymbol{J}^T \cdot \mathbf{M}^{-1} \cdot \boldsymbol{J}/4\alpha} \sqrt{\frac{\pi^n}{\alpha^n \det \mathbf{M}}}. \tag{1.110}$$

If $n = 1$, then $\mathbf{M} = 1$, $\mathbf{M}^{-1} = 1$, and $\det \mathbf{M} = 1$, so that equation (1.110) reduces to equation (1.99).

Another important Gaussian integral involves integration over complex variables, denoted λ. The simplest form for this integral is given by

$$I_c = \int_{-\infty}^{\infty} \mathrm{d}\lambda^* \, \mathrm{d}\lambda \exp\left(-\alpha\lambda^*\lambda + J^*\lambda + J\lambda^*\right). \tag{1.111}$$

where α is a complex number whose real part is positive. The integral (1.111) is understood in the following way. The variable λ lies in the complex plane and the integration $\mathrm{d}^2\lambda = \mathrm{d}\lambda^* \, \mathrm{d}\lambda$ covers the *entire complex plane*. There are several ways to parameterize this integration. The first is to write $\lambda = \lambda_R + i\lambda_I$ and $\mathrm{d}^2\lambda = \mathrm{d}\lambda_R \, \mathrm{d}\lambda_I$, where the limits on both λ_R and λ_I are $(-\infty, \infty)$. The second is to write $\lambda = re^{i\theta}$ and $\mathrm{d}^2\lambda = r \, \mathrm{d}r \, \mathrm{d}\theta$, where the limits on r are $(0, \infty)$ and the limits on θ are $(0, 2\pi)$. Using the first parameterization gives

$$I_c = \int_{-\infty}^{\infty} \mathrm{d}\lambda_R \; e^{-\alpha\lambda_R^2+\lambda_R(J+J^*)} \int_{-\infty}^{\infty} \mathrm{d}\lambda_I \; e^{-\alpha\lambda_I^2+i\lambda_I(J^*-J)}, \tag{1.112}$$

which is the product of two copies of equation (1.99), so that

$$I_c = \frac{\pi}{\alpha}\, e^{J^*J/\alpha}. \tag{1.113}$$

The second parameterization is used in equation (2.80) to demonstrate coherent state completeness. The extension of equation (1.111) to n complex variables λ_j with a *symmetric invertible matrix* $\mathbf{M}$ reduces to the product of two copies of equation (1.110), so that

$$\begin{aligned} I_{cn} &= \int \prod_{j=1}^{n} \mathrm{d}^2\lambda_j \exp\left(-\alpha \sum_{j,k=1}^{n} \lambda_j^* M_{jk} \lambda_k + \sum_{k=1}^{n} (J_k^* \lambda_k + J_k \lambda_k^*)\right) \\ &= \frac{\pi^n}{\alpha^n \det \mathbf{M}} \exp\left(\frac{1}{\alpha} \mathbf{J}^* {\cdot} \mathbf{M}^{-1} \cdot \mathbf{J}\right) \end{aligned} \tag{1.114}$$

where $\mathbf{J}^*{\cdot}\mathbf{M}^{-1} \cdot \mathbf{J}$ stands for the matrix product.

Gaussian Grassmann integrals are approached similarly. However, just as equation (1.72) gave the inverse of the c-number result, it will now be shown that a change of coordinates for a Grassmann multivariable integral is accompanied by the appearance of the inverse Jacobian. The form of the Grassmann Jacobian J_G follows from demanding invariance of the multivariable version of equation (1.68) under a linear change of coordinates from $\boldsymbol{\eta}$ to $\boldsymbol{\theta}$, where $\eta_j = \sum_{k=1}^{n} M_{jk}\theta_k$. The c-number coefficients M_{jk} form the elements of a matrix $\mathbf{M}$, so that the coordinate transformation can be written $\boldsymbol{\eta} = \mathbf{M} \cdot \boldsymbol{\theta}$ or $\boldsymbol{\theta} = \mathbf{M}^{-1} \cdot \boldsymbol{\eta}$. The Jacobian J_G of this transformation, $\mathrm{d}^n\boldsymbol{\eta} = \mathrm{d}^n\boldsymbol{\theta}\, J_G$, is determined by demanding the new Grassmann integral coincides with the old Grassmann integral when $\eta_j = \sum_{k=1}^{n} M_{jk}\, \theta_k$ is inserted. Using property (1.84) and definition (1.102) gives

$$\begin{aligned} \int \mathrm{d}\eta_1 \cdots \mathrm{d}\eta_n \, \eta_1 \cdots \eta_n &= \int \mathrm{d}\theta_1 \cdots \mathrm{d}\theta_n \, J_G \, \eta_1(\boldsymbol{\theta}) \cdots \eta_n(\boldsymbol{\theta}) \\ &= \int \mathrm{d}\theta_1 \cdots \mathrm{d}\theta_n \, J_G \sum_{a_1,\ldots,a_n=1}^{n} M_{1a_1} \cdots M_{na_n} \theta_{a_1} \cdots \theta_{a_n} \\ &= \int \mathrm{d}\theta_1 \cdots \mathrm{d}\theta_n \, \theta_1 \cdots \theta_n \, J_G \sum_{a_1,\ldots,a_n=1}^{n} \varepsilon^{a_1\cdots a_n} M_{1a_1} \cdots M_{na_n} \\ &= \int \mathrm{d}\theta_1 \cdots \mathrm{d}\theta_n \, \theta_1 \cdots \theta_n \, J_G \det \mathbf{M}. \end{aligned} \tag{1.115}$$

Result (1.115) immediately gives

$$J_G = \frac{1}{\det \mathbf{M}} = \det(\mathbf{M}^{-1}) = \det \frac{\partial \boldsymbol{\theta}}{\partial \boldsymbol{\eta}}. \tag{1.116}$$

The Grassmann Jacobian (1.116) is therefore the inverse of the c-number Jacobian of equation (1.103), reproducing the earlier result (1.72) for scaling a single variable. Although equation (1.116) was derived for a linear change of variables, it can be generalized to the case that the transformation includes odd powers of the new Grassmann variables.

The simplest real Grassmann variable Gaussian is defined by introducing two Grassmann variables of integration, η_1 and η_2, and two additional Grassmann variables, K_1 and K_2, that play the role of J in equation (1.99), to give

$$I = \int \mathrm{d}\eta_1 \, \mathrm{d}\eta_2 \, e^{-\alpha\eta_1\eta_2 + K_1\eta_1 + K_2\eta_2}, \tag{1.117}$$

where α is an arbitrary complex number. It is important to note that $\eta_1^2 = \eta_2^2 = 0$ prevents a nontrivial Gaussian involving a single Grassmann variable, and instead two real Grassmann variables are required. It is instructive to evaluate the integral (1.117) by first performing the Grassmann equivalent of completing the square and then using equation (1.62) to factorize the result,

$$I = e^{K_1K_2/\alpha} \int \mathrm{d}\eta_1 \, \mathrm{d}\eta_2 \, e^{-\alpha(\eta_1 - K_2/\alpha)(\eta_2 + K_1/\alpha)}. \tag{1.118}$$

The sign differences in equation (1.118) result from the Grassmann nature of the variables. Coupling the translation invariance property (1.70) with result (1.71) gives

$$I = e^{K_1K_2/\alpha} \int \mathrm{d}\eta_1' \, \mathrm{d}\eta_2' \, e^{-\alpha\eta_1'\eta_2'} = \alpha \, e^{K_1K_2/\alpha} = \alpha + K_1K_2. \tag{1.119}$$

In the process of generalizing equation (1.117) it is useful to note that the argument of the exponential will always behave as a c-number since it is quadratic in Grassmann variables. If an odd number of Grassmann variables are being integrated, this requires the integral to yield a Grassmann number, and the only possibility is zero. As a result, a nontrivial generalization of equation (1.117) requires an even number of real Grassmann variables, but takes a form similar to equation (1.105),

$$\begin{aligned} I_G &= \int \mathrm{d}\eta_1 \cdots \mathrm{d}\eta_n \exp\left(-\alpha \sum_{j,k=1}^{n} \eta_j M_{jk} \eta_k + \sum_{j=1}^{n} K_j \eta_j \right) \\ &= \int \mathrm{d}^n\boldsymbol{\eta} \exp\left(-\alpha \, \boldsymbol{\eta}^T \cdot \mathbf{M} \cdot \boldsymbol{\eta} + \boldsymbol{K}^T \cdot \boldsymbol{\eta}\right). \end{aligned} \tag{1.120}$$

where n is an even number, the M_{jk} are assumed to be real c-number elements of the matrix $\mathbf{M}$, and the η_j and K_j are the components of the Grassmann vectors $\boldsymbol{\eta}$ and $\boldsymbol{K}$. Using an argument similar to equation (1.106), the Grassmann nature of the variables shows that only the antisymmetric part of $\mathbf{M}$ contributes to equation (1.120).

Evaluating equation (1.120) begins by completing the square, written in matrix notation as $\boldsymbol{\eta} \to \boldsymbol{\eta} + \frac{1}{2}\alpha^{-1}\mathbf{M}^{-1} \cdot \boldsymbol{K}$. This is another indication that n must be even, since $\det \mathbf{M} = 0$ for an $n \times n$ antisymmetric matrix if n is odd, a fact easily proven using the definition of the determinant (1.102). This would prevent the existence of $\mathbf{M}^{-1}$, causing equation (1.120) to be undefined. Because $\mathbf{M}$ is antisymmetric, $\mathbf{M}^{-1}$ is also antisymmetric. This means that $\boldsymbol{\eta}^T \to \boldsymbol{\eta}^T - \frac{1}{2}\alpha^{-1}\boldsymbol{K}^T \cdot \mathbf{M}^{-1}$. Since the Grassmann integrals are invariant under translation, the integral (1.120) becomes

$$I_G = \exp\left(-\frac{1}{4\alpha}\boldsymbol{K}^T \cdot \mathbf{M}^{-1} \cdot \boldsymbol{K}\right)\int \mathrm{d}^n\boldsymbol{\eta}\, \exp\left(-\alpha\, \boldsymbol{\eta}^T \cdot \mathbf{M} \cdot \boldsymbol{\eta}\right). \tag{1.121}$$

The remaining Grassmann integral in equation (1.121), denoted G, can be evaluated by first squaring it. Since the arguments of the exponentials are quadratic, they commute with the Grassmann volume element $\mathrm{d}^n\boldsymbol{\xi}$ and each other. This gives

$$G^2 = \int \mathrm{d}^n\boldsymbol{\eta}\, \mathrm{d}^n\boldsymbol{\xi}\, \exp\left\{-\alpha(\boldsymbol{\eta}^T \cdot \mathbf{M} \cdot \boldsymbol{\eta} + \boldsymbol{\xi}^T \cdot \mathbf{M} \cdot \boldsymbol{\xi})\right\}. \tag{1.122}$$

Introducing n complex Grassmann variables $\boldsymbol{\zeta} = (\boldsymbol{\eta} + i\boldsymbol{\xi})/\sqrt{2}$ allows G^2 to be written

$$G^2 = \int \mathrm{d}\zeta_1^*\, \mathrm{d}\zeta_1 \cdots \mathrm{d}\zeta_n^*\, \mathrm{d}\zeta_n\, J_G \exp\left\{2i\alpha\, \boldsymbol{\zeta}^{*T} \cdot (i\mathbf{M}) \cdot \boldsymbol{\zeta}\right\}. \tag{1.123}$$

A key step in demonstrating equation (1.123) is the identity $\boldsymbol{\eta}^T \cdot \mathbf{M} \cdot \boldsymbol{\xi} - \boldsymbol{\xi}^T \cdot \mathbf{M} \cdot \boldsymbol{\eta} = 0$, which holds since $\boldsymbol{\eta}$ and $\boldsymbol{\xi}$ anticommute and $\mathbf{M}$ is antisymmetric, so that $\boldsymbol{\eta}^T \cdot \boldsymbol{\xi} = -\boldsymbol{\xi}^T \cdot \boldsymbol{\eta}$ and $\mathbf{M}^T = -\mathbf{M}$. The Jacobian for the variable change can be found by first arranging the differentials as $\mathrm{d}\eta_1 \cdots \mathrm{d}\eta_n\, \mathrm{d}\xi_1 \cdots \mathrm{d}\xi_n = (-1)^{n/2}\mathrm{d}\eta_1\, \mathrm{d}\xi_1 \cdots \mathrm{d}\eta_n\, \mathrm{d}\xi_n$, where the fact that n must be an even number was used. This gives n factors of the Jacobian for each individual change to complex variables $\mathrm{d}\eta_j\, \mathrm{d}\xi_j = \mathrm{d}\zeta_j\, \mathrm{d}\zeta_j^*\, J$. These n Jacobians are identical and easily computed to be

$$J = \det\begin{pmatrix} 1/\sqrt{2} & i/\sqrt{2} \\ 1/\sqrt{2} & -i/\sqrt{2} \end{pmatrix} = -i, \tag{1.124}$$

so that the total Jacobian factor in equation (1.122) is $J_G = (-1)^{n/2}J^n = i^n(-i)^n = 1$. By virtue of the matrix $\mathbf{M}$ being antisymmetric and real, the matrix $\mathbf{H} = i\mathbf{M}$ satisfies $\mathbf{H}^\dagger = -i\mathbf{M}^\dagger = -i\mathbf{M}^T = i\mathbf{M} = \mathbf{H}$ and is therefore Hermitian. As a result, the same technique used to obtain equation (1.110) can be employed to evaluate equation (1.123). The matrix of eigenvectors $\mathbf{U}$ is used to define the new complex Grassmann variables $\boldsymbol{\zeta} = \mathbf{U} \cdot \boldsymbol{\theta}$, where $\det \mathbf{U} = 1$. If the $\lambda^{(j)}$ are the n eigenvalues of $\mathbf{H}$, then equation (1.123) becomes

$$\begin{aligned} G^2 &= \prod_{j=1}^{n} \int \mathrm{d}\theta_j \mathrm{d}\theta_j^* e^{2i\alpha\lambda^{(j)}\theta_j^*\theta_j} = \prod_{j=1}^{n} (-2i\alpha\lambda^{(j)}) = (-2i\alpha)^n \prod_{j=1}^{n} \lambda^{(j)} \\ &= (-2i\alpha)^n \det \mathbf{H} = (-2i\alpha)^n i^n \det \mathbf{M} = (2\alpha)^n \det \mathbf{M}. \end{aligned} \tag{1.125}$$

Returning the square root of equation (1.125) to (1.121) shows that equation (1.120) is given by

$$I_G = \exp\left(-\frac{1}{4\alpha}\boldsymbol{K}^T \cdot \mathbf{M}^{-1} \cdot \boldsymbol{K}\right)(2\alpha)^{n/2}\sqrt{\det \mathbf{M}}. \tag{1.126}$$

The square root of the determinant is sometimes referred to as the *Pfaffian* of the associated matrix. Choosing $n = 2$ and $M_{ab} = \frac{1}{2}\varepsilon^{ab}$, so that $M_{ab}^{-1} = -2\varepsilon^{ab}$ and $\det \mathbf{M} = 1/4$, immediately reproduces result (1.119) obtained by evaluating equation (1.117).

The methods used to obtain result (1.125) can be adapted to evaluate a variant of the Grassmann Gaussian integral involving a c-number $n \times n$ matrix $\mathbf{M}$ that is assumed to be invertible. The variables of integration are two sets of n real Grassmann variables, ζ and η, coupled to two arbitrary Grassmann n-vectors, K and J,

$$I_c = \int d\zeta_1\, d\eta_1 \cdots d\zeta_n\, d\eta_n \exp\{-\zeta^T \cdot \mathbf{M} \cdot \eta + K^T \cdot \eta + \zeta^T \cdot J\}, \tag{1.127}$$

As before, the variables of integration are translated to $\zeta_j \to \zeta_j + \sum_{k=1}^{n} K_k M_{kj}^{-1}$ and $\eta_j \to \eta_j + \sum_{k=1}^{n} M_{jk}^{-1} J_k$. The integral (1.127) becomes

$$I_c = \exp\{K^T \cdot \mathbf{M}^{-1} \cdot J\} \int d\zeta_1\, d\eta_1 \cdots d\zeta_n\, d\eta_n \exp\{-\zeta^T \cdot \mathbf{M} \cdot \eta\}. \tag{1.128}$$

The remaining Grassmann integral can be evaluated by the change of coordinates $\eta = \mathbf{U} \cdot \eta'$ and $\zeta^T = \zeta^{T'} \cdot \mathbf{U}^T$, where the orthogonal matrix $\mathbf{U}$ is chosen to place $\boldsymbol{M}$ into diagonal form with its eigenvalues as entries. Since $\det(\mathbf{U} \cdot \mathbf{U}^T) = 1$ the measure of equation (1.128) picks up no additional factors. Dropping the primes and denoting the n eigenvalues of $\mathbf{M}$ as $\lambda^{(k)}$, the rules of Grassmann integration give

$$\begin{aligned} I_c &= \exp\{K^T \cdot \mathbf{M}^{-1} \cdot J\} \int d\zeta_1\, d\eta_1 \cdots d\zeta_n d\eta_n \exp\left\{\sum_{k=1}^{n} \lambda^{(k)} \eta_k \zeta_k\right\} \\ &= \exp\{K^T \cdot \mathbf{M}^{-1} \cdot J\} \prod_{k=1}^{n} \int d\zeta_k\, d\eta_k \exp\{\lambda^{(k)} \eta_k \zeta_k\} \\ &= \exp\{K^T \cdot \mathbf{M}^{-1} \cdot J\} \prod_{k=1}^{n} \lambda^{(k)} = \exp\{K^T \cdot \mathbf{M}^{-1} \cdot J\} \det \mathbf{M} \end{aligned} \tag{1.129}$$

where the identity $\prod_{k=1}^{n} \lambda^{(k)} = \det \mathbf{M}$ was once again used.

1.6 Poisson resummation

Quantum mechanical systems often give rise to functions $F(x)$ of the form

$$F(x) = \sum_{n=-\infty}^{\infty} f(x + nL), \tag{1.130}$$

where L is a constant length. The function $f(x)$ is assumed to be piecewise continuous, so that it possesses a Fourier series or transform. The function $F(x)$

is periodic with period L, so that $F(x + Lj) = F(x)$, where j is an arbitrary integer. This follows from the form of $F(x)$, since

$$F(x + Lj) = \sum_{n=-\infty}^{\infty} f(x + L(n + j)) = \sum_{k=-\infty}^{\infty} f(x + Lk) = F(x). \tag{1.131}$$

The key step follows from defining the new integer $k = n + j$ and noting that its range is the same as the original integer n.

Because $F(x)$ is periodic with period L it follows that it has a Fourier series representation like that of equation (1.20),

$$F(x) = \sum_{n=-\infty}^{\infty} a_n \exp\left(\frac{i2\pi nx}{L}\right), \tag{1.132}$$

where the Fourier coefficients are given by

$$a_n = \frac{1}{L}\int_{-\frac{1}{2}L}^{\frac{1}{2}L} \mathrm{d}y\ F(y)\exp\left(-\frac{i2\pi ny}{L}\right). \tag{1.133}$$

Inserting the definition of $F(y)$ into equation (1.133) and interchanging sum and integration, the Fourier coefficient becomes

$$a_n = \frac{1}{L}\sum_{j=-\infty}^{\infty}\int_{-\frac{1}{2}L}^{\frac{1}{2}L} \mathrm{d}y\, f(y + Lj)\exp\left(-\frac{i2\pi ny}{L}\right). \tag{1.134}$$

The range on the integration is important, since it can be now be combined with the sum. For each integer j, the variable of integration can be changed to $z = y + Lj$, so that $\mathrm{d}y = \mathrm{d}z$. The range of integration for this change becomes $L(j - \frac{1}{2})$ to $L(j + \frac{1}{2})$, while the integrand becomes

$$\begin{aligned} f(y + Lj)\exp\left(-\frac{i2\pi ny}{L}\right) &= f(z)\exp\left(-\frac{i2\pi n(z - Lj)}{L}\right) \\ &= f(z)\exp\left(-\frac{i2\pi nz}{L}\right), \end{aligned} \tag{1.135}$$

which follows from $e^{i2\pi nj} = 1$ by virtue of nj being an integer. The expression for the Fourier coefficient now becomes

$$a_n = \frac{1}{L}\sum_{j=-\infty}^{\infty}\int_{(j-\frac{1}{2})L}^{(j+\frac{1}{2})L} \mathrm{d}z\, f(z)\exp\left(-\frac{i2\pi nz}{L}\right). \tag{1.136}$$

Because the integrand is identical for all the integrals, the sum of integrals becomes the integral over the sum of ranges,

$$\sum_{j=-\infty}^{\infty}\int_{(j-\frac{1}{2})L}^{(j+\frac{1}{2})L} \to \int_{-\infty}^{\infty}, \tag{1.137}$$

with the result that the Fourier coefficient is given by

$$a_n = \frac{1}{L}\int_{-\infty}^{\infty} \mathrm{d}z\, f(z)\exp\left(-\frac{i2\pi nz}{L}\right). \tag{1.138}$$

The final step is to return this result to the original Fourier series, yielding

$$\sum_{n=-\infty}^{\infty} f(x+Ln) = \frac{1}{L}\sum_{n=-\infty}^{\infty}\int_{-\infty}^{\infty} \mathrm{d}z\, f(z)\exp\left(\frac{i2\pi n(x-z)}{L}\right). \tag{1.139}$$

Result (1.139) is known as the *Poisson resummation theorem.*

It is worth noting that the property of a summation,

$$\sum_{n=-\infty}^{\infty} f(n) = \sum_{n=\infty}^{-\infty} f(n) = \sum_{n=-\infty}^{\infty} f(-n), \tag{1.140}$$

allows the Poisson resummation formula to be written

$$\sum_{n=-\infty}^{\infty} f(x+Ln) = \frac{1}{L}\sum_{n=-\infty}^{\infty}\int_{-\infty}^{\infty} \mathrm{d}z\, f(z)\exp\left(-\frac{i2\pi n(x-z)}{L}\right), \tag{1.141}$$

by simply letting $n \to -n$ on the right-hand side of equation (1.139). It is common to consider the $x = 0$ case of the theorem, which gives

$$\sum_{n=-\infty}^{\infty} f(Ln) = \frac{1}{L}\sum_{n=-\infty}^{\infty}\int_{-\infty}^{\infty} \mathrm{d}z\, f(z)\exp\left(\frac{i2\pi nz}{L}\right). \tag{1.142}$$

Setting $L = 1$ and relabeling both the dummy variables on the right-hand side from n to j and z to n gives the often used but mildly misleading form of the Poisson resummation theorem,

$$\sum_{n=-\infty}^{\infty} f(n) = \sum_{j=-\infty}^{\infty}\int_{-\infty}^{\infty} \mathrm{d}n\, f(n)\, e^{i2\pi nj}. \tag{1.143}$$

In effect, this result states that the sum over the original function $f(n)$ is the same as the sum over the Fourier transform of that function.

As an example, the function $e^{-\alpha n^2}$ can be resummed using equation (1.143) and the variant of the Gaussian integral (1.99) where j is replaced by $i2\pi j$,

$$\int_{-\infty}^{\infty} \mathrm{d}n\, e^{-\alpha n^2 - i2\pi nj} = \sqrt{\frac{\pi}{\alpha}}\, e^{-\pi^2 j^2/\alpha}. \tag{1.144}$$

This yields the equality of two sums,

$$\sum_{n=-\infty}^{\infty} e^{-\alpha n^2} = \sqrt{\frac{\pi}{\alpha}} \sum_{n=-\infty}^{\infty} e^{-\pi^2 n^2/\alpha}. \tag{1.145}$$

Further reading

Functional techniques are presented in

- V Volterra 1959 *Theory of Functionals* (New York: Dover)
- R Courant and D Hilbert 1953 *Methods of Mathematical Physics* vol 1 (New York: Wiley-Interscience)
- H Fried 1972 *Functional Methods and Models in Quantum Field Theory* (Cambridge, MA: MIT Press)

Classical mechanics in variational form is presented in

- J R Taylor 2005 *Classical Mechanics* (Mill Valley, CA: University Science Books)
- H Goldstein, C Poole Jr and J Safko 2001 *Classical Mechanics, Third Edition* (Reading, MA: Addison-Wesley)

Fourier series and integrals, linear algebra and matrix theory, and basic complex analysis are presented in

- G Arfken, H Weber and F Harris 2012 *Mathematical Methods for Physicists* (India: Elsevier)
- B Kusse and E Westwig 2006 *Mathematical Physics* 2nd edn (New York: Wiley)
- F Byron and R Fuller 1992 *Mathematics of Classical and Quantum Physics* (New York: Dover)

Basic quantum mechanics is presented in

- C Cohen-Tannoudji, B Diu and F Laloë 1977 *Quantum Mechanics, Volumes I and II* (New York: Wiley)
- R Feynman, R Leighton, and M Sands 1965 *The Feynman Lectures on Physics, Volume III* (Reading, MA: Addison-Wesley)
- W Greiner 1994 *Quantum Mechanics: An Introduction* 3rd edn (Berlin: Springer)
- A Messiah 2017 *Quantum Mechanics, Volumes I and II* (New York: Dover)
- L Schiff and J Bandhyopadhyay 2014 *Quantum Mechanics* 4th edn (New York: McGraw-Hill)
- F Schwabl 1995 *Quantum Mechanics* 2nd edn (Berlin: Springer)

The Baker–Campbell–Hausdorff theorem is discussed in detail in

- W Greiner and J Reinhardt 1996 *Field Quantization* (Berlin: Springer)

and the Trotter product formula is derived in

- B Simon 2005 *Functional Integration and Quantum Physics* 2nd edn (New York: AMS Chelsea Publishing)

Grassmann variables are discussed extensively in

- B DeWitt 1992 *Supermanifolds* 2nd edn (Cambridge: Cambridge University Press)

IOP Publishing

Path Integral Quantization

Mark S Swanson

Chapter 2

The quantum mechanical path integral

In this chapter the quantum mechanical path integral for the transition amplitude of several illustrative systems will be derived from the Dirac formulation of quantum mechanics. Although Feynman's original definition of the path integral was more intuitive, the derivation presented here has the advantage that it yields a form of the path integral that is exactly equivalent to the standard form of quantum mechanics.

2.1 The infinitesimal transition element

The starting point is an observation, first made by Dirac, regarding the nature of the quantum mechanical transition amplitude for the case of an infinitesimal time interval. For simplicity of notation initial attention will be restricted to the configuration space transition element in one dimension for a time-independent Hamiltonian $\hat{H}(\hat{P}, \hat{X})$. The infinitesimal transition element $G(x', x, \epsilon)$ is defined as the configuration space matrix element of the evolution operator (1.46) for an infinitesimal time interval $\epsilon = t' - t \approx 0$, and is defined as

$$G(x', x, \epsilon) = \langle x'|e^{-i\epsilon\hat{H}(\hat{P}, \hat{X})/\hbar}|x\rangle. \tag{2.1}$$

In the Copenhagen interpretation of quantum mechanics (2.1) gives the probability amplitude for a particle initially at the position x to be observed at the position x' after the infinitesimal time interval ϵ has elapsed.

A common form for the Hamiltonian is $\hat{H}(\hat{P}, \hat{X}) = \hat{P}^2/2\,m + V(\hat{X})$. For the purposes of evaluating equation (2.1) the BCH theorem (1.55) allows the separation of the time evolution operator into exponential factors, which gives

$$e^{-i\epsilon\hat{H}/\hbar} = e^{-i\epsilon\hat{P}^2/2m\hbar}e^{-i\epsilon V(\hat{X})/\hbar}e^{-\epsilon^2[\hat{P}^2,\, V(\hat{X})]/2m\hbar^2}e^{O(\epsilon^3)}\ldots \tag{2.2}$$

The arguments of the higher order exponentials in the BCH theorem, including the third exponential factor in equation (2.2), are suppressed since the commutators are $O(\epsilon^2)$ and higher, assuming the potential V is *not singular*. Discarding these higher order exponentials as unity in the limit $\epsilon \to 0$ allows equation (2.1) to be evaluated by using the unit momentum projection operator of equation (1.33) and the inner product (1.37), giving

$$\begin{aligned} G(x', x, \epsilon) &= \int_{-\infty}^{\infty} \mathrm{d}p \, \langle x'|e^{-i\epsilon\hat{P}^2/2m\hbar}|p\rangle\langle p|e^{-i\epsilon V(\hat{X})/\hbar}|x\rangle \\ &= \int_{-\infty}^{\infty} \mathrm{d}p \, e^{-i\epsilon(p^2/2m+V(x))/\hbar}\langle x'|p\rangle\langle p|x\rangle \\ &= \int_{-\infty}^{\infty} \frac{\mathrm{d}p}{2\pi\hbar} \, e^{-i\epsilon(p^2/2m+V(x))/\hbar}e^{ip(x'-x)/\hbar} \\ &= \int_{-\infty}^{\infty} \frac{\mathrm{d}p}{2\pi\hbar} \exp\left\{\frac{i}{\hbar}\epsilon\left(p\frac{(x'-x)}{\epsilon} - \frac{p^2}{2m} - V(x)\right)\right\}. \end{aligned} \tag{2.3}$$

If the particle, initially at x, is observed at x' after the time interval ϵ, then the *average speed* over that time interval would have been $v = (x' - x)/\epsilon$, and it is therefore extremely tempting to write $\dot{x} = (x' - x)/\epsilon$. Using this formal notation the argument of the exponential becomes $i\epsilon(p\dot{x} - H(p, x))/\hbar = i\epsilon\, \mathcal{L}(p, x)/\hbar$, where the *classical* Lagrangian (1.10), expressed in terms of p and x, has apparently emerged in the context of the infinitesimal quantum mechanical transition element.

The integral over p is a Gaussian with the form (1.101), so that the Fresnel theorem can be used to evaluate it. However, it is often the case that Gaussian integrals with imaginary arguments are evaluated after performing what is called the *Wick rotation*. The Wick rotation consists of continuing the integral to imaginary time using the recipe $\epsilon \to -i\epsilon$, performing the integral over p, and then returning to real time with $\epsilon \to i\epsilon$. The Wick rotation typically removes the oscillatory nature of the quadratic part of the integrand, rendering the integral manifestly convergent. In the case of equation (2.3) the Wick rotation procedure gives the same result as equation (1.99), so that the infinitesimal transition element becomes

$$G(x', x, \epsilon) = \sqrt{\frac{m}{2\pi i\epsilon\hbar}} \exp\left\{\frac{i\epsilon}{\hbar}\left(\frac{1}{2}m\dot{x}^2 - V(x)\right)\right\}, \tag{2.4}$$

where, once again, $\dot{x}$ stands for the *purely formal identification* $\dot{x} = (x' - x)/\epsilon$. The infinitesimal transition element is proportional to $\exp(i\epsilon\mathcal{L}(x, \dot{x})/\hbar)$, where the Lagrangian (1.1) in terms of x and $\dot{x}$ has formally appeared.

The identification of $\mathcal{L}(x, \dot{x})$ in equation (2.4) is purely *formal* in the sense that x' and x are *not* related through a time-dependent function $x(t)$. Instead, they are arbitrary initial and final possible positions for the particle. Quantum mechanics makes no mention of classical trajectories in its assumptions, instead presenting a method to calculate the probability that the particle is observed at the two positions over the time interval. Nevertheless, result (2.4) shows that the particle's *quantum motion* is related to *classical mechanics*, in particular the value of the classical action $\epsilon\, \mathcal{L}(x, \dot{x})$ associated with the time interval. This association is most apparent for infinitesimal time intervals, since for that case the average velocity of the particle is inferred to have been $v \approx (x' - x)/\epsilon$ in order for the particle to be observed at the two positions. The appearance of the classical action is arguably the feature of path integrals most often put to use.

2.2 The basic quantum mechanical path integral

For the case of a time-dependent potential $V(\hat{X},\, t)$ and a *finite* time interval, the path integral is derived by replacing the integral appearing in the evolution operator (1.46) with a Riemann sum. Denoting $T = t_f - t_i$, the infinitesimal time interval ϵ is given by partitioning T into N intervals, so that $\epsilon = T/N$, where N is an arbitrarily large integer. The evolution operator $\hat{U}(t_f,\, t_i)$ becomes

$$\hat{U}(t_f,\, t_i) = \lim_{N\to\infty} \mathcal{T}\left\{\exp\left(-\frac{i}{\hbar}\sum_{j=0}^{N-1} \epsilon \hat{H}(t_j)\right)\right\}. \tag{2.5}$$

The time argument of the Hamiltonian has been indexed by the integer j, so that $t_j = t_i + j\epsilon$. Expression (2.5) can be simplified using result (1.60) for time-ordering the exponential of a sum of operators. Applying equation (1.60) allows the evolution operator to be written

$$\begin{aligned}\hat{U}(t_f,\, t_i) &= \lim_{N\to\infty} \mathcal{T}\left\{e^{-i\epsilon[\hat{H}(t_0)+\hat{H}(t_1)+\ldots+\hat{H}(t_{N-1})]/\hbar}\right\}\\ &= \lim_{N\to\infty} e^{-i\epsilon\hat{H}(t_{N-1})/\hbar}e^{-i\epsilon\hat{H}(t_{N-2})/\hbar}\ldots e^{-i\epsilon\hat{H}(t_0)/\hbar}.\end{aligned} \tag{2.6}$$

Result (2.6) shows that the evolution operator for finite time intervals can be viewed as the product of the N infinitesimal evolution operators. This is consistent with the Schrödinger equation, which gives the evolution of the wave function as

$$e^{-i\epsilon\hat{H}(t_j)/\hbar}\psi(x,\, t_j) = e^{\epsilon\partial/\partial t_j}\psi(x,\, t_j) = \psi(x,\, t_j + \epsilon). \tag{2.7}$$

The process of dividing the finite time interval into a sequence of N infinitesimal time intervals is referred to as *time slicing* in the path integral literature.

The matrix elements of $\hat{U}(t_f,\, t_i)$ are typically evaluated between initial and final position or momentum states. For the case of position states, the matrix element of the evolution operator is referred to as the *kernel* or more commonly as the *propagator*. It is given explicitly by

$$G(x_f,\, x_i,\, t_f,\, t_i) = \langle x_f|\hat{U}(t_f,\, t_i)|x_i\rangle. \tag{2.8}$$

The propagator obeys the Schrödinger equation since $i\hbar\, \partial\hat{U}(t,\, t_i)/\partial t = \hat{H}\hat{U}(t,\, t_i)$. It is interpreted as the probability amplitude that the particle, initially observed at x_i, is observed at x_f after the time $T = t_f - t_i$ has elapsed. Each of the N infinitesimal evolution operators, denoted $\hat{U}_j = e^{-i\epsilon\hat{H}(t_j)/\hbar}$, can be analyzed by inserting $N - 1$ position projection operators,

$$G(x_f, x_i, t_f, t_i) = \int_{-\infty}^{\infty} \mathrm{d}x_1 \cdots \mathrm{d}x_{N-1} \, \langle x_f | \hat{U}_{N-1} | x_{N-1} \rangle \langle x_{N-1} | \hat{U}_{N-2} | x_{N-2} \rangle \cdots$$
$$\cdots \langle x_2 | \hat{U}_1 | x_1 \rangle \langle x_1 | \hat{U}_0 | x_i \rangle \tag{2.9}$$
$$= \int_{-\infty}^{\infty} \mathrm{d}x_1 \cdots \mathrm{d}x_{N-1} \prod_{j=0}^{N-1} \langle x_{j+1} | \hat{U}_j | x_j \rangle,$$

where the identifications $x_N = x_f$ and $x_0 = x_i$ are in effect. The subscripts on the variables of integration in equation (2.9) have been chosen to match the index of t_j for notational convenience. Each x_j is then paired with the time t_j, but the x_j have no intrinsic time-dependence. Depending on the form of $\hat{H}(t)$, each of the individual infinitesimal evolution operators in equation (2.6) can be further factorized. If attention is restricted to a one-dimensional system and the Hamiltonian takes the standard form $\hat{H}(t) = \hat{P}^2/2\,m + V(\hat{X}, t)$, with the possibility of a time-dependent potential, the individual infinitesimal evolution operators can once again be factorized and analyzed identically to equation (2.2). Using the identifications $x_0 = x_i$ and $x_N = x_f$ and inserting a complete set of momentum states, each infinitesimal element becomes

$$\langle x_{j+1} | \hat{U}_j | x_j \rangle = \int_{-\infty}^{\infty} \mathrm{d}p_j \, \langle x_{j+1} | p_j \rangle \langle p_j | \exp\left\{ -\frac{i}{\hbar} \epsilon \left(\frac{\hat{P}^2}{2m} + V(\hat{X}, t_j) \right) \right\} | x_j \rangle$$
$$= \int_{-\infty}^{\infty} \frac{\mathrm{d}p_j}{2\pi\hbar} \exp\left\{ \frac{i}{\hbar} \epsilon \left(p_j \dot{x}_j - \frac{p_j^2}{2m} - V(x_j, t_j) \right) \right\}, \tag{2.10}$$

where the notation $\dot{x}_j = (x_{j+1} - x_j)/\epsilon$ has again been used.

It is worth noting that the momentum states could have been inserted to the right of the evolution operator, and this would appear to change the result. However, the possible choices for the order can be shown to yield equivalent path integrals. It is also possible that the Hamiltonian may contain terms involving powers of both $\hat{P}$ and $\hat{X}$. For such a case, the order of the operators becomes critical to evaluating them. It is typical to place the operators in what is known as *Weyl order*, which is the average of the set of all possible permutations of the operators. For brevity, these issues will not be discussed in detail, and the interested reader should consult the references.

Result (2.10) may be returned to equation (2.9) to obtain the *path integral representation of the propagator*,

$$G(x_f, x_i, t_f, t_i) = \int [\mathcal{D}p][\mathcal{D}x] \exp\left\{ \frac{i}{\hbar} \sum_{j=0}^{N-1} \epsilon \left(p_j \dot{x}_j - \frac{p_j^2}{2m} - V(x_j, t_j) \right) \right\}, \tag{2.11}$$

where the *path integral measure* is given by

$$[\mathcal{D}p][\mathcal{D}x] = \frac{\mathrm{d}p_0}{2\pi\hbar}\cdots\frac{\mathrm{d}p_{N-1}}{2\pi\hbar}\,\mathrm{d}x_1\cdots\mathrm{d}x_{N-1}. \tag{2.12}$$

Of course, the limit $N \to \infty$ is implicit in expressions (2.11) and (2.12). It is worth noting that the product of the N exponentials in equation (2.9) simply becomes the exponential of the sum of the N arguments in equation (2.11). This is a virtue of the path integral, which is composed solely of commuting or anticommuting variables with no reference to their operator antecedents. There is no need to apply the BCH theorem since it was implicitly invoked in the derivation of the infinitesimal elements. Because there are integrations over both momentum and position, the path integral of equation (2.11) is often referred to as the *phase space path integral.* It is also important to note that throughout the derivation of equation (2.11) it has been assumed that the range on the integrations in the position and momentum projection operators is $(-\infty, \infty)$ the case since many physical systems are assumed to occupy an infinite volume. However, boundary conditions may dictate a finite range. Such a possibility will be considered in the next section.

Performing the Gaussian momentum integrals in equation (2.11) using either the Wick rotation or Fresnel's theorem (1.99) yields N copies of equation (2.4), so that

$$G(x_f, x_i, t_f, t_i) = \int [\tilde{\mathcal{D}}x] \exp\left\{\frac{i}{\hbar}\sum_{j=0}^{N-1} \epsilon\left(\frac{1}{2}m\dot{x}_j^2 - V(x_j, t_j)\right)\right\}, \tag{2.13}$$

where the measure subsequent to the momentum integrations has become

$$[\tilde{\mathcal{D}}x] = \left(\frac{m}{2\pi i\epsilon\hbar}\right)^{N/2} \mathrm{d}x_1\cdots\mathrm{d}x_{N-1}. \tag{2.14}$$

The more commonly used form for the Lagrangian, $\mathcal{L}(x, \dot{x}, t)$, is formally present in the action of equation (1.1). For that reason, the path integral of equation (2.13) is referred to as the *configuration space path integral.* The new measure (2.14) appears to be badly divergent in the limit $\epsilon \to 0$. However, this divergence disappears if the variables of integration are scaled according to $x_j \to \sqrt{\epsilon}\, x_j$. This gives $\epsilon\, \dot{x}_j^2 \to (x_{j+1} - x_j)^2$ in the action sum, while all but one of the N factors of $1/\sqrt{\epsilon}$ are canceled from the measure. If the potential is composed of terms quadratic and higher in x_j it develops an overall scale factor of ϵ^2 and does not reintroduce any divergences. In what follows the Lagrangian forms appearing in equations (2.11) and (2.13) will be called *time-sliced Lagrangians,* referring to the formal identification of time derivatives that have resulted from the time slicing approach.

In some applications it is necessary to have matching numbers of momentum and position integrations in the phase space path integral, which is not the case in equation (2.12). This can be remedied by considering a mixed transition element. For example, $G(x_f, p_i, t_f, t_i) = \langle x_f|\hat{U}(t_f, t_i)|p_i\rangle$ is the probability amplitude for a particle with the initial momentum p_i to be observed at the position x_f after the time $T = t_f - t_i$. It can be found from equation (2.11) by using the definition of $G(x_f, x_i, t_f, t_i)$ and the unit position projection operator (1.33),

$$G(x_f, p_i, t_f, t_i) = \int_{-\infty}^{\infty} \mathrm{d}x_0 \, \langle x_f|\hat{U}(t_f, t_i)|x_0\rangle\langle x_0|p_i\rangle$$

$$= \int [\bar{\mathcal{D}}p][\bar{\mathcal{D}}x] \exp\left\{\frac{i}{\hbar}p_i x_0 + \frac{i}{\hbar}\sum_{j=0}^{N-1} \epsilon\left(p_j \dot{x}_j - \frac{p_j^2}{2m} - V(x_j, t_j)\right)\right\}, \tag{2.15}$$

where again $\dot{x}_j = (x_{j+1} - x_j)/\epsilon$ and $x_N = x_f$, but now x_0 is an integration variable. The phase space measure has become symmetric and is given by

$$\int [\bar{\mathcal{D}}p][\bar{\mathcal{D}}x] = \sqrt{\frac{1}{2\pi\hbar}} \frac{\mathrm{d}p_0}{2\pi\hbar} \cdots \frac{\mathrm{d}p_{N-1}}{2\pi\hbar} \mathrm{d}x_0 \cdots \mathrm{d}x_{N-1}. \tag{2.16}$$

The price paid for a symmetric measure is the additional term $ip_i x_0/\hbar$ in the action sum in the exponential. The action sum can be written

$$p_i x_0 + \sum_{j=0}^{N-1} p_j(x_{j+1} - x_j) = p_i x_i + \sum_{j=-1}^{N-1} p_j(x_{j+1} - x_j), \tag{2.17}$$

where the identifications $p_{-1} = p_i$ and $x_{-1} = x_i$ have been made. The quantity x_i is neither a variable of integration nor is it defined by the transition element. It is tempting to define x_i by solving the classical equations of motion consistent with the boundary conditions $x_c(t_f) = x_f$ and $p_c(t_i) = m\dot{x}_c(t_i) = p_i$ and then identifying $x_i = x_c(t_i)$. Doing so defines x_i in terms of x_f and p_i, but there is no *a priori* justification for this. It is important to note that the two terms in equation (2.17) involving x_i cancel each other, and in that sense x_i is a *pseudovariable* since equation (2.17) is independent of its value. These two terms vanish individually if $x_i = 0$, and so any manipulations performed on equation (2.17) can be evaluated at $x_i = 0$ to achieve consistency. In analyzing equation (2.17) it will be useful to employ the path integral equivalent of integration by parts. It is straightforward to verify the following equality,

$$p_i x_0 + \sum_{j=0}^{N-1} p_j(x_{j+1} - x_j) = p_f x_f - \sum_{j=-1}^{N-1} x_{j+1}(p_{j+1} - p_j), \tag{2.18}$$

where the identifications $p_{-1} = p_i$ and $p_N = p_f$ have been made. As with x_i in equation (2.17), p_f is a pseudovariable and the two terms in equation (2.18) involving p_f cancel.

2.3 The path in path integrals

It is natural to ask why equations (2.11) and (2.13) are referred to as *path integrals*. It will be seen in the next chapter that the most commonly used method to evaluate path integrals relies on the classical trajectories or *paths* obtained from Hamilton's equations (1.11) or the Euler–Lagrange equations (1.7). Subsequent to this discovery, it was considered possible that the quantum transition element of the particle between x_i and x_f could be understood as the *sum over all possible paths* between the initial and final state, with each path from x_i to x_f weighted by the phase given by the classical action for that path, $\exp iS_c/\hbar$. This viewpoint treats the sum appearing in equation (2.11) as an integral in the $\epsilon \to 0$ limit. For the specific form of the Hamiltonian used in equation (2.11), making the *continuum limit* identifications $x_j \to x(t_j)$, $p_j \to p(t_j)$, and $\epsilon \to \mathrm{d}t$ causes the action sum in the path integral to become a true integral, and the path integral could then be written

$$G(x_f, x_i, t_f, t_i) = \int [\mathcal{D}p][\mathcal{D}x] \exp\left\{\frac{i}{\hbar}\int_{t_i}^{t_f} \mathrm{d}t \left(p\dot{x} - \frac{p^2}{2m} - V(x, t)\right)\right\}. \tag{2.19}$$

It must be stressed that this is an *assumption*, and in some cases a mistake. The variables p_j and x_j appearing in equation (2.11) are *not* functions of time associated with some particle trajectory, just as $\dot{x}_j = (x_{j+1} - x_j)/\epsilon$ is *not* the time derivative of a trajectory. Instead, x_j and p_j are variables of integration and $\dot{x}_j$ is simply a notational convenience. Analysis presented in the next chapter will show that the use of differentiable paths is not sufficient to derive the correct quantum behavior *if the measure appearing in* equation (2.19) *is used.* As a result, taking the continuum limit and using differentiable paths in this manner requires a reinterpretation of the measure appearing in equation (2.19). This reinterpretation is presented in the next chapter in the section on continuum methods.

In the path integral of equation (2.11), the integrations over all possible intermediate momenta and positions are often interpreted as generating all possible *paths* between x_i and x_f. Since taking the continuum limit in the action sum treats these paths as trajectories or paths parameterized by time, it would therefore also need to be accompanied by revising the measure to a *sum over paths*. The continuum version of the path integral would then be *symbolically* written

$$G(x_f, x_i, T) = \sum_{\text{paths}} e^{iS_c(\text{path})/\hbar}. \tag{2.20}$$

Statement (2.20) is understood as taking an arbitrary path $x(t)$ from x_i to x_f in the time interval T, computing the value of the classical action S_c along that path using (1.1), and attaching the weighting phase $\exp(iS_c/\hbar)$ to that path. Summing over *all paths* from x_i to x_f using this recipe *was anticipated* to yield the correct quantum mechanical transition element. While this is an intuitively appealing interpretation of quantum processes, it has proved exceedingly difficult to provide a mathematical definition of what a *sum over all paths* means. The efforts to quantify a *path measure* in a functional formalism date back to Wiener's treatment of Brownian motion in diffusive processes and remains a difficult mathematical endeavor. Further discussion of this is presented in the next chapter where Fourier techniques are combined with the differentiable paths approach.

The forms given by equations (2.11) and (2.13) will therefore be treated in the remainder of this text as the correct path integral expression for the quantum mechanical transition element in the sense that they are equivalent to the standard formulation of quantum mechanics. There is an important *caveat*, which is that a singular potential will cause the Trotter product formula to fail in the neighborhood of the singularity, rendering the naïve forms (2.11) and (2.13) invalid. This is discussed in a later section for the case of polar coordinates. Nevertheless, expression (2.19) is often used to infer the path integral for an arbitrary quantum mechanical system. This approach uses the classical Lagrangian $\mathcal{L}(p, x, t)$ for the system under consideration. The action integral becomes a Riemann sum by replacing $\mathrm{d}t$ with ϵ, $p(t_j)$ with p_j, $x(t_j)$ with x_j, and $\dot{x}(t)$ with $(x_{j+1} - x_j)/\epsilon$. All the time-sliced intermediate variables are integrated to obtain the quantum mechanical path integral (2.11). This is *assumed* to be correct even when the Hamiltonian takes a different form than the one which led to equation (2.11). This procedure, referred to as *the time slicing recipe*, does *not* always yield the correct path integral. For example, this recipe can be followed with a *classical* Lagrangian whose quantum mechanical interpretation is unknown or unclear, such as unstable systems, and so there is no way to verify the quantum mechanical correctness of the path integral in such a case. It will be seen that the simple classical to quantum time-slicing recipe breaks down for singular potentials and for coordinate systems with singularities. Inferring the correct form for the path integral by starting with the action using $\mathcal{L}(x, \dot{x}, t)$ is even more difficult because of the factors appearing in the configuration measure of equation (2.14). It will be shown that these factors are necessary to normalize the path integral. It is sometimes possible to determine these factors from the requirement that $\lim_{t_f \to t_i} G(x_f, x_i, t_f, t_i) = \delta(x_f - x_i)$ or simply to cancel them by forming *normalized ratios*. Such an approach to developing the path integral representation of the transition element is discussed further in the remaining chapters.

2.4 Discrete systems and path integrals

An important aspect of expression (2.11) and the related expression (2.13) is that the range of intermediate integrations in the path integral were assumed to be both infinite and continuous. There are many quantum mechanical systems whose volumes are finite and whose momentum and energy spectrums are discrete rather than continuous. It is important to understand how and in what way path integrals can be applied to the quantum transition elements in such a case.

It is instructive to investigate the path integral for the relatively simple case of a mass m point particle confined to an infinitely deep one-dimensional box of width L occupying the interval $(0, L)$. The particle is free in the interior, so that $V(x) = 0$ for $0 < x < L$ and the Hamiltonian is given by $\hat{H} = \hat{P}^2/2m$. The boundary conditions on the energy eigenfunctions require that they vanish at the walls, and the solutions to the Schrödinger equation in position representation are given by $\langle x|E_n\rangle = \sqrt{2/L}\,\sin(n\pi x/L)$. The eigenfunctions are associated with the discrete set of energy eigenvalues $E_n = n^2(\pi^2\hbar^2/2\,mL^2)$, where n is an integer. It is important to note that the energy eigenfunctions are *not* momentum eigenfunctions. Instead, they are a superposition of momentum eigenfunctions, $|E_n\rangle = (|p_n\rangle - |p_{-n}\rangle)/i\sqrt{2}$, where $\hat{P}|p_n\rangle = p_n|p_n\rangle = (n\pi\hbar/L)|p_n\rangle$ with n an integer. The energy eigenstates can be viewed as the quantum mechanical equivalent of standing waves and by construction satisfy $|E_{-n}\rangle = -|E_n\rangle$. In general, the momentum eigenstates are not orthogonal since

$\langle p_j | p_k \rangle \neq \delta_{jk}$ if $j - k$ is an odd integer. This is easily verified by using $\langle x | p_n \rangle = \exp(ip_n x/\hbar)$ and integrating the momentum inner product $\langle p_j | x \rangle \langle x | p_k \rangle$ over the interval $(0, L)$. However, for $j \neq k$ the inner products satisfy $\langle p_j | p_k \rangle = -\langle p_{-j} | p_{-k} \rangle$ and $\langle p_j | p_{-k} \rangle = -\langle p_{-j} | p_k \rangle$, and so the energy eigenstates are orthonormal.

For the purpose of developing the path integral, the unit position and energy projection operators are given by

$$\hat{\mathbf{1}}_x = \int_0^L \mathrm{d}x \, |x\rangle\langle x| \quad \text{and} \quad \hat{\mathbf{1}}_E = \sum_{n=1}^{\infty} |E_n\rangle\langle E_n|. \tag{2.21}$$

It follows that

$$\langle x|y\rangle = \langle x|\mathbf{1}_E|y\rangle = \frac{2}{L}\sum_{n=1}^{\infty} \sin\left(\frac{n\pi x}{L}\right)\sin\left(\frac{n\pi y}{L}\right) = \delta(x - y), \tag{2.22}$$

where the Fourier series representation of the Dirac delta (1.22) consistent with the boundary conditions has appeared, i.e. $\delta(x) = \delta(-y) = \delta(x - L) = \delta(L - y) = 0$.

As with the path integral for an infinite volume, the starting point is the infinitesimal transition element for motion in the interior of the box where $\hat{H} = \hat{P}^2/2m$. The BCH theorem is irrelevant since there is no potential in the interior of the box. Using completeness of the energy states and $\langle x|E_n\rangle = \sqrt{2/L}\,\sin(p_n x/\hbar)$ gives

$$\begin{aligned} \langle x|e^{-i\epsilon\hat{H}/\hbar}|y\rangle &= \sum_{n=1}^{\infty} \exp\left(-\frac{i}{\hbar}\epsilon\frac{p_n^2}{2m}\right)\langle x|E_n\rangle\langle E_n|y\rangle \\ &= \frac{1}{L}\sum_{n=-\infty}^{\infty} \exp\left(-\frac{i}{\hbar}\epsilon\frac{p_n^2}{2m}\right)\sin\left(\frac{p_n x}{\hbar}\right)\sin\left(\frac{p_n y}{\hbar}\right), \end{aligned} \tag{2.23}$$

where the range of the sum has been extended for later convenience. Expressing the sine functions as exponentials in the sum of equation (2.23) gives

$$G(x, y, \epsilon) = \frac{1}{2L}\sum_{n=-\infty}^{\infty} \exp\left(-\frac{i}{\hbar}\epsilon\frac{p_n^2}{2m}\right)\left(e^{ip_n(x-y)/\hbar} - e^{ip_n(x+y)/\hbar}\right) \tag{2.24}$$

where the range on the sum allowed consolidating the four crossterms into two. The unusual aspect of equation (2.24) is the appearance of $x + y$ in the second exponential. This is a direct outgrowth of the absence of *translational invariance* for the box, which simply means that the system and therefore result (2.24) are not invariant under the change in coordinates given by $x \to x + a$ and $y \to y + a$. The two sums appearing in equation (2.24), designated $G = G_- - G_+$ for the sign of y in the exponential, can be rewritten using the Poisson resummation formula of equation (1.143),

$$G_\pm(x, y, \epsilon) = \frac{1}{2L} \sum_{n=-\infty}^{\infty} \exp\left\{\frac{i}{\hbar}\epsilon\left(p_n \frac{(x \pm y)}{\epsilon} - \frac{p_n^2}{2m}\right)\right\}$$
$$= \frac{1}{2L} \sum_{j=-\infty}^{\infty} \int_{-\infty}^{\infty} \mathrm{d}n\, e^{i2\pi nj} \exp\left\{\frac{i}{\hbar}\epsilon\left(p_n \frac{(x \pm y)}{\epsilon} - \frac{p_n^2}{2m}\right)\right\} \tag{2.25}$$
$$= \sum_{j=-\infty}^{\infty} \int_{-\infty}^{\infty} \frac{\mathrm{d}p}{2\pi\hbar} \exp\left\{\frac{i}{\hbar}\epsilon\left(p\, \frac{(x \pm y + 2jL)}{\epsilon} - \frac{p^2}{2m}\right)\right\},$$

where the last step in equation (2.25) introduced the continuous momentum variable $p = n\pi\hbar/L$. The momentum integration is the standard Gaussian of (1.99) and it yields

$$G_\pm(x, y, \epsilon) = \sqrt{\frac{m}{2\pi i\epsilon\hbar}} \sum_{j=-\infty}^{\infty} \exp\left\{\frac{i}{\hbar}\epsilon\, \frac{1}{2}m\left(\frac{x \pm y + 2jL}{\epsilon}\right)^2\right\}. \tag{2.26}$$

Result (2.26) has an important property related to the Poisson resummation theorem that is revealed by considering the fact that the time independence of the Hamiltonian gives

$$\int_0^L \mathrm{d}y\, G(x, y, \epsilon)G(y, z, \epsilon) = \int_0^L \mathrm{d}y\, \langle x|e^{-i\epsilon\hat{H}/\hbar}|y\rangle\langle y|e^{-i\epsilon\hat{H}/\hbar}|z\rangle$$
$$= \langle x|e^{-i2\epsilon\hat{H}/\hbar}|z\rangle = G(x, z, 2\epsilon). \tag{2.27}$$

The details of how result (2.27) holds will be important in deriving the full path integral for this system. The two infinitesimal propagators on the left-hand side of equation (2.27) can be expressed using equation (2.26) in the combination $G = G_+ - G_-$. The specific form of equation (2.26) establishes the following two identities,

$$\int_0^L \mathrm{d}y\, G_+(x, y, \epsilon)\, G_+(y, z, \epsilon) = \int_{-L}^0 \mathrm{d}y\, G_-(x, y, \epsilon)G_-(y, z, \epsilon), \tag{2.28}$$

$$\int_0^L \mathrm{d}y\, G_+(x, y, \epsilon)\, G_-(y, z, \epsilon) = \int_{-L}^0 \mathrm{d}y\, G_-(x, y, \epsilon)G_+(y, z, \epsilon). \tag{2.29}$$

These identities are proven by simply changing the variable of integration from y to $-y$ and using the freedom in the infinite sums to let either $j \to -j$ or $j' \to -j'$. As a result, the left-hand side of (2.27) can be written

$$\int_0^L \mathrm{d}y\, G(x, y, \epsilon)G(y, z, \epsilon) = \int_{-L}^L \mathrm{d}y\, G_-(x, y, \epsilon)$$
$$\{G_-(y, z, \epsilon) - G_+(y, z, \epsilon)\} \tag{2.30}$$
$$= \int_{-L}^L \mathrm{d}y\, G_-(x, y, \epsilon)G(y, z, \epsilon).$$

Next, one of the sums in equation (2.30) is absorbed into extending the limits on the integral by using the same technique employed in equation (1.137). Using the forms for $G_\pm$ given by equation (2.26), the steps in this demonstration are given by

$$\begin{aligned}
&\int_{-L}^{L} \mathrm{d}y\; G_-(x, y, \epsilon) G_\pm(y, z, \epsilon) \\
&= \frac{m}{2\pi i \epsilon \hbar} \sum_{j,j'=-\infty}^{\infty} \int_{-L}^{L} \mathrm{d}y \\
&\quad \exp\left\{\frac{i}{\hbar}\epsilon\left(\frac{1}{2}m\frac{(x-y+2jL)^2}{\epsilon^2} + \frac{1}{2}m\frac{(y \pm z + 2j'L)^2}{\epsilon^2}\right)\right\} \\
&= \frac{m}{2\pi i \epsilon \hbar} \sum_{j,j'=-\infty}^{\infty} \int_{-L+2jL}^{L+2jL} \mathrm{d}y \\
&\quad \exp\left\{\frac{i}{\hbar}\epsilon\left(\frac{1}{2}m\frac{(x-y)^2}{\epsilon^2} + \frac{1}{2}m\frac{(y \pm z + 2(j'+j)L)^2}{\epsilon^2}\right)\right\} \\
&= \frac{m}{2\pi i \epsilon \hbar} \sum_{j'=-\infty}^{\infty} \int_{-\infty}^{\infty} \mathrm{d}y \\
&\quad \exp\left\{\frac{i}{\hbar}\epsilon\left(\frac{1}{2}m\frac{(x-y)^2}{\epsilon^2} + \frac{1}{2}m\frac{(y \pm z + 2j'L)^2}{\epsilon^2}\right)\right\},
\end{aligned} \tag{2.31}$$

where the last step in equation (2.31) follows from the ability to redefine $j' \to j' - j$ with no effect on the infinite sum over j'. In this case the integration can be performed using a variant of the Gaussian (1.99) obtained from completing the square,

$$\int_{-\infty}^{\infty} \mathrm{d}y\; e^{-\alpha(x-y)^2 - \beta(y \pm z)^2} = \sqrt{\frac{\pi}{\alpha+\beta}} \exp\left\{-\left(\frac{\alpha\beta}{\alpha+\beta}\right)(x \pm z)^2\right\}. \tag{2.32}$$

Identifying $\alpha = \beta = -im/2\hbar\epsilon$ and using equation (2.32) reduces equation (2.31) to

$$\begin{aligned}
&\int_{-L}^{L} \mathrm{d}y\; G_-(x, y, \epsilon) G_\pm(y, z, \epsilon) \\
&= \sqrt{\frac{m}{2\pi i (2\epsilon)\hbar}} \sum_{j=-\infty}^{\infty} \exp\left\{\frac{i}{\hbar}(2\epsilon)\left(\frac{1}{2}m\frac{(x \pm z + 2Lj)^2}{(2\epsilon)^2}\right)\right\}.
\end{aligned} \tag{2.33}$$

Returning equation (2.33) to (2.30) immediately verifies that property (2.27) holds. This is an important result because it shows that the summations associated with the Poisson resummation theorem disappear by changing the limits of the integrations over the intermediate position variables.

It is now straightforward to combine the same approach of equation (2.9) with the general form for the infinitesimal transition amplitude to obtain the time-sliced finite transition element applicable to the particle in a box,

$$G(x_f, x_i, T) = \prod_{j=1}^{N-1} \left(\int_0^L \mathrm{d}x_j \; G(x_{j+1}, x_j, \epsilon) \right) G(x_1, x_0, \epsilon), \tag{2.34}$$

where the identifications $x_N = x_f$ and $x_0 = x_i$ are currently in place. Using result (2.30) the limits on the integrations over all $N - 1$ of the intermediate variables x_j can be extended, and equation (2.34) takes the form

$$G(x_f, x_i, T) = \prod_{j=1}^{N-1} \left(\int_{-L}^L \mathrm{d}x_j \; G_-(x_{j+1}, x_j, \epsilon) \right) G(x_1, x_0, \epsilon). \tag{2.35}$$

It is important to note that the rightmost infinitesimal propagator remains the original infinitesimal propagator $G(x_1, x_0, \epsilon) = G_-(x_1, x_0, \epsilon) - G_+(x_1, x_0, \epsilon)$ since there are N infinitesimal propagators in the original expression (2.34) and only $N - 1$ intermediate integrations. There are also summations of the type present in equation (2.31) in all N factors due to the Poisson resummation of equation (2.25) that occurred in all N of the infinitesimal propagators. All but one of these summations can be absorbed by extending the range of the $N - 1$ intermediate integrations. The notation is streamlined by redefining x_0 to $x_0 = (-1)^r x_i + 2kL$, where $r = 0$ or 1. Using this identification and result (2.31) allows equation (2.35) to be written

$$G(x_f, x_i, T) = \sum_{k=-\infty}^{\infty} \sum_{r=0}^{1} (-1)^r \int_{-\infty}^{\infty} [\tilde{\mathcal{D}}x] \exp\left(\frac{i}{\hbar} \sum_{j=0}^{N-1} \epsilon \, \frac{1}{2} m \dot{x}_j^2 \right), \tag{2.36}$$

where $\dot{x}_j = (x_{j+1} - x_j)/\epsilon$ as before and the measure $[\tilde{\mathcal{D}}x]$ is given by equation (2.14). Result (2.36) is the configuration space path integral for a free particle in a box.

Remarkably, result (2.36) has both the same action and measure as the path integral for the free particle in an infinite volume given by equation (2.13). The entire effect of being constrained to the box is encapsulated in the sum over k in the initial point, $x_0 = (-1)^r x_i + 2kL$, and the overall prefactor $(-1)^r$. It is useful to note that the magnitude $|x_f - x_0| = |x_f - (-1)^r x_i - 2kL|$ coincides with the distance traveled when the *classical particle* bounces off the walls of the box going from x_i to x_f at a constant speed. For example, if the particle bounces off the right wall once and then travels back to x_f, the total distance travelled is $(L - x_i) + (L - x_f) = 2L - (x_i + x_f)$. If the particle bounces off both the left wall and the right wall before traveling to x_f, the total distance traveled is $x_i + L + (L - x_f) = 2L - (x_f - x_i) = |x_f - x_i - 2L|$. If $r = 0$ the distances $|x_f - x_0|$ correspond to an even number of bounces, while if $r = 1$ the distances correspond to an odd number of bounces. This shows that the correct configuration space path integral (2.36) includes not just $x_f - x_i$, but all possible *classical distances* traveled by the point particle as it moves from x_i to x_f.

A related simple system is a point mass m constrained to move in a circle of radius R. It is effectively a one-dimensional system with the position parameterized by the angle θ familiar from polar coordinates. The Hamiltonian can be derived from the two-dimensional classical free particle Lagrangian in polar coordinates $\mathcal{L} = \frac{1}{2}m\dot{r}^2 + \frac{1}{2}mr^2\dot{\theta}^2$ by fixing $r = R$ and $\dot{r} = 0$, and is given by $\hat{H} = \hat{L}^2/2\,mR^2$, where $\hat{L}$ the angular momentum of the particle. The orthonormal energy eigenfunctions are given in configuration representation by $\langle\theta|E_n\rangle = e^{in\theta}/\sqrt{2\pi}$. The energy eigenfunctions coincide with the angular momentum eigenfunctions, so that $\hat{L}|E_n\rangle = n\hbar|E_n\rangle$, and this yields the energy eigenvalues $E_n = n^2\hbar^2/2\,mR^2$. The unit projection operators $\hat{\mathbf{1}}_E$ and $\hat{\mathbf{1}}_\theta$ satisfy

$$\langle\theta'|\hat{\mathbf{1}}_E|\theta\rangle = \sum_{n=-\infty}^{\infty}\langle\theta'|E_n\rangle\langle E_n|\theta\rangle = \frac{1}{2\pi}\sum_{n=-\infty}^{\infty} e^{in(\theta-\theta')} = \delta(\theta-\theta'), \tag{2.37}$$

$$\langle E_n|\hat{\mathbf{1}}_\theta|E_m\rangle = \int_0^{2\pi} \mathrm{d}\theta\,\langle E_n|\theta\rangle\langle\theta|E_m\rangle = \int_0^{2\pi}\frac{\mathrm{d}\theta}{2\pi}\,e^{-i(n-m)\theta} = \delta_{nm}. \tag{2.38}$$

In effect, the circle is a boundaryless version of the particle in a box, and so the derivation of its path integral is quite similar to that of the box.

The infinitesimal transition element for the free particle is given by

$$\begin{aligned}
\langle\theta'|e^{-i\epsilon\hat{H}/\hbar}|\theta\rangle &= \sum_{n=-\infty}^{\infty}\langle\theta'|e^{-i\epsilon\hat{H}/\hbar}|E_n\rangle\langle E_n|\theta\rangle \\
&= \sum_{n=-\infty}^{\infty}\frac{1}{2\pi}\exp\left\{\frac{i}{\hbar}\epsilon\left(n\hbar\frac{(\theta-\theta')}{\epsilon} - \frac{n^2\hbar^2}{2\,mR^2}\right)\right\} \\
&= \sum_{j=-\infty}^{\infty}\int_{-\infty}^{\infty}\frac{\mathrm{d}n}{2\pi}\,e^{i2\pi jn}\exp\left\{\frac{i}{\hbar}\epsilon\left(n\hbar\frac{(\theta-\theta')}{\epsilon} - \frac{n^2\hbar^2}{2\,mR^2}\right)\right\} \\
&= \sum_{j=-\infty}^{\infty}\int_{-\infty}^{\infty}\frac{\mathrm{d}p_\theta}{2\pi\hbar}\exp\left\{\frac{i}{\hbar}\epsilon\left(p_\theta\frac{(\theta-\theta'+2\pi j)}{\epsilon} - \frac{p_\theta^2}{2\,mR^2}\right)\right\},
\end{aligned} \tag{2.39}$$

where the Poisson resummation was applied in the third step and the continuous angular momentum $p_\theta = n\hbar$ was introduced in the fourth step.

Result (2.39) can be used in analyzing the finite transition element given by the time slicing procedure of equation (2.9),

$$G(\theta_f, \theta_i, t_f, t_i) = \int_0^{2\pi}\mathrm{d}\theta_1\cdots\mathrm{d}\theta_{N-1}\prod_{j=0}^{N-1}\langle\theta_{j+1}|e^{-i\epsilon\hat{H}/\hbar}|\theta_j\rangle, \tag{2.40}$$

where the initial identifications $\theta_N = \theta_f$ and $\theta_0 = \theta_i$ have been made. As with the particle in a box, the sums appearing in equation (2.39) can be absorbed into the integration limits in the same manner as equation (2.31), and this gives

$$\int_0^{2\pi} \mathrm{d}\theta_j \, \langle \theta_{j+1} | e^{-i\epsilon \hat{H}/\hbar} | \theta_j \rangle$$

$$= \sum_{k=-\infty}^{\infty} \int_{-\infty}^{\infty} \frac{\mathrm{d}p_j}{2\pi\hbar} \int_0^{2\pi} \mathrm{d}\theta_j \exp\left\{ \frac{i}{\hbar}\epsilon \left(p_j \frac{(\theta_{j+1} - \theta_j + 2\pi k)}{\epsilon} - \frac{p_j^2}{2\,mR^2} \right) \right\} \tag{2.41}$$

$$= \int_{-\infty}^{\infty} \frac{\mathrm{d}p_j}{2\pi\hbar} \int_{-\infty}^{\infty} \mathrm{d}\theta_j \exp\left\{ \frac{i}{\hbar}\epsilon \left(p_j \frac{(\theta_{j+1} - \theta_j)}{\epsilon} - \frac{p_j^2}{2\,mR^2} \right) \right\}.$$

This can be done for $N - 1$ of the N sums present in equation (2.40), leaving only the sum associated with the last infinitesimal element. The final result of returning equation (2.41) back to the factors in equation (2.40) gives the phase space path integral for the free particle constrained to lie on a circle,

$$G(\theta_f, \theta_i, T)$$

$$= \sum_{k=-\infty}^{\infty} \int_{-\infty}^{\infty} [\mathcal{D}p \, \mathcal{D}\theta] \exp\left\{ \frac{i}{\hbar} \sum_{j=0}^{N-1} \epsilon \left(p_j \frac{(\theta_{j+1} - \theta_j)}{\epsilon} - \frac{p_j^2}{2\,mR^2} \right) \right\}. \tag{2.42}$$

Result (2.42) still has $\theta_N = \theta_f$, but now $\theta_0 = \theta_i + 2\pi k$. The measure is given by

$$[\mathcal{D}p \, \mathcal{D}\theta] = \frac{\mathrm{d}p_0}{2\pi\hbar} \cdots \frac{\mathrm{d}p_{N-1}}{2\pi\hbar} \, \mathrm{d}\theta_1 \cdots \mathrm{d}\theta_{N-1}. \tag{2.43}$$

The intermediate momenta in equation (2.42) can be integrated using the Gaussian integral result (1.99). The result is

$$G(\theta_f, \theta_i, T) = \sum_{k=-\infty}^{\infty} \int_{-\infty}^{\infty} [\tilde{\mathcal{D}}\theta] \exp\left\{ \frac{i}{\hbar} \sum_{j=0}^{N-1} \epsilon \, \frac{1}{2} mR^2 \left(\frac{\theta_{j+1} - \theta_j}{\epsilon} \right)^2 \right\}, \tag{2.44}$$

where

$$[\tilde{\mathcal{D}}\theta] = \left(\frac{mR^2}{2\pi i \epsilon \hbar} \right)^{N/2} \mathrm{d}\theta_1 \cdots \mathrm{d}\theta_{N-1}. \tag{2.45}$$

Since the initial angle of equation (2.44) is given by $\theta_0 = \theta_i + 2\pi k$, the sum over k creates all possible angles that are equivalent to θ_i on the circle. This integer corresponds to what is called the *winding number* of the classical path from θ_i to θ_f and represents how many times the path winds around the circle starting from θ_i before reaching θ_f. In this regard, it is similar to the integers in equation (2.36), which corresponded to the number of wall bounces for the particle in the box. The winding number is present because the *topology* of a circle is nontrivial. It is often stated that the path integral measure must include a sum over the distinct *topological sectors* of the underlying manifold for the system, and the derivation for the circle quantifies the meaning of this statement.

2.5 Path integrals with singular coordinates and potentials

At first glance it would seem to be a fairly straightforward task to extend the path integral of equation (2.11) or (2.13) to higher dimensions by simply replacing the intermediate integrations with the appropriate higher dimensional volumes, $\mathrm{d}^n\boldsymbol{x}_j$ or $\mathrm{d}^n\boldsymbol{p}_j$, and using the time-sliced Lagrangian written in terms of the particular coordinate system in use. In some cases this is a viable procedure. However, there are obstacles to doing this for an arbitrary potential *as well as in an arbitrary coordinate system.*

It was stated earlier that the separation of the kinetic and potential energies of the evolution operator in equation (2.2) relied on the suppression of commutators in the BCH expansion. Singular potentials, such as the physically important Coulomb potential $V(r) = -k/r$, cause the commutator $\epsilon^2[\hat{P}^2, \hat{V}(\hat{X})]$ to diverge in the neighborhood of the singularity, and this prevents discarding the contribution of higher order commutators in the limit $\epsilon \to 0$. For such a case, the naïve time slicing approach used to derive equation (2.11) is inapplicable since the Trotter product formula fails. This problem even appears in the two-dimensional path integral for a free particle when polar coordinates are chosen, since the centrifugal barrier that appears in polar coordinates creates a coordinate singularity in the Hamiltonian. The details of this failure are important to understand, since they help determine when and how the path integral approach, in particular the time slicing recipe, can be misleading. The free particle path integral in polar coordinates provides some relatively easy insights into this surprisingly subtle problem.

The path integral for the transition element of a free particle moving in two dimensions is straightforward to derive in Cartesian coordinates. The procedure is identical to the time slicing that led to equation (2.11) with the exception that the position and momentum unit projection operators are given by

$$\hat{\mathbf{1}}_x = \int_{-\infty}^{\infty} \mathrm{d}x\, \mathrm{d}y\, |x, y\rangle\langle x, y| \quad \text{and} \quad \hat{\mathbf{1}}_p = \int_{-\infty}^{\infty} \mathrm{d}p_x\, \mathrm{d}p_y\, |p_x, p_y\rangle\langle p_x, p_y|, \tag{2.46}$$

where $|x, y\rangle = |x\rangle \otimes |y\rangle$ and $|p_x, p_y\rangle = |p_x\rangle \otimes |p_y\rangle$ are the usual tensor products of the one-dimensional states. As a result, the inner product is the product of two copies of equation (1.37) and given by $\langle x, y|p_x, p_y\rangle = \exp(i\boldsymbol{p}\cdot\boldsymbol{r}/\hbar)/2\pi\hbar$, where $\boldsymbol{p}\cdot\boldsymbol{r} = p_x x + p_y y$ is the usual scalar product of two vectors. Using $\mathbf{1}_p$ and the inner product gives the infinitesimal transition element for $\hat{H} = \hat{\boldsymbol{P}}^2/2\,m = (\hat{P}_x^2 + \hat{P}_y^2)/2m$,

$$\langle x', y'|e^{-i\epsilon\hat{H}/\hbar}|x, y\rangle = \int_{-\infty}^{\infty} \frac{\mathrm{d}^2\boldsymbol{p}}{(2\pi\hbar)^2} \exp\left\{\frac{i}{\hbar}\epsilon\left(\boldsymbol{p}\cdot\frac{(\boldsymbol{r}' - \boldsymbol{r})}{\epsilon} - \frac{\boldsymbol{p}^2}{2m}\right)\right\}, \tag{2.47}$$

where $\boldsymbol{p}^2 = p_x^2 + p_y^2$ and $\mathrm{d}^2\boldsymbol{p} = \mathrm{d}p_x\,\mathrm{d}p_y$. The action appearing in equation (2.47) is the time-sliced version of the classical phase space Lagrangian $\boldsymbol{p} \cdot \dot{\boldsymbol{r}} - \boldsymbol{p}^2/2m$. The two momentum integrations are Gaussian, and the result is the configuration space infinitesimal transition element,

$$\langle x', y'|e^{-i\epsilon\hat{H}/\hbar}|x, y\rangle = \left(\frac{m}{2\pi i\hbar\epsilon}\right)\exp\left\{\frac{i}{\hbar}\epsilon\left(\frac{1}{2}m\frac{(\boldsymbol{r}' - \boldsymbol{r})^2}{\epsilon^2}\right)\right\}, \tag{2.48}$$

where $(\boldsymbol{r}' - \boldsymbol{r})^2 = (x' - x)^2 + (y' - y)^2$. The action appearing in equation (2.48) is the time-sliced version of the classical action $\mathcal{L} = \frac{1}{2}m\dot{\boldsymbol{r}}^2 = \frac{1}{2}m(\dot{x}^2 + \dot{y}^2)$. Using Cartesian coordinates generates no ambiguities and yields two expressions, equations (2.47) and (2.48), that are simply the product of two copies of the one-dimensional Cartesian results (2.3) and (2.4). Adding a potential that has no singularities is straightforward since the separation (2.2) is valid.

Expressing the free particle path integral using polar coordinates r and θ starts by considering the classical Lagrangian $\mathcal{L} = \frac{1}{2}m\dot{r}^2 + \frac{1}{2}mr^2\dot{\theta}^2$ previously used with the particle moving in a circle. The *classical Hamiltonian* in polar coordinates is easily found using (1.8), and is given by

$$H(\boldsymbol{p}, \boldsymbol{r}) = \frac{p_r^2}{2m} + \frac{p_\theta^2}{2\,mr^2}, \tag{2.49}$$

where p_r is the radial momentum and p_θ is the angular momentum. The singularity at $r = 0$ is a signal that the BCH expansion (2.2) will be problematic, but the full demonstration of this requires further effort. Using equation (2.49), the associated classical phase space Lagrangian is found to be $\mathcal{L} = p_r\dot{r} + p_\theta\dot{\theta} - p_r^2/2\,m - p_\theta^2/2\,mr^2$, where $p_r = m\dot{r}$ and $p_\theta = mr^2\dot{\theta}$. With no detailed knowledge of the underlying quantum mechanical theory in polar coordinates, the time slicing recipe discussed in the previous section would replace the phase space Lagrangian appearing in equation (2.47) with the polar coordinate version, $\mathcal{L} = p_r(r' - r)/\epsilon + p_\theta(\theta' - \theta)/\epsilon - p_r^2/2\,m - p_\theta^2/2\,mr^2$. It is assumed that the integration over the polar coordinate momenta, p_r and p_θ, would result in the polar coordinate time-sliced Lagrangian $\mathcal{L} = \frac{1}{2}m(r' - r)^2/\epsilon^2 + \frac{1}{2}mr^2(\theta' - \theta)^2/\epsilon^2$ appearing in the action, along with the overall factor $m/2\pi i\epsilon\hbar$. It is not difficult to show that this result occurs only if the Cartesian momentum integrations of equation (2.47) are replaced with the polar momentum integrations $\mathrm{d}p_x\,\mathrm{d}p_y/(2\pi\hbar)^2 = \mathrm{d}p_r\,\mathrm{d}p_\theta/r(2\pi\hbar)^2$. This is not the standard polar coordinate momentum space measure, $\mathrm{d}^2\boldsymbol{p}/(2\pi\hbar)^2 = p_r\,\mathrm{d}p_r\,\mathrm{d}\theta_p/(2\pi\hbar)^2$, signaling problems with the time slicing recipe.

In order to reveal the correct time-sliced matrix element it is necessary to consider some of the underlying quantum properties of the system. This starts by introducing the polar configuration states $|r, \theta\rangle$. These states must span the same Hilbert space as the Cartesian states, $|x, y\rangle$, in such a way that the polar projection operator is given by

$$\hat{\mathbf{1}}_r = \int \mathrm{d}^2\boldsymbol{r}\, |r,\theta\rangle\langle r,\theta| = \int_0^\infty \mathrm{d}r\, r \int_0^{2\pi} \mathrm{d}\theta\, |r,\theta\rangle\langle r,\theta|. \tag{2.50}$$

Orthonormality of the polar coordinate position states is stated as

$$\begin{aligned}\langle r',\theta'|r,\theta\rangle = \delta^2(\boldsymbol{r}'-\boldsymbol{r}) &= \frac{1}{r}\delta(r'-r)\,\delta(\theta'-\theta)\\ &= \delta(r\cos\theta - r'\cos\theta')\,\delta(r\sin\theta - r'\sin\theta').\end{aligned} \tag{2.51}$$

The second form for the polar coordinate Dirac delta appearing in equation (2.51) can be proven to be equivalent to the first more commonly used version from property (1.19) of the Dirac delta. If equation (2.50) is a unit projection operator in the Hilbert space, the two sets of states must have the inner product

$$\langle x, y|r,\theta\rangle = \delta(x - r\cos\theta)\,\delta(y - r\sin\theta), \tag{2.52}$$

which is consistent with the standard relationship between polar and Cartesian coordinates. Using equations (2.52) and (2.51) then gives

$$\begin{aligned}\langle r',\theta'|r,\theta\rangle = \langle r',\theta'|\hat{\mathbf{1}}_x|r,\theta\rangle &= \int \mathrm{d}x\,\mathrm{d}y\,\langle r',\theta'|x,y\rangle\langle x,y|r,\theta\rangle\\ &= \delta(r\cos\theta - r'\cos\theta')\,\delta(r\sin\theta - r'\sin\theta') = \delta^2(\boldsymbol{r}'-\boldsymbol{r}).\end{aligned} \tag{2.53}$$

A similar result, $\langle x', y'|\hat{\mathbf{1}}_r|x, y\rangle = \delta^2(\boldsymbol{x}'-\boldsymbol{x})$, holds for the polar projection operator of equation (2.50), although the demonstration requires using equation (1.19).

Using the inner product (2.52) and the projection operator $\hat{\mathbf{1}}_x$ allows the polar coordinate version of the infinitesimal transition element to be found from equation (2.48),

$$\begin{aligned}G(\boldsymbol{r}',\boldsymbol{r},\epsilon) &= \langle r',\theta'|e^{-i\epsilon\hat{H}/\hbar}|r,\theta\rangle\\ &= \int \mathrm{d}^2\boldsymbol{x}\,\mathrm{d}^2\boldsymbol{x}'\,\langle r',\theta'|x',y'\rangle\langle x',y'|e^{-i\epsilon\hat{H}/\hbar}|x,y\rangle\langle x,y|r,\theta\rangle\\ &= \left(\frac{m}{2\pi i\hbar\epsilon}\right)\exp\left\{\frac{i}{\hbar}\epsilon\left(\frac{1}{2}m\frac{(r'\cos\theta' - r\cos\theta)^2}{\epsilon^2} + \frac{1}{2}m\frac{(r'\sin\theta' - r\sin\theta)^2}{\epsilon^2}\right)\right\}\\ &= \left(\frac{m}{2\pi i\hbar\epsilon}\right)\exp\left\{\frac{i}{\hbar}\left(\frac{1}{2}\frac{m}{\epsilon}(r'^2 + r^2) - m\frac{r'r}{\epsilon}\cos(\theta'-\theta)\right)\right\}\\ &= \left(\frac{m}{2\pi i\hbar\epsilon}\right)\exp\left\{\frac{i}{\hbar}\epsilon\left(\frac{1}{2}m\frac{(r'-r)^2}{\epsilon^2} + 2\,m\frac{r'r}{\epsilon^2}\sin^2\frac{1}{2}(\theta'-\theta)\right)\right\}.\end{aligned} \tag{2.54}$$

The final line of equation (2.54) is reminiscent of the time slicing recipe for the polar coordinate Lagrangian if $\sin^2\frac{1}{2}(\theta'-\theta) \to \frac{1}{4}(\theta'-\theta)^2$, which would hold if $\theta'-\theta \to 0$. However, as has been pointed out earlier, this is not the case since both θ' and θ are arbitrary angles. As a result, the correct time-sliced action appearing in the transition element differs significantly from the classical form.

In order to understand this difference, it is useful to examine the correct quantum mechanical Hamiltonian expressed entirely in terms of self-adjoint operators. The free Hamiltonian in wave mechanics is given by

$$\begin{aligned}\hat{H} &= -\frac{\hbar^2}{2m}\left(\frac{\partial^2}{\partial x^2} + \frac{\partial^2}{\partial x^2}\right) = -\frac{\hbar^2}{2m}\left(\frac{\partial^2}{\partial r^2} + \frac{1}{r}\frac{\partial}{\partial r} + \frac{1}{r^2}\frac{\partial^2}{\partial \theta^2}\right)\\ &= -\frac{\hbar^2}{2m}\left(\frac{\partial}{\partial r} + \frac{1}{2r}\right)^2 - \frac{\hbar^2}{2mr^2}\left(\frac{\partial^2}{\partial \theta^2} + \frac{1}{4}\right).\end{aligned} \tag{2.55}$$

Both of the polar coordinate forms of the Hamiltonian in equation (2.55) are self-adjoint and the angular momentum operator, $\hat{P}_\theta = -i\hbar\, \partial/\partial\theta$, familiar from wave mechanics is clearly present in both. The second polar form reflects the fact that the radial momentum *cannot* be identified as $-i\hbar\, \partial/\partial r$ since it would *not be self-adjoint*. This is due to the presence of the factor of r in the integral measure $\mathrm{d}^2\boldsymbol{r} = r\, \mathrm{d}r\, \mathrm{d}\theta$ of polar coordinates, so that integrating by parts will create *imaginary terms* in the expectation value. The correct self-adjoint form for the radial momentum in the wave mechanical context is given by $\hat{P}_r = -i\hbar\, (\partial/\partial r + 1/2r)$ since it allows integration by parts in the expectation value. Since $[r, 1/r] = 0$, this form for the radial momentum also satisfies the required quantization condition $[r, \hat{P}_r] = i\hbar$. The momentum states associated with $\hat{P}_r$ and $\hat{P}_\theta$ are written $|p_r, \nu\rangle$ and satisfy $\hat{P}_r|p_r, \nu\rangle = p_r|p_r, \nu\rangle$ and $\hat{P}_\theta|p_r, \nu\rangle = \nu\hbar|p_r, \nu\rangle$. The eigenvalue p_r is continuous, while ν is an integer. In position representation they are given by

$$\langle r, \theta|p_r, \nu\rangle = \frac{1}{2\pi\sqrt{r\hbar}}\, e^{ip_r r/\hbar} e^{i\nu\theta} \implies \hat{\mathbf{1}}_p = \int_{-\infty}^{\infty} \mathrm{d}p_r \sum_{\nu=-\infty}^{\infty} |p_r, \nu\rangle\langle p_r, \nu|. \tag{2.56}$$

The correct quantum mechanical Hamiltonian written in terms of self-adjoint operators for polar coordinates is therefore obtained from the second form in equation (2.55),

$$\hat{H}(\hat{P}_r, \hat{P}_\theta, \hat{R}, \hat{\theta}) = \frac{1}{2m}\hat{P}_r^2 + \frac{1}{2m\hat{R}^2}\left(\hat{P}_\theta^2 - \frac{1}{4}\hbar^2\right). \tag{2.57}$$

Result (2.57) contains an attractive potential due to quantum effects that is singular at the origin, just as the centrifugal barrier is.

The goal is to determine how results (2.56) and (2.57) appear in the infinitesimal propagator (2.54). Adapting an approach developed by Kleinert, the second to last line of equation (2.54) can be rewritten using an identity involving the modified Bessel functions $I_\nu(z)$ of integer order,

$$e^{z\cos(\theta'-\theta)} = \sum_{\nu=-\infty}^{\infty} I_\nu(z) e^{i\nu(\theta'-\theta)}, \tag{2.58}$$

which gives

$$G(\mathbf{r}', \mathbf{r}, \epsilon) = \left(\frac{m}{2\pi i\hbar\epsilon}\right) \sum_{\nu=-\infty}^{\infty} I_\nu\left(\frac{mr'r}{i\hbar\epsilon}\right) \exp\left\{\frac{i}{\hbar}\frac{1}{2}\frac{m}{\epsilon}(r'^2 + r^2)\right\} e^{i\nu(\theta'-\theta)}. \tag{2.59}$$

It is important to verify that equation (2.59) satisfies the requirement (2.27) when integrated over the intermediate polar coordinates. Integrating $G(\mathbf{r}', \mathbf{r}, \epsilon)\, G(\mathbf{r}, \mathbf{r}'', \epsilon)$ over θ generates the Kronecker delta $2\pi\, \delta_{\nu\nu'}$ in the order indices of the modified Bessel functions in the same manner as the angular integral in equation (2.38). Summing over the second Bessel function index then results in

$$\begin{aligned}\int d^2r\, G(\mathbf{r}', \mathbf{r}, \epsilon)\, G(\mathbf{r}, \mathbf{r}'', \epsilon) = &\left(\frac{i^2m^2}{2\pi\hbar^2\epsilon^2}\right) \sum_{\nu=-\infty}^{\infty} \int_0^\infty dr\, r\, I_\nu\left(\frac{mr'r}{i\hbar\epsilon}\right) I_\nu\left(\frac{mrr''}{i\hbar\epsilon}\right) \\ &\times \exp\left\{\frac{i}{\hbar}\frac{1}{2}\frac{m}{\epsilon}\left(r'^2 + 2r^2 + r''^2\right)\right\} e^{i\nu(\theta'-\theta'')}.\end{aligned} \tag{2.60}$$

The integral over r follows from a second Bessel function identity,

$$\int_0^\infty dr\, r\, e^{-r^2/\alpha} I_\nu(\beta r) I_\nu(\gamma r) = \frac{1}{2}\, \alpha\, e^{-\frac{1}{4}(\beta^2+\gamma^2)\alpha} I_\nu\left(\frac{1}{2}\alpha\beta\gamma\right). \tag{2.61}$$

Setting $\alpha = i\epsilon\hbar/m$, $\beta = mr'/i\epsilon\hbar$, and $\gamma = mr''/i\epsilon\hbar$ turns the left-hand side of equation (2.61) into the integral of equation (2.60), so that

$$\begin{aligned}&\int d^2r\, G(\mathbf{r}', \mathbf{r}, \epsilon)\, G(\mathbf{r}, \mathbf{r}'', \epsilon) \\ &= \left(\frac{m}{2\pi i\hbar(2\epsilon)}\right) \sum_{\nu=-\infty}^{\infty} I_\nu\left(\frac{mr'r''}{i\hbar(2\epsilon)}\right) \exp\left\{\frac{i}{\hbar}\frac{1}{2}\frac{m}{(2\epsilon)}(r'^2 + r''^2)\right\} e^{i\nu(\theta'-\theta'')} \\ &= G(\mathbf{r}', \mathbf{r}'', 2\epsilon).\end{aligned} \tag{2.62}$$

Result (2.62) verifies that equation (2.59) is the correct polar coordinate form for the infinitesimal transition element.

Result (2.59) can now be used to identify the failure of the time slicing recipe by looking at two different limits. The first limit holds r' and r finite and lets $\epsilon \to 0$. In that limit the Bessel function becomes, to $O(\epsilon)$,

$$\lim_{\epsilon\to 0} I_\nu\left(\frac{mr'r}{i\hbar\epsilon}\right) \approx \sqrt{\frac{i\hbar\epsilon}{2\pi mrr'}} \exp\left(-\frac{i}{\hbar}\frac{mr'r}{\epsilon} - \frac{i\hbar\epsilon}{2\, mr'r}\left(\nu^2 - \frac{1}{4}\right)\right). \tag{2.63}$$

Using equation (2.63) in equation (2.59) gives the $\epsilon \to 0$ limit for fixed r and r',

$$
\begin{aligned}
G(\mathbf{r}', \mathbf{r}, \epsilon) &= \frac{1}{2\pi}\sqrt{\frac{m}{2\pi i\epsilon\hbar r' r}} \exp\left\{\frac{i}{\hbar}\epsilon\frac{1}{2}m\frac{(r'-r)^2}{\epsilon^2}\right\} \\
&\quad\times \sum_{\nu=-\infty}^{\infty} \exp\left\{-\frac{i}{\hbar}\epsilon\frac{1}{2\,mr'r}\left(\hbar^2\nu^2 - \frac{\hbar^2}{4}\right)\right\} e^{i\nu(\theta'-\theta)} \\
&= \frac{1}{\sqrt{r'r}}\int_{-\infty}^{\infty}\frac{dp_r}{2\pi\hbar}\exp\left\{-\frac{i}{\hbar}\epsilon\frac{p_r^2}{2m}\right\}e^{ip_r(r'-r)/\hbar} \\
&\quad\times \sum_{\nu=-\infty}^{\infty}\frac{1}{2\pi}\exp\left\{-\frac{i}{\hbar}\epsilon\frac{1}{2\,mr'r}\left(\hbar^2\nu^2 - \frac{\hbar^2}{4}\right)\right\}e^{i\nu(\theta'-\theta)}.
\end{aligned}
\tag{2.64}
$$

Using equation (2.56) shows that equation (2.64) is identical to the infinitesimal transition element

$$
\langle r', \theta'|e^{-i\epsilon\hat{H}/\hbar}|r, \theta\rangle = \int_{-\infty}^{\infty} dp_r \sum_{\nu=-\infty}^{\infty} \langle r', \theta'|e^{-i\epsilon\hat{H}/\hbar}|p_r, \nu\rangle\langle p_r, \nu|r, \theta\rangle, \tag{2.65}
$$

where the commutators in equation (2.2) *have been discarded.* The Hamiltonian appearing in equation (2.65) is the correct polar coordinate form (2.57). Therefore, *if* r and r' are finite in the limit that $\epsilon \to 0$, the Trotter product theorem is valid and the correct result (2.64) for the infinitesimal transition element is obtained.

On the other hand, the limit $r \to 0$ or $r' \to 0$ before $\epsilon \to 0$ yields a form inconsistent with the Trotter product theorem being applied to the Hamiltonian (2.57). For such a case, the Bessel function in (2.59) becomes

$$
\lim_{r\to 0} I_\nu\left(\frac{mr'r}{i\hbar\epsilon}\right) \approx \frac{1}{|\nu|!}\left(\frac{mr'r}{2i\hbar\epsilon}\right)^{|\nu|}, \tag{2.66}
$$

and this limiting form gives

$$
\begin{aligned}
\lim_{r\to 0}\sum_{\nu=0}^{\infty} I_\nu\left(\frac{mr'r}{i\hbar\epsilon}\right)e^{i\nu(\theta'-\theta)} &\approx \sum_{\nu=0}^{\infty}\frac{1}{\nu!}\left(\frac{mr'r}{2i\hbar\epsilon}e^{i(\theta'-\theta)}\right)^{\nu} \\
&= \exp\left\{-\frac{i}{\hbar}\frac{mr'r}{2\epsilon}e^{i(\theta'-\theta)}\right\}.
\end{aligned}
\tag{2.67}
$$

As a result, in the neighborhood of $r = 0$, the sum in equation (2.59) becomes

$$
\lim_{r\to 0}\sum_{\nu=-\infty}^{\infty} I_\nu\left(\frac{mr'r}{i\hbar\epsilon}\right)e^{i\nu(\theta'-\theta)} = \lim_{r\to 0}\exp\left\{-\frac{i}{\hbar}\frac{mr'r}{\epsilon}\cos(\theta'-\theta)\right\}. \tag{2.68}
$$

This shows that the deviation of the correct polar coordinate transition element (2.68) from the expected time slicing recipe (2.65) is brought about by the failure of the Trotter product formula in the neighborhood of the singularity at $r = 0$.

The failure of the time slicing recipe in the limit $r \to 0$ *cannot be avoided* in the full polar coordinate path integral, since the intermediate integrations over r will include those values. As a result, the correct path integral in polar coordinates for *all* r and r' must use the form (2.59) for the infinitesimal transition element, and is given in configuration space form by

$$G(\boldsymbol{r}_f, \boldsymbol{r}_i, T)$$
$$= \int [\tilde{\mathcal{D}}^2\boldsymbol{r}] \prod_{j=0}^{N-1} \left(\sum_{\nu_j=-\infty}^{\infty} I_{\nu_j}\left(\frac{m r_{j+1} r_j}{i\hbar\epsilon}\right) \exp\left\{\frac{i}{\hbar}\frac{1}{2}\frac{m}{\epsilon}\left(r_{j+1}^2 + r_j^2\right)\right\} e^{i\nu_j(\theta_{j+1}-\theta_j)} \right), \tag{2.69}$$
$$[\tilde{\mathcal{D}}^2\boldsymbol{r}] = \left(\frac{m}{2\pi i\hbar\epsilon}\right)^N \mathrm{d}^2\boldsymbol{r}_1 \cdots \mathrm{d}^2\boldsymbol{r}_{N-1}.$$

where the usual identifications $\boldsymbol{r}_N = \boldsymbol{r}_f$ and $\boldsymbol{r}_0 = \boldsymbol{r}_i$ have been made and $\mathrm{d}^2\boldsymbol{r}_j = r_j\, \mathrm{d}r_j\, \mathrm{d}\theta_j$. It is straightforward to add a potential energy $V(r_j, \theta_j)$ in the first exponential as long as it is not singular.

Singular potentials, such as the critically important Coulomb potential $V(r) = -Ke^2/r$, will suffer from the same difficulty as polar coordinates in the vicinity of the singularity. Although the naïve time slicing approach fails, such potentials can be given a path integral representation using a method developed by Duru and Kleinert. The key step in their approach is modifying the time slicing recipe in the vicinity of the singularity by making the recipe position dependent. In its simplest form this is accomplished by rewriting the evolution operator as

$$\exp\left\{-\frac{i}{\hbar}\epsilon\, H(p_j, x_j)\right\} = \exp\left\{-\frac{i}{\hbar}\mathrm{d}s_j\, \zeta(x_j) H(p_j, x_j)\right\}, \tag{2.70}$$

where $\mathrm{d}s_j = \epsilon/\zeta(x_j)$ is a reparameterization of the time variable. The parameter s_j replaces the time parameter t, and $\mathrm{d}s_j$ replaces ϵ in the Trotter product formula, allowing $\zeta(x_j)$ to be chosen to suppress the singularities in the potential. This approach requires revising the transition element into a *fixed energy transition element*, which allows time reparameterization. While the fixed energy transition element is discussed in the context of tunneling in the next chapter, a full explication of the Duru–Kleinert approach is beyond the scope of this brief introductory text. The reader is recommended to the complete description of the Duru–Kleinert method contained in the very thorough and valuable monograph by Kleinert, where, among many applications, the singular Coulomb system is transformed into the well behaved harmonic oscillator system.

2.6 The harmonic oscillator and coherent state path integrals

The simple harmonic oscillator is an extremely important exactly solvable quantum mechanical system. The energy states and levels explain a great deal about the interaction of matter and light. In addition, they are often used as a basis for a *perturbative* approach to analyzing polynomial potentials sketched in the next chapter. In one dimension, the Hamiltonian is parameterized by the mass m and the angular frequency ω and is given by $\hat{H} = \frac{1}{2m}\hat{P}^2 + \frac{1}{2}m\omega^2\hat{X}^2$. Since $V(\hat{X}) = \frac{1}{2}m\omega^2\hat{X}^2$ is not singular and the space of interest is not compact, the path integral for the transition amplitude is given by equation (2.11) or (2.13). This path integral will be evaluated in the next chapter.

However, there is a very useful alternative formulation of the harmonic oscillator due to Dirac, and it will be of great value in formulating the path integral for quantum field theory. Using the self-adjoint operators $\hat{P}$ and $\hat{X}$, the dimensionless *annihilation operator* $\hat{a}$ and its adjoint, the *creation operator* $\hat{a}^\dagger$, are defined as

$$\hat{a} = \sqrt{\frac{m\omega}{2\hbar}}\left(\hat{X} + \frac{i}{m\omega}\hat{P}\right) \implies \hat{a}^\dagger = \sqrt{\frac{m\omega}{2\hbar}}\left(\hat{X} - \frac{i}{m\omega}\hat{P}\right). \tag{2.71}$$

The commutation relation (1.35) for $\hat{X}$ and $\hat{P}$ immediately gives

$$[\hat{a}, \hat{a}^\dagger] = 1. \tag{2.72}$$

The Hamiltonian can be written in terms of $\hat{a}$ and $\hat{a}^\dagger$ by noting that

$$\hat{X} = \sqrt{\frac{\hbar}{2\,m\omega}}(\hat{a}^\dagger + \hat{a}), \quad \hat{P} = i\sqrt{\frac{m\omega\hbar}{2}}(\hat{a}^\dagger - \hat{a}). \tag{2.73}$$

Inserting equation (2.73) into the Hamiltonian gives

$$\hat{H} = \frac{1}{2}\hbar\omega(\hat{a}^\dagger\hat{a} + \hat{a}\hat{a}^\dagger) = \hbar\omega\left(\hat{a}^\dagger\hat{a} + \frac{1}{2}[\hat{a}, \hat{a}^\dagger]\right) = \hbar\omega\left(\hat{a}^\dagger\hat{a} + \frac{1}{2}\right). \tag{2.74}$$

The ground state of the harmonic oscillator is denoted $|0\rangle$ and has the property that it is annihilated by $\hat{a}$, meaning that $\hat{a}|0\rangle = 0$. This is easy to see in position representation, where $\langle x|0\rangle = N\exp(-m\omega x^2/2\hbar)$ and $\hat{a} = \alpha(m\omega x + \hbar\,\partial/\partial x)$. The constant N is chosen so that $\langle 0|0\rangle = 1$. It follows that $\hat{H}|0\rangle = \frac{1}{2}\hbar\omega|0\rangle$, the usual ground state energy of the harmonic oscillator. Using $\langle 0|\hat{a}^\dagger = (\hat{a}|0\rangle)^\dagger = 0$, it follows that $|1\rangle = \hat{a}^\dagger|0\rangle$ satisfies $\langle 1|0\rangle = 0$ and $\langle 1|1\rangle = \langle 0|\hat{a}\hat{a}^\dagger|0\rangle = \langle 0|\hat{a}^\dagger\hat{a}|0\rangle + \langle 0|[\hat{a}, \hat{a}^\dagger]|0\rangle = \langle 0|0\rangle = 1$. The harmonic oscillator energy spectrum is generated by applying $\hat{a}^\dagger$ to the ground state to create the sequence of orthonormal energy eigenstates, so that $|n\rangle = (\hat{a}^\dagger)^n|0\rangle/\sqrt{n!}$ has the energy eigenvalue $E_n = (n + \frac{1}{2})\hbar\omega$ and satisfies $\langle n|m\rangle = \delta_{nm}$. The operator $\hat{N} = \hat{a}^\dagger\hat{a}$ is referred to as the *number operator*, since it has the property that $[\hat{N}, (\hat{a}^\dagger)^n] = n(\hat{a}^\dagger)^n$, which gives $\hat{N}|n\rangle = n|n\rangle$. The Hamiltonian can therefore be written $\hat{H} = \hbar\omega(\hat{a}^\dagger\hat{a} + \frac{1}{2}) = \hbar\omega(\hat{N} + \frac{1}{2})$.

Using the Dirac formulation, Schrödinger and later Glauber defined the *coherent state* $|\lambda\rangle$ using a *complex dimensionless number* λ as

$$|\lambda\rangle = e^{\lambda\hat{a}^\dagger - \lambda^*\hat{a}}|0\rangle \implies \langle\lambda| = \left(e^{\lambda\hat{a}^\dagger - \lambda^*\hat{a}}|0\rangle\right)^\dagger = \langle 0|e^{-(\lambda\hat{a}^\dagger - \lambda^*\hat{a})}. \tag{2.75}$$

It follows from equation (2.75) that $\langle\lambda|\lambda\rangle = \langle 0|0\rangle = 1$. Using the commutator (2.72) and $\hat{a}|0\rangle = 0$, the BCH theorem (1.55) factorizes the exponential operator, yielding

$$|\lambda\rangle = e^{\lambda\hat{a}^\dagger} e^{-\lambda^*\hat{a}}|0\rangle e^{-\frac{1}{2}\lambda^*\lambda} = e^{\lambda\hat{a}^\dagger}|0\rangle e^{-\frac{1}{2}\lambda^*\lambda} = \sum_{n=0}^{\infty} \frac{\lambda^n}{\sqrt{n!}}|n\rangle e^{-\frac{1}{2}\lambda^*\lambda}, \tag{2.76}$$

so that the coherent state is a mixture of energy eigenstates. Using the orthonormality of the energy eigenstates, the inner product of two different coherent states written in the form (2.76) is given by

$$\langle\sigma|\lambda\rangle = e^{-\frac{1}{2}\sigma^*\sigma} e^{-\frac{1}{2}\lambda^*\lambda} \sum_{n=0}^{\infty} \frac{(\sigma^*\lambda)^n}{n!} = e^{-\frac{1}{2}\sigma^*\sigma - \frac{1}{2}\lambda^*\lambda + \sigma^*\lambda}. \tag{2.77}$$

The coherent states are complete over the space of harmonic oscillator states, possessing the unit projection operator

$$\hat{\mathbf{1}}_\lambda = \int \frac{\mathrm{d}\lambda^R\, \mathrm{d}\lambda^I}{\pi} |\lambda\rangle\langle\lambda| = \int \frac{\mathrm{d}\lambda^*\, \mathrm{d}\lambda}{2\pi i} |\lambda\rangle\langle\lambda| \equiv \int \frac{\mathrm{d}^2\lambda}{\pi} |\lambda\rangle\langle\lambda|. \tag{2.78}$$

The measure in equation (2.78) is understood as an integration over the entire complex plane, so that it is an integration over both the real and the imaginary parts of $\lambda = \lambda^R + i\lambda^I$. The second form of the measure in equation (2.78), $\mathrm{d}\lambda^*\, \mathrm{d}\lambda/2\pi i$, is obtained from the Jacobian,

$$\frac{\mathrm{d}\lambda^*\, \mathrm{d}\lambda}{2\pi i} = \left(\frac{\partial\lambda^*}{\partial\lambda^R}\frac{\partial\lambda}{\partial\lambda^I} - \frac{\partial\lambda^*}{\partial\lambda^I}\frac{\partial\lambda}{\partial\lambda^R}\right)\frac{\mathrm{d}\lambda^R\, \mathrm{d}\lambda^I}{2\pi i} = 2i\frac{\mathrm{d}\lambda^R\, \mathrm{d}\lambda^I}{2\pi i} = \frac{\mathrm{d}\lambda^R\, \mathrm{d}\lambda^I}{\pi}. \tag{2.79}$$

While one form appears to be real and the other imaginary, it should be remembered that the minus sign of the Jacobian is always discarded. The measure can also be written in complex polar coordinates $\lambda = re^{i\theta}$, and this change of coordinates gives the positive measure $\mathrm{d}\lambda^R\, \mathrm{d}\lambda^I = r\, \mathrm{d}r\, \mathrm{d}\theta$. Using $\lambda^*\lambda = r^2$, $\lambda^n = r^n e^{in\theta}$, and $\lambda^{*m} = r^m e^{-im\theta}$ in expression (2.76), it follows from $\int_0^{2\pi} \mathrm{d}\theta\, e^{i(n-m)} = 2\pi\, \delta_{nm}$ that

$$\begin{aligned}\int \frac{\mathrm{d}^2\lambda}{\pi}|\lambda\rangle\langle\lambda| &= \sum_{n,m=0}^{\infty} \int_0^\infty \mathrm{d}r\, r\, \frac{r^{n+m}}{\sqrt{n!\, m!}} e^{-r^2}|n\rangle\langle m| \int_0^{2\pi} \frac{\mathrm{d}\theta}{\pi} e^{i(n-m)\theta} \\ &= \sum_{n=0}^{\infty}\left(\int_0^\infty \mathrm{d}r\, \frac{2r^{2n+1}}{n!} e^{-r^2}\right)|n\rangle\langle n| = \sum_{n=0}^{\infty}|n\rangle\langle n| = \hat{\mathbf{1}}.\end{aligned} \tag{2.80}$$

The coherent states are therefore complete but not orthogonal, a property referred to as *overcompleteness*.

One of the valuable properties of coherent states is found by using the operator identity (1.54), which gives

$$e^{-(\lambda\hat{a}^\dagger - \lambda^*\hat{a})}\hat{a}\, e^{\lambda\hat{a}^\dagger - \lambda^*\hat{a}} = \hat{a} + \lambda \implies \hat{a}\, e^{\lambda\hat{a}^\dagger - \lambda^*\hat{a}} = e^{\lambda\hat{a}^\dagger - \lambda^*\hat{a}}(\hat{a} + \lambda). \tag{2.81}$$

Result (2.81) gives

$$\hat{a}|\lambda\rangle = \lambda|\lambda\rangle \implies (\hat{a}|\lambda\rangle)^\dagger = \langle\lambda|\hat{a}^\dagger = \lambda^*\langle\lambda|. \tag{2.82}$$

The coherent state is an eigenstate of the annihilation operator, but it is *not* an eigenstate of $\hat{a}^\dagger$, $\hat{X}$, or $\hat{P}$. Operators written in terms of $\hat{a}$ and $\hat{a}^\dagger$ have matrix elements found from equation (2.82). For example, the number operator gives

$$\langle\sigma|\hat{N}|\lambda\rangle = \langle\sigma|\hat{a}^\dagger\hat{a}|\lambda\rangle = \sigma^*\lambda\,\langle\sigma|\lambda\rangle. \tag{2.83}$$

Using this result to evaluate arbitrary products of annihilation and creation operators requires placing the product into *normal order*, which moves all of the creation operators to the left and the annihilation operators to the right while observing the commutation relation (2.72). Both $\hat{X}$ and $\hat{P}$ have well defined matrix elements for coherent states. Using equations (2.73) and (2.82) gives

$$\langle\sigma|\hat{X}|\lambda\rangle = \sqrt{\frac{\hbar}{2\,m\omega}}\,(\sigma^*+\lambda)\langle\sigma|\lambda\rangle,\quad \langle\sigma|\hat{P}|\lambda\rangle = i\sqrt{\frac{m\omega\hbar}{2}}\,(\sigma^*-\lambda)\langle\sigma|\lambda\rangle. \tag{2.84}$$

Result (2.84) allows a semiclassical interpretation for coherent states.

Using equations (2.77), (2.78), and (2.83), Klauder derived the harmonic oscillator transition element between two coherent states in terms of a path integral. The path integral is developed using the familiar time slicing argument of equation (2.9) for the time-ordered evolution operator $\hat{U}(T)$ defined by equation (2.5). Once again, denoting $\hat{U}_j = \exp(-i\epsilon\hat{H}(t_j)/\hbar)$, the coherent state path integral is defined by

$$\begin{aligned} G(\lambda_f, \lambda_i, t_f, t_i) &= \langle\lambda_f|\hat{U}(t_i, t_f)|\lambda_i\rangle \\ &= \int\frac{d^2\lambda_1}{\pi}\cdots\frac{d^2\lambda_{N-1}}{\pi}\,\langle\lambda_f|\hat{U}_{N-1}|\lambda_{n-1}\rangle\langle\lambda_{N-1}|\hat{U}_{N-2}\cdots\hat{U}_1|\lambda_1\rangle\langle\lambda_1|\hat{U}_0|\lambda_i\rangle \\ &= \int\frac{d^2\lambda_1}{\pi}\cdots\frac{d^2\lambda_{N-1}}{\pi}\left(\prod_{j=0}^{N-1}\langle\lambda_{j+1}|\hat{U}_j|\lambda_j\rangle\right), \end{aligned} \tag{2.85}$$

where the notation $\lambda_0 = \lambda_i$ and $\lambda_N = \lambda_f$ is in use, along with their complex conjugates. Each of the infinitesimal transition elements in equation (2.9) can be evaluated using the properties (2.77), (2.83), and (2.84), as well as the Trotter product formula (1.58), yielding

$$\begin{aligned} \langle\lambda_{j+1}|e^{-i\epsilon\hat{H}(t_j)/\hbar}|\lambda_j\rangle &= \langle\lambda_{j+1}|\lambda_j\rangle\exp\left\{-\frac{i}{\hbar}\epsilon\left[\hbar\omega(\lambda^*_{j+1}\lambda_j + \frac{1}{2})\right]\right\} \\ &= \exp\left\{\frac{i}{\hbar}\left[\frac{1}{2}i\hbar\lambda^*_{j+1}(\lambda_{j+1} - \lambda_j) - \frac{1}{2}i\hbar(\lambda^*_{j+1} - \lambda^*_j)\lambda_j - \hbar\omega\epsilon(\lambda^*_{j+1}\lambda_j + \frac{1}{2})\right]\right\} \\ &= \exp\left\{\frac{i}{\hbar}\epsilon\left[\frac{1}{2}i\hbar\lambda^*_{j+1}\dot{\lambda}_j - \frac{1}{2}i\hbar\dot{\lambda}^*_j\lambda_j - \hbar\omega(\lambda^*_{j+1}\lambda_j + \frac{1}{2})\right]\right\}, \end{aligned} \tag{2.86}$$

where the notation $\dot{\lambda}_j = (\lambda_{j+1} - \lambda_j)/\epsilon$ is again in use. Using equation (2.86) gives the final result for equation (2.85),

$$G(\lambda_f, \lambda_i, t_f, t_i) = \int [\mathcal{D}^2\lambda] \exp\left\{\frac{i}{\hbar}\sum_{j=0}^{N-1}\epsilon\left(\frac{1}{2}i\hbar\lambda^*_{j+1}\dot{\lambda}_j - \frac{1}{2}i\hbar\dot{\lambda}^*_j\lambda_j - \hbar\omega\left(\lambda^*_{j+1}\lambda_j + \frac{1}{2}\right)\right)\right\}, \tag{2.87}$$

where the measure is given by

$$[\mathcal{D}^2\lambda] = \frac{d^2\lambda_1}{\pi}\cdots\frac{d^2\lambda_{N-1}}{\pi}. \tag{2.88}$$

The time-sliced action appearing in equation (2.87) corresponds to the classical Lagrangian $\mathcal{L} = \frac{1}{2}i\hbar\lambda^*\dot{\lambda} - \frac{1}{2}i\hbar\dot{\lambda}^*\lambda - \hbar\omega(\lambda^*\lambda + \frac{1}{2})$. A full identification of this classical Lagrangian from the time-sliced action of equation (2.87) requires treating $\lambda^*_{j+1} = \lambda^*_j + \epsilon(\lambda^*_{j+1} - \lambda^*_j)/\epsilon \approx \lambda^*_j + O(\epsilon)$. This reveals another flaw in the time slicing recipe, which is that it replaces the correct variable of integration λ^*_{j+1} in equation (2.87) with the time-sliced variable λ^*_j. This is possible only if λ^*_{j+1} can be treated as $\lambda^*_j + O(\epsilon)$ within the context of the path integral. The validity of this is discussed in the next chapter. It is worth noting that the classical Euler–Lagrange equation yields two *first order* differential equations for λ and λ^*, given by $i\dot{\lambda} - \omega\lambda = 0$ and $i\dot{\lambda}^* + \omega\lambda^* = 0$. These equations of motion admit only one *complex* boundary condition, either an initial or final value for λ or λ^*, and therefore cannot match the transition element boundary conditions.

The Hamiltonian is often generalized to include time-dependent external source terms $-J(t)\,\hat{a}^\dagger - J^*(t)\,\hat{a}$. The complex valued time-dependent function $J(t)$, assumed to be piecewise continuous with no quantum mechanical properties, has the units of $\hbar$ and represents externally induced energy changing processes. The path integral with the source term will be designated as $Z_{if}[J]$ since it is a *functional* of $J(t)$. The benefit of adding the source term is that the path integral becomes a *generating functional* for time-ordered products of the *Heisenberg operators* $\hat{a}(t)$ and $\hat{a}^\dagger(t)$, much in the same way the exponential e^{Jx} in equation (1.99) generated all possible Gaussian integrals by taking the derivative with respect to J. This is most easily seen by considering the factorized form (2.6) for the evolution operator $\hat{U}_J(t_i, t_f)$ used to define the path integral transition element in the presence of the source function. The functional derivatives with respect to $J(t_1)$ and $J^*(t_2)$ yield

$$\begin{aligned}(-i\hbar)^2\frac{\delta^2 Z_{if}[J]}{\delta J(t_1)\,\delta J^*(t_2)} &= \theta(t_1 - t_2)\hat{U}_J(t_f, t_1)\,\hat{a}^\dagger\,\hat{U}_J(t_1, t_2)\,\hat{a}\,\hat{U}_J(t_2, t_i) \\ &+ \theta(t_2 - t_1)\hat{U}_J(t_f, t_2)\,\hat{a}\,\hat{U}_J(t_2, t_1)\,\hat{a}^\dagger\,\hat{U}_J(t_1, t_i) = \hat{U}_J(t_f, t_i)\mathcal{T}\{\hat{a}^\dagger(t_1)\,\hat{a}(t_2)\},\end{aligned} \tag{2.89}$$

which follows from relation (1.50). Using equation (2.82) shows that the source terms appear in the time-sliced Lagrangian in equation (2.87) with the form

$$\mathcal{L} = \frac{1}{2}i\hbar\dot{\lambda}^*_{j+1}\lambda_j - \frac{1}{2}i\hbar\dot{\lambda}^*_j\lambda_j - \hbar\omega(\lambda^*_{j+1}\lambda_j + \frac{1}{2}) + J(t_j)\lambda^*_{j+1} + J^*(t_j)\lambda_j. \quad (2.90)$$

It will be shown later that the functional derivatives of the path integral with the action (2.90) give time-ordered products of the creation and annihilation operators.

2.7 Grassmann quantum mechanics and path integrals

Using the classical Grassmann Lagrangians (1.92) and (1.97), the quantum mechanical path integral for anticommuting variables can be formulated. In keeping with the goal of this text, it is important to understand the relation of the path integral to the underlying Grassmann quantum mechanics. The transition from Grassmann classical mechanics to Grassmann quantum mechanics is complicated by the underlying anticommuting algebra, and often leads to unexpected differences from standard quantum mechanics.

In developing the Dirac version of Grassmann quantum mechanics the starting point is the Grassmann configuration state $|\eta\rangle$. This state is assumed to be an eigenstate of the Hermitian Grassmann operator $\hat{\mathcal{N}}$, so that $\hat{\mathcal{N}}|\eta\rangle = \eta|\eta\rangle$, where η is a real Grassmann variable. The Grassmann operator $\hat{\mathcal{N}}$ must anticommute with Grassmann variables in order to preserve the Grassmann algebra of its eigenvalues. This follows by considering two Grassmann operators, $\hat{\mathcal{N}}_1$ and $\hat{\mathcal{N}}_2$, acting on the tensor product of their eigenstates, $\hat{\mathcal{N}}_1\hat{\mathcal{N}}_2|\eta_1, \eta_2\rangle = \hat{\mathcal{N}}_1\eta_2|\eta_1, \eta_2\rangle = -\eta_2\hat{\mathcal{N}}_1|\eta_1, \eta_2\rangle = -\eta_2\eta_1|\eta_1, \eta_2\rangle = \eta_1\eta_2|\eta_1, \eta_2\rangle$. In turn, this is consistent with $\{\hat{\mathcal{N}}_1, \hat{\mathcal{N}}_2\} = 0$, so that the two Grassmann position operators anticommute as well. Given the Grassmann nature of η, this immediately gives $\hat{\mathcal{N}}^2|\eta\rangle = \hat{\mathcal{N}}\eta|\eta\rangle = -\eta\hat{\mathcal{N}}|\eta\rangle = -\eta^2|\eta\rangle = 0$, so that $\hat{\mathcal{N}}$ is also a *nilpotent operator*. Assuming self-adjointness for $\hat{\mathcal{N}}$ and, importantly, *that* η commutes with the states, it follows that $\eta\langle\eta'|\eta\rangle = \langle\eta'|\hat{\mathcal{N}}|\eta\rangle = \eta'\langle\eta'|\eta\rangle$, so that $(\eta - \eta')\langle\eta'|\eta\rangle = 0$. The states are therefore orthogonal in the Dirac sense, since this result is consistent with $\langle\eta'|\eta\rangle = i\delta(\eta - \eta') = i(\eta - \eta')$, where the Grassmann Dirac delta (1.86) was used. The factor of i has appeared in the inner product in order to satisfy the requirement that $\langle\eta'|\eta\rangle^*=\langle\eta|\eta'\rangle = -\langle\eta'|\eta\rangle$. Using the rules of Grassmann integration (1.68) and (1.69) shows that

$$-i\int d\eta\, \langle\eta'|\eta\rangle\langle\eta|\eta''\rangle = i\int d\eta\, (\eta - \eta')(\eta'' - \eta) = i(\eta'' - \eta') = \langle\eta'|\eta''\rangle, \quad (2.91)$$

so that the states form a unit projection operator $\hat{\mathbf{1}}_\eta = -i\int d\eta\, |\eta\rangle\langle\eta|$, as long as the overall factor of $-i$ is included. This factor is also required to make the projection operator self-adjoint, since it was shown in the previous chapter that $(d\eta)^*=-d\eta$ for real Grassmann variables. However, the states have the unusual property that $\langle\eta|\eta\rangle = 0$. Such states are referred to as *ghost states* and cannot be physically observable since they are not normalizable. Similar results follow for the Grassmann momentum operator $\hat{\mathcal{P}}$, whose eigenstates are $|\rho\rangle$. It is assumed that $\{\eta, \rho\} = 0$, since both ρ and η are Grassmann variables.

Grassmann quantum mechanics for a single pair of real Grassmann variables is defined by enforcing the *anticommutation relation* $\{\hat{\mathcal{N}}, \hat{\mathcal{P}}\} = \hbar$. The anticommutator is required due to the nilpotency of both operators, since multiplying the commutator $[\hat{\mathcal{N}}, \hat{\mathcal{P}}] = i\hbar$ from the right with $\hat{\mathcal{P}}$ or $\hat{\mathcal{N}}$ immediately reveals the commutation relation is inconsistent with nilpotency. Self-adjointness gives $\{\hat{\mathcal{N}}, \hat{\mathcal{P}}\}^\dagger = \{\hat{\mathcal{N}}, \hat{\mathcal{P}}\}$, so that the anticommutator must be real. Using the anticommutation relation between Grassmann operators and Grassmann variables and the unit projection operator of equation (2.91) gives

$$i\hbar(\eta' - \eta) = \hbar\langle\eta|\eta'\rangle = \langle\eta|\{\hat{\mathcal{N}}, \hat{\mathcal{P}}\}|\eta'\rangle = (\eta - \eta')\langle\eta|\hat{\mathcal{P}}|\eta'\rangle. \tag{2.92}$$

Result (2.92) gives the configuration representation of the Grassmann momentum operator, $\langle\eta|\hat{\mathcal{P}}|\eta'\rangle = -i\hbar = \hbar\,(\partial/\partial\eta)\langle\eta|\eta'\rangle = -\hbar\,(\partial/\partial\eta')\langle\eta|\eta'\rangle$. This representation for the momentum satisfies the anticommutation relation since (1.66) gives $\{\eta, \partial/\partial\eta\} = 1$. Using this result for the momentum shows that

$$\begin{aligned}
&\hbar\langle\eta|\rho\rangle = \langle\eta|\{\hat{\mathcal{N}}, \hat{\mathcal{P}}\}|\rho\rangle = \eta\rho\langle\eta|\rho\rangle + i\hbar\int d\eta'\left(\frac{\partial}{\partial\eta'}\langle\eta|\eta'\rangle\right)\eta'\langle\eta'|\rho\rangle \\
&\Longrightarrow \hbar\frac{\partial}{\partial\eta}\langle\eta|\rho\rangle - \rho\langle\eta|\rho\rangle = 0 \Longrightarrow \langle\eta|\rho\rangle = e^{\eta\rho/\hbar},
\end{aligned} \tag{2.93}$$

where the first order differential equation of the second line results from integration by parts for the last term in the first line. The result for $\langle\eta|\rho\rangle$ is formally similar to the c-number result (1.37). It follows that $\langle\rho|\eta\rangle = \langle\eta|\rho\rangle^* = (e^{\eta\rho/\hbar})^* = e^{\rho\eta/\hbar} = e^{-\eta\rho/\hbar}$, where the rules of complex conjugation for real Grassmann variables has been used. Using these results and the rules for Grassmann integration reveals that the momentum states satisfy

$$i\int\hbar\,d\rho\,\langle\eta'|\rho\rangle\langle\rho|\eta\rangle = i\int\hbar\,d\rho\,e^{\eta'\rho/\hbar}e^{-\eta\rho/\hbar} = i(\eta - \eta') = \langle\eta'|\eta\rangle, \tag{2.94}$$

so that $i\int\hbar\,d\rho\,|\rho\rangle\langle\rho| = \hat{\mathbf{1}}_\rho$ is a unit projection operator.

The results so far are sufficient to develop the path integral for the single Grassmann variable Lagrangian (1.92) by the time slicing approach of equation (2.9). The time interval T is partitioned into N subintervals and $N - 1$ complete sets of Grassmann position states are inserted. However, since $\hat{H} = 0$ for the Lagrangian (1.92), the dynamics are trivial and each of the infinitesimal transition elements in the path integral reduces to a Grassmann Dirac delta. Using equation (2.94) this can be written

$$\langle\eta_{j+1}|e^{-i\epsilon\hat{H}/\hbar}|\eta_j\rangle = \int\hbar\,d\rho_j\,\exp\left\{\frac{i}{\hbar}\epsilon\left(\frac{\eta_{j+1} - \eta_j}{\epsilon}\rho_j\right)\right\}, \tag{2.95}$$

where the factor of i has been absorbed into the exponential by the scaling $\rho_j \to i\rho_j$, which gives $d\rho_j \to -i\,d\rho_j$. Returning equation (2.95) to the time-sliced expression gives

$$G(\eta_f, \eta_i, T) = \langle \eta_f | e^{-i\hat{H}T/\hbar} | \eta_i \rangle = \int [\mathcal{D}\rho\ \mathcal{D}\eta] \exp\left\{ \frac{i}{\hbar} \sum_{j=0}^{N-1} \epsilon\ \dot{\eta}_j \rho_j \right\}, \tag{2.96}$$

where the usual notation, $\dot{\eta}_j = (\eta_{j+1} - \eta_j)/\epsilon$, $\eta_0 = \eta_i$, and $\eta_N = \eta_f$ has been used. The measure in equation (2.96) is given by

$$[\mathcal{D}\rho\ \mathcal{D}\eta] = \hbar(-i\hbar\ \mathrm{d}\eta_1)\cdots(-i\hbar\ \mathrm{d}\eta_{N-1})\ \mathrm{d}\rho_0 \cdots \mathrm{d}\rho_{N-1}. \tag{2.97}$$

It is worth noting that both η and ρ have the dimensions of $\sqrt{\hbar}$, while the dimensions of $\mathrm{d}\eta$ and $\mathrm{d}\rho$ are $1/\sqrt{\hbar}$. The time-sliced Lagrangian appearing in equation (2.96) is $\mathcal{L} = \dot{\eta}\rho$, which is identical to the Lagrangian (1.92) with the identification $\rho = -\frac{1}{2}i\eta$. Result (2.96) follows the usual time slicing recipe despite its Grassmann nature.

The Grassmann harmonic oscillator is more complicated but critical to developing Fermionic field theory. In that regard, the complex variable Lagrangian (1.97) is more useful in developing the quantum mechanical version. The standard quantization process is ambiguous since the momentum canonically conjugate to ζ is $-\frac{1}{2}i\zeta^*$, while the momentum canonically conjugate to ζ^* is $-\frac{1}{2}i\zeta$. It is unclear how to replace these momenta with Grassmann derivatives since they are simultaneously the Grassmann position variables. Fortunately, the quantization process can be implemented by enforcing the Heisenberg picture time evolution (1.49) using the Hamiltonian (1.98). The complex variables become operators, and the quantum mechanical Hamiltonian is $\hat{H} = \frac{1}{2}\omega(\hat{\zeta}^\dagger\hat{\zeta} - \hat{\zeta}\hat{\zeta}^\dagger)$. Treating $\hat{\zeta}$ and $\hat{\zeta}^\dagger$ as nilpotent Heisenberg operators, so that $\hat{\zeta}^2 = \hat{\zeta}^{\dagger 2} = 0$, it follows that equation (1.49) requires

$$\begin{aligned} \frac{\mathrm{d}}{\mathrm{d}t}\hat{\zeta}^\dagger &= \frac{i}{\hbar}[\hat{H}, \hat{\zeta}^\dagger] = \frac{i\omega}{2\hbar}[\hat{\zeta}^\dagger\hat{\zeta} - \hat{\zeta}\hat{\zeta}^\dagger, \hat{\zeta}^\dagger] = \frac{i\omega}{\hbar}\hat{\zeta}^\dagger\hat{\zeta}\hat{\zeta}^\dagger, \\ \frac{\mathrm{d}}{\mathrm{d}t}\hat{\zeta} &= \frac{i}{\hbar}[\hat{H}, \hat{\zeta}] = \frac{i\omega}{2\hbar}[\hat{\zeta}^\dagger\hat{\zeta} - \hat{\zeta}\hat{\zeta}^\dagger, \hat{\zeta}] = -\frac{i\omega}{\hbar}\hat{\zeta}\hat{\zeta}^\dagger\hat{\zeta}. \end{aligned} \tag{2.98}$$

The time dependent classical solutions associated with equation (1.97) satisfy $\dot{\zeta}^* = i\omega\zeta^*$ and $\dot{\zeta} = -i\omega\zeta$. In order to be consistent with this result, the two equations of equation (2.98) require both nilpotency *and the anticommutation relation*

$$\{\hat{\zeta}^\dagger, \hat{\zeta}\} = \hat{\zeta}^\dagger\hat{\zeta} + \hat{\zeta}\hat{\zeta}^\dagger = \hbar. \tag{2.99}$$

Relation (2.99) gives both $\zeta\zeta^\dagger\zeta = -\zeta\zeta\zeta^\dagger + \hbar\zeta = \hbar\zeta$ and $\zeta^\dagger\zeta\zeta^\dagger = -\zeta\zeta^\dagger\zeta^\dagger + \hbar\zeta^\dagger = \hbar\zeta^\dagger$, so that both relations of equation (2.98) are equivalent to the equations of motion. It should be clear that a commutation relation similar to equation (2.73) would not yield consistency with the classical equations of motion.

The anticommutation relation (2.99) is the Grassmann version of the c-number oscillator relation (2.72). For convenience, the operator $\hat{\zeta}$ is scaled to the new operator $\hat{a}$ defined by $\hat{\zeta} = \sqrt{\hbar}\ \hat{a}$, so that the anticommutator of equation (2.99) becomes

$$\{\hat{\alpha}^\dagger, \hat{\alpha}\} = 1 \implies \hat{H} = \frac{1}{2}\hbar\omega(\hat{\alpha}^\dagger\hat{\alpha} - \hat{\alpha}\hat{\alpha}^\dagger) = \hbar\omega(\hat{\alpha}^\dagger\hat{\alpha} - \frac{1}{2}). \tag{2.100}$$

The ground state of the Grassmann oscillator is denoted $|0\rangle$ and is defined by $\hat{\alpha}|0\rangle = 0$ and $\langle 0|\hat{\alpha}^\dagger = 0$. By its definition it satisfies $\hat{H}|0\rangle = -\frac{1}{2}\hbar\omega|0\rangle$, so that the Grassmann oscillator has a ground state energy that is exactly opposite to the c-number oscillator. The state is normalized, so that $\langle 0|0\rangle = 1$. The excited state is then given by $|1\rangle = \hat{\alpha}^\dagger|0\rangle$, so that $\langle 1| = \langle 0|\hat{\alpha}$. The definition of the ground state $|0\rangle$ and relation (2.99) yield the following important properties of $|1\rangle$,

$$\begin{aligned} \hat{H}|1\rangle &= \hbar\omega\,(\hat{\alpha}^\dagger\hat{\alpha} - \frac{1}{2})\hat{\alpha}^\dagger|0\rangle = \frac{1}{2}\hbar\omega|1\rangle, \\ \langle 1|1\rangle &= \langle 0|\hat{\alpha}\hat{\alpha}^\dagger|0\rangle = -\langle 0|\hat{\alpha}^\dagger\hat{\alpha}|0\rangle + \langle 0|0\rangle = \langle 0|0\rangle = 1, \\ \langle 1|0\rangle &= \langle 0|\hat{\alpha}|0\rangle = 0. \end{aligned} \tag{2.101}$$

There are no higher excited states since $\hat{\alpha}^{\dagger 2} = 0$. There are therefore only two states available to the Grassmann oscillator, the ground state $|0\rangle$ and the excited state $|1\rangle$. The excited state can be viewed as *fully occupied* by the quantum of energy $\hbar\omega$, implementing what is known as *Fermi–Dirac statistics.* This is revisited in the next chapter where the thermodynamic partition function is analyzed using a path integral representation. The states $|0\rangle$ and $|1\rangle$ can be given a Grassmann position representation as superfunctions, but that will not be presented here.

Introducing a dimensionless complex Grassmann variable ξ that obeys both $\{\hat{\alpha}, \xi\} = 0$ and $(\hat{\alpha}^\dagger\xi)^\dagger = \xi^*\hat{\alpha}$, the Grassmann coherent state $|\xi\rangle$ is defined as

$$|\xi\rangle = \exp(\hat{\alpha}^\dagger\xi - \xi^*\hat{\alpha})|0\rangle \implies \langle\xi| = \langle 0|\exp(-\hat{\alpha}^\dagger\xi + \xi^*\hat{\alpha}). \tag{2.102}$$

The order in equation (2.102) is important because of the Grassmann algebra. Expanding the exponential using nilpotency, equation (2.100), and $\hat{\alpha}^\dagger|0\rangle = |1\rangle$ gives

$$|\xi\rangle = \left(1 - \frac{1}{2}\xi^*\xi\right)|0\rangle - \xi|1\rangle \implies \xi|\xi\rangle = \xi|0\rangle. \tag{2.103}$$

Using equation (2.103), the orthonormality of $|0\rangle$ and $|1\rangle$, and the rules of Grassmann algebra, the inner product of two coherent states can be written

$$\langle\xi|\chi\rangle = (1 - \frac{1}{2}\xi^*\xi)(1 - \frac{1}{2}\chi^*\chi) + \xi^*\chi = \exp\left(-\frac{1}{2}\xi^*\xi - \frac{1}{2}\chi^*\chi + \xi^*\chi\right). \tag{2.104}$$

Using equations (2.103) and (2.100) shows that the coherent states have the property

$$\hat{\alpha}|\xi\rangle = -\hat{\alpha}\,\xi|1\rangle = -\hat{\alpha}\,\xi\hat{\alpha}^\dagger|0\rangle = \xi\hat{\alpha}\hat{\alpha}^\dagger|0\rangle = \xi|0\rangle = \xi|\xi\rangle. \tag{2.105}$$

Result (2.105) gives

$$\langle\chi|\hat{\alpha}|\xi\rangle = \xi\langle\chi|\xi\rangle \implies \langle\chi|\hat{\alpha}^\dagger|\xi\rangle = \langle\xi|\hat{\alpha}|\chi\rangle^* = \chi^*\langle\chi|\xi\rangle. \tag{2.106}$$

The coherent states are complete since equations (2.103), (1.78), (1.79), and nilpotency give

$$\int d^2\xi\ |\xi\rangle\langle\xi| = \int d\xi\ d\xi^*\{(1 - \xi^*\xi)|0\rangle\langle 0| + \xi^*|0\rangle\langle 1| + \xi|1\rangle\langle 0| + \xi\xi^*|1\rangle\langle 1|\} \\ = |0\rangle\langle 0| + |1\rangle\langle 1| = \hat{\mathbf{1}}_E. \tag{2.107}$$

These properties are similar to those of the c-number coherent states (2.75) with one important exception. Using equation (2.103) and anticommutativity, it follows that

$$\sum_{E=0}^{1}\langle E|\xi'\rangle\langle\xi|E\rangle = (1 - \frac{1}{2}\xi^*\xi)(1 - \frac{1}{2}\xi^{*'}\xi') + \xi'\xi^* \\ = (1 - \frac{1}{2}\xi^*\xi)(1 - \frac{1}{2}\xi^{*'}\xi') - \xi^*\xi' = \sum_{E=0}^{1}\langle -\xi|E\rangle\langle E|\xi'\rangle = \langle -\xi|\xi'\rangle. \tag{2.108}$$

The last step shows that the sign change is consistent with $\sum_E|E\rangle\langle E| = \mathbf{1}_E$.

The coherent state transition amplitude $G(\xi_f, \xi_i, T) = \langle\xi_f|\exp(-i\hat{H}T/\hbar)|\xi_i\rangle$ for the Grassmann oscillator can now be given a path integral representation. As with all the previous transition elements, the time interval T is partitioned into N subintervals of duration $\epsilon = T/N$, and $N - 1$ complete sets of coherent states are inserted as part of the time slicing process. As in equation (2.9), the result is the product of N infinitesimal transition elements of the form

$$G(\xi_{j+1}, \xi_j, \epsilon) = \langle\xi_{j+1}|e^{-i\epsilon\hat{H}/\hbar}|\xi_j\rangle, \tag{2.109}$$

where the identifications $\xi_N = \xi_f$ and $\xi_0 = \xi_i$ are made. Combining the form of the Hamiltonian (2.100) and properties (2.104) and (2.106) along with the Grassmann nature of the variables, the infinitesimal transition element is given by

$$G(\xi_{j+1}, \xi_j, \epsilon) = \langle\xi_{j+1}|\xi_j\rangle \exp\left(-i\epsilon\omega(\xi^*_{j+1}\xi_j - \frac{1}{2})\right) \\ = \exp\left\{-\frac{1}{2}\xi^*_{j+1}\xi_{j+1} - \frac{1}{2}\xi^*_j\xi_j + \xi^*_{j+1}\xi_j - i\epsilon\omega\left(\xi^*_{j+1}\xi_j - \frac{1}{2}\right)\right\} \\ = \exp\left\{i\left(\frac{1}{2}i\xi^*_{j+1}(\xi_{j+1} - \xi_j) - \frac{1}{2}i(\xi^*_{j+1} - \xi^*_j)\xi_j - \epsilon\omega\left(\xi^*_{j+1}\xi_j - \frac{1}{2}\right)\right)\right\} \\ = \exp\left\{i\epsilon\left(\frac{1}{2}i\xi^*_{j+1}\dot{\xi}_j - \frac{1}{2}i\dot{\xi}^*_j\xi_j - \omega\left(\xi^*_{j+1}\xi_j - \frac{1}{2}\right)\right)\right\}, \tag{2.110}$$

where the notation $\dot{\xi}_j = (\xi_{j+1} - \xi_j)/\epsilon$ is once again in use. Throughout the manipulations of equation (2.110) the Grassmann algebra could be ignored since pairs of Grassmann variables commute with other pairs of Grassmann variables. The path integral for the Grassmann oscillator coherent state transition element is then given by the product of all N infinitesimal elements of the form (2.110) along with the respective coherent state measures,

$$G(\xi_f, \xi_i, t_f, t_i)$$
$$= \int [\mathcal{D}^2\xi]\exp\left\{\frac{i}{\hbar}\sum_{j=0}^{N-1}\epsilon\left(\frac{1}{2}i\hbar\xi_{j+1}^*\,\dot{\xi}_j - \frac{1}{2}i\hbar\dot{\xi}_j^*\xi_j - \hbar\omega\left(\xi_{j+1}^*\xi_j - \frac{1}{2}\right)\right)\right\}, \quad (2.111)$$

where the measure is given by

$$[\mathcal{D}^2\xi] = \mathrm{d}^2\xi_1\cdots\mathrm{d}^2\xi_{N-1}. \quad (2.112)$$

Apart from the overall factor of $\hbar$ the time-sliced version of the classical Grassmann oscillator Lagrangian (1.97) has appeared in the path integral. The Grassmann oscillator path integral (2.111) is identical in form to the c-number coherent state path integral of equation (2.87). In that regard it has the same variable combination $\xi_{j+1}^*\,\xi_j$ that appeared in the c-number oscillator path integral (2.87). Applying the time slicing recipe would have resulted in replacing ξ_{j+1}^* with $\xi_{j+1}^* = \xi_j^* + O(\epsilon)$. This difference is critical to evaluating the path integral, which is done in the next chapter.

As with the c-number coherent state path integral, the Grassmann Hamiltonian is often generalized to include the terms $-K^*(t)\,\hat{\alpha} - \hat{\alpha}^\dagger K(t)$, where $K(t)$ is an anticommuting Grassmann variable source function. The time-sliced Lagrangian in the Grassmann path integral of equation (2.111) is then given by

$$\mathcal{L}_j = \frac{1}{2}i\hbar\,\xi_{j+1}^*\,\dot{\xi}_j - \frac{1}{2}i\hbar\,\dot{\xi}_j^*\xi_j - \hbar\omega(\xi_{j+1}^*\xi_j - \frac{1}{2}) + K^*(t_j)\,\xi_j + \xi_{j+1}^*K(t_j). \quad (2.113)$$

Functional derivatives with respect to $K(t)$ and $K^*(t)$ give time-ordered products of the anticommuting Heisenberg picture annihilation and creation operators respectively in a manner similar to equation (1.50). This is more complicated than the c-number result (1.50) due to the anticommuting nature of the operators and the Grassmann variables. First, it is important to note that the order of the source term and operator in the product $\hat{\alpha}^\dagger K(t)$ gives an additional minus sign for the functional derivative since $\delta/\delta K(t_1)$ anticommutes with $\hat{\alpha}^\dagger$. Denoting the path integral with source terms as $G_K(\xi_f, \xi_i, t_f, t_i)$, it follows that

$$i\hbar\frac{\delta G_K(\xi_f, \xi_i, t_f, t_i)}{\delta K(t_1)} = \langle\xi_f|\,\hat{U}(t_f, t_1)\hat{\alpha}^\dagger\hat{U}(t_1, t_i)|\xi_i\rangle. \quad (2.114)$$

However, the second functional derivative $\delta/\delta K^*(t_2)$ must be anticommuted through the factor of $\hat{\alpha}^\dagger$, *but only if* $t_1 > t_2$. This gives

$$\begin{aligned}\hbar^2\frac{\delta^2 G_K(\xi_f, \xi_i, t_f, t_i)}{\delta K^*(t_2)\,\delta K(t_1)} &= \theta(t_2 - t_1)\langle\xi_f|\,\hat{U}(t_f, t_2)\hat{\alpha}\hat{U}(t_2, t_1)\hat{\alpha}^\dagger\hat{U}(t_1, t_i)|\xi_i\rangle\\ &\quad - \theta(t_1 - t_2)\langle\xi_f|\,\hat{U}(t_f, t_1)\hat{\alpha}^\dagger\hat{U}(t_1, t_2)\hat{\alpha}\hat{U}(t_2, t_i)|\xi_i\rangle\\ &= \langle\xi_f|\,\hat{U}(t_f, t_i)\mathcal{T}\left\{\hat{\alpha}_H(t_2)\,\hat{\alpha}_H^\dagger(t_1)\right\}|\xi_i\rangle.\end{aligned} \quad (2.115)$$

Therefore, if two Heisenberg operators, $\hat{A}(t_1)$ and $\hat{B}(t_2)$, obey an anticommutation relation, then their time-ordered product is defined as

$$\mathcal{T}\left\{\hat{A}(t_1)\hat{B}(t_2)\right\} = \theta(t_1 - t_2)\hat{A}(t_1)\hat{B}(t_2) - \theta(t_2 - t_1)\hat{B}(t_2)\hat{A}(t_1). \tag{2.116}$$

This definition gives $\mathcal{T}\{\hat{B}(t_2)\hat{A}(t_1)\} = -\mathcal{T}\{\hat{A}(t_1)\hat{B}(t_2)\}$, which is consistent with the underlying anticommutation relations and reflected in the functional derivatives,

$$\frac{\delta^2 G_K(\xi_f, \xi_i, t_f, t_i)}{\delta K(t_1)\, \delta K^*(t_2)} = -\frac{\delta^2 G_K(\xi_f, \xi_i, t_f, t_i)}{\delta K^*(t_2)\, \delta K(t_1)}. \tag{2.117}$$

This will be discussed again in the next chapter where generating functionals for time-ordered products are analyzed.

Further reading

There have been many texts and monographs that focus on the derivation and the nature of quantum mechanical path integrals. In order of original publication date these include

- R P Feynman, A R Hibbs, and D Styer 2010 *Quantum Mechanics and Path Integrals* (New York: Dover)
- L S Schulman 2005 *Techniques and Applications of Path Integration* (New York: Dover)
- R J Rivers 1987 *Path Integral Methods in Quantum Field Theory* (Cambridge: Cambridge University Press)
- H Kleinert 2004 *Path Integrals in Quantum Mechanics, Statistics, Polymer Physics, and Financial Markets* (Singapore: World Scientific)
- M S Swanson 2014 *Path Integrals and Quantum Processes* (New York: Dover)
- C Grosche and F Steiner 1998 *Handbook of Feynman Path Integrals* (Berlin: Springer)
- U Mosel 2004 *Path Integrals in Field Theory: An Introduction* (Berlin: Springer)
- K Fujikawa and H Suzuki 2004 *Path Integrals and Quantum Anomalies* (Oxford: Clarendon Press)
- J Zinn-Justin 2005 *Path Integrals in Quantum Mechanics* (Oxford: Oxford University Press)
- J R Klauder 2011 *A Modern Approach to Functional Integration* (Boston, MA: Birkhäuser)
- A Das 2012 *Field Theory: A Path Integral Approach* 3rd edn (Singapore: World Scientific)
- R Shankar 2017 *Quantum Field Theory and Condensed Matter: An Introduction* (Cambridge: Cambridge University Press)

In addition, there are online reviews and articles that develop many of the introductory concepts of the path integral. These include

- C Grosche 1993 *An Introduction Into the Feynman Path Integral* (arXiv:hep-th/9302097v1)

- R Mackenzie 2000 *Path Integral Methods and Applications* (arXiv:quant-ph/0004090v1)

A mathematical definition of path measure for Brownian motion is discussed in
- N Wiener 1930 *Acta Math.* **55** 117

and in
- M Kac 1959 *Probability and Related Topics in the Physical Sciences* (New York: Interscience)

The Duru–Kleinert method for dealing with singular potentials in the context of the path integral stems from
- H Duru and H Kleinert 1979 *Phys. Lett. B* **84** 185

and is discussed in detail in the monograph on path integrals by Kleinert. The use of coherent states in quantum optics stems from
- R Glauber 1963 *Phys. Rev.* **130** 2529

The initial application of coherent states in path integrals is made in
- J R Klauder 1960 *Ann. Phys.* **11** 123
- J R Klauder 1979 *Phys. Rev. D* **19** 2349

Chapter 3

Evaluating the path integral

In this chapter various techniques for evaluating the quantum mechanical path integral are presented. The related partition function as well as source functionals are also defined and analyzed.

3.1 Performing the intermediate integrations

In the event that the path integral action is quadratic and linear in the integration variables, the path integral can be evaluated exactly by simply performing the intermediate integrations. A simple example of this is the path integral for a free particle moving in one dimension, which is given in configuration space form by setting $V = 0$ in equation (2.13),

$$G(x_f, x_i, t_f, t_i) = \lim_{n\to\infty} \int [\tilde{\mathcal{D}}x] \exp\left\{\frac{i}{\hbar}\sum_{j=0}^{N-1} \epsilon\left(\frac{1}{2}m\dot{x}_j^2\right)\right\}, \tag{3.1}$$

where $\epsilon\, \dot{x}_j^2 = (x_{j+1} - x_j)^2/\epsilon$, $x_0 = x_i$, $x_N = x_f$, and the measure is

$$[\tilde{\mathcal{D}}x] = \left(\frac{m}{2\pi i\epsilon\hbar}\right)^{N/2} \mathrm{d}x_1 \cdots \mathrm{d}x_{N-1}. \tag{3.2}$$

Evaluation starts by setting $\alpha = -im/2\hbar\epsilon$ and $\beta = -im/2\hbar j\epsilon$ in equation (2.32) to find

$$\begin{aligned}&\int_{-\infty}^{\infty} \mathrm{d}y \exp\left\{\frac{im}{2\hbar\epsilon}\left((x-y)^2 + \frac{1}{j}(y-z)^2\right)\right\}\\ &= \sqrt{\frac{2\pi i\hbar j\epsilon}{m(j+1)}} \exp\left\{\frac{im}{2\hbar(j+1)\epsilon}(x-z)^2\right\}.\end{aligned} \tag{3.3}$$

Using equation (3.3) the $N - 1$ integrations in equation (3.1) can now be performed sequentially, starting with x_1. Performing the x_1 integration yields

$$\begin{aligned}&\sqrt{\frac{m}{2\pi i\hbar\epsilon}}\int_{-\infty}^{\infty}\mathrm{d}x_1\exp\left\{\frac{im}{2\hbar\epsilon}\left((x_2 - x_1)^2 + (x_1 - x_i)^2\right)\right\}\\ &= \sqrt{\frac{1}{2}}\exp\left\{\frac{im}{(2\cdot 2)\hbar\epsilon}(x_2 - x_i)^2\right\}.\end{aligned}\tag{3.4}$$

Result (3.4) now contributes to the x_2 integration, which equation (3.3) shows is

$$\begin{aligned}&\sqrt{\frac{1}{2}}\sqrt{\frac{m}{2\pi i\hbar\epsilon}}\int_{-\infty}^{\infty}\mathrm{d}x_2\exp\left\{\frac{im}{2\hbar\epsilon}\left((x_3 - x_2)^2 + \frac{1}{2}(x_2 - x_i)^2\right)\right\}\\ &= \sqrt{\frac{1}{3}}\exp\left\{\frac{im}{(3\cdot 2)\hbar\epsilon}(x_3 - x_i)^2\right\}.\end{aligned}\tag{3.5}$$

This process continues through all $N - 1$ integrations, with the final result for the path integral given by

$$\begin{aligned}G(x_f, x_i, t_f, t_i) &= \lim_{N\to\infty}\sqrt{\frac{1}{N}}\sqrt{\frac{m}{2\pi i\hbar\epsilon}}\exp\left\{\frac{im}{2\hbar N\epsilon}(x_f - x_i)^2\right\}\\ &= \sqrt{\frac{m}{2\pi i\hbar T}}\exp\left\{\frac{i}{\hbar}\left(\frac{1}{2}m\frac{(x_f - x_i)^2}{T}\right)\right\},\end{aligned}\tag{3.6}$$

where $\epsilon = T/N$ was used. Result (3.6) coincides with the operator evaluation,

$$\begin{aligned}\langle x_f|e^{-i\hat{H}T/\hbar}|x_i\rangle &= \int_{-\infty}^{\infty}\mathrm{d}p\,\langle x_f|e^{-i(\hat{P}^2T/2m\hbar)}|p\rangle\langle p|x_i\rangle\\ &= \int_{-\infty}^{\infty}\frac{\mathrm{d}p}{2\pi\hbar}\exp\left\{\frac{i}{\hbar}\left(p(x_f - x_i) - \frac{p^2}{2m}T\right)\right\}\\ &= \sqrt{\frac{m}{2\pi i\hbar T}}\exp\left\{\frac{i}{\hbar}\left(\frac{1}{2}m\frac{(x_f - x_i)^2}{T}\right)\right\},\end{aligned}\tag{3.7}$$

where the Gaussian integral (1.99) was used. It is straightforward to show that equation (3.6) also satisfies the free Schrödinger equation,

$$\left(-\frac{\hbar^2}{2m}\frac{\partial^2}{\partial x^2} - i\hbar\frac{\partial}{\partial t}\right)G(x, x_i, t, t_i) = 0.\tag{3.8}$$

It also obeys $\lim_{t\to t_i} G(x_f, x_i, t_f, t_i) = \delta(x_f - x_i)$. This can be proven by integrating it over x_i against an arbitrary function of x_i and then taking the limit.

In this case, the path integral provided no calculational simplification compared to standard techniques, but the mechanisms by which it yields the correct answer were revealed. The path integral approach is put to better use in the case of a free particle moving one-dimensionally under the influence of an external force $J(t)$. This is modeled by adding the term $V(x, t) = -xJ(t)$ to the classical Hamiltonian. It is important to note that $J(t)$ is restricted to the class of functions *with well-defined derivatives and integrals.* The classical equation of motion is given by $\dot{p}(t) = J(t)$, which is solved by integration to give $p(t) = p_i + \int_{t_o}^{t} \mathrm{d}\tau\, J(\tau)$, where $p_i = p(t_o)$.

The phase space path integral (2.11) with an external source is given by

$$Z_{if}[J] = \int_{-\infty}^{\infty} [\mathcal{D}p][\mathcal{D}x] \exp\left\{\frac{i}{\hbar} \sum_{j=0}^{N-1} \epsilon \left(p_j \dot{x}_j - \frac{p_j^2}{2m} + x_j J(t_j)\right)\right\}, \tag{3.9}$$

where $\epsilon\, p_j \dot{x}_j = p_j(x_{j+1} - x_j)$, $x_N = x_f$, $x_0 = x_i$, and the measure is given by

$$[\mathcal{D}p][\mathcal{D}x] = \frac{\mathrm{d}p_0}{2\pi\hbar} \cdots \frac{\mathrm{d}p_{N-1}}{2\pi\hbar}\, \mathrm{d}x_1 \cdots \mathrm{d}x_{N-1}. \tag{3.10}$$

The path integral (3.9) is denoted $Z_{if}[J]$ since it is a *functional* of the source $J(t)$. It is possible to integrate the momenta p_j and then proceed to the variables x_j. However, it is far easier to integrate the x_j variables first. This process starts by rewriting the action in equation (3.9) using a variant of the integration by parts introduced in equation (2.18). This gives

$$\sum_{j=0}^{N-1} \epsilon\, p_j \dot{x}_j = \sum_{j=0}^{N-1} p_j(x_{j+1} - x_j) = x_f p_{N-1} - x_i p_0 - \sum_{j=1}^{N-1} x_j(p_j - p_{j-1}). \tag{3.11}$$

Using equation (3.11) in equation (3.9) shows that each of the integrations over the x_j are of the form (1.25), which gives

$$\begin{aligned} &\int_{-\infty}^{\infty} \mathrm{d}x_1 \cdots \mathrm{d}x_{N-1} \exp\left\{\frac{i}{\hbar} \sum_{j=1}^{N-1} x_j(p_j - p_{j-1} - \epsilon J(t_j))\right\} \\ &= (2\pi\hbar)^{N-1} \prod_{j=1}^{N-1} \delta(p_j - p_{j-1} - \epsilon J(t_j)). \end{aligned} \tag{3.12}$$

Each of the Dirac deltas appearing in equation (3.12) corresponds to the classical equation of motion $\dot{p}(t) = J(t)$. Using equation (3.12) it is now straightforward to perform $N - 1$ of the momentum integrations. Starting with p_0, the first Dirac delta $\delta(p_1 - p_0 - \epsilon J(t_1))$ replaces p_0 with $p_1 - \epsilon J(t_1)$ in the action. The next Dirac delta replaces p_1 with $p_2 - \epsilon J(t_2)$, so that the previous expression becomes $p_2 - \epsilon J(t_2) - \epsilon J(t_1)$. This process repeats until p_{N-2} is integrated, becoming $p_{N-1} - \epsilon J(t_{N-1})$ in all previously integrated momenta. The final outcome of integrating over all the momenta except for p_{N-1} is to replace the original momentum variables by

$$p_j = p_{N-1} - \sum_{k=j+1}^{N-1} \epsilon J(t_k) = p_{N-1} - \int_{t_j}^{t_f} \mathrm{d}\tau\, J(\tau) = p_{N-1} - I(t_j), \tag{3.13}$$

where the integral occurs as the limit of the Riemann sum. The integral $I(t_j)$ is referred to as the *impulse function*, and can be written using the Heaviside step function (1.16) as

$$I(t_j) = \int_{t_i}^{t_f} \mathrm{d}\tau\, \theta(\tau - t_j) J(\tau). \tag{3.14}$$

It is worth noting that the impulse function of equation (3.14) obeys $\dot{I}(t) = -J(t)$. In that sense, the impulse function integral contains the Green's function $\Delta(t, t') = \theta(t - t')$ for the differential operator $\mathrm{d}/\mathrm{d}t$, since the step function obeys $(\mathrm{d}/\mathrm{d}t)\Delta(t, t') = \delta(t - t')$. Denoting p_{N-1} as p, equation (3.9) becomes a Gaussian integral over p, given by

$$\begin{aligned} Z_{if}[J] &= \int \frac{\mathrm{d}p}{2\pi\hbar} \exp\left\{\frac{i}{\hbar}\left(+x_i I(t_i) + (x_f - x_i)p - \sum_{j=1}^{N-1} \epsilon \frac{(p - I(t_j))^2}{2m}\right)\right\} \\ &= \int \frac{\mathrm{d}p}{2\pi\hbar} \exp\left\{\frac{i}{\hbar}\left(x_i I(t_i) + \left(x_f - x_i + \frac{Q_J}{m}\right)p - \frac{T}{2m}p^2 - \frac{R_J}{2m}\right)\right\}, \end{aligned} \tag{3.15}$$

where the functionals Q_J and R_J are given by

$$Q_J = \sum_{j=1}^{N-1} \epsilon\, I(t_j) \to \int_{t_i}^{t_f} \mathrm{d}t\, I(t), \quad R_J = \sum_{j=1}^{N-1} \epsilon\, I^2(t_j) \to \int_{t_i}^{t_f} \mathrm{d}t\, I^2(t). \tag{3.16}$$

The Gaussian integral over p is given by equation (1.99), and the final result is

$$\begin{aligned} Z_{if}[J] &= \sqrt{\frac{m}{2\pi i\hbar T}} \exp\left\{\frac{i}{\hbar} S[J]\right\}, \\ S[J] &= x_i I(t_i) + \frac{m(x_f - x_i)^2}{2T} + \frac{(x_f - x_i)Q_J}{T} + \frac{Q_J^{\,2}}{2\,mT} - \frac{R_J}{2m}. \end{aligned} \tag{3.17}$$

In the limit $J \to 0$ result (3.17) is identical to equation (3.6) since I, Q_J, and R_J vanish.

Result (3.17) can be adapted to evaluate the path integral for the case of the linear potential $V(x) = -fx$ by setting $J(t) = f$, where f is a constant. For this case, $I(t) = (T - t)f$, $R_J = \frac{1}{2}fT^2$, and $S_J = \frac{1}{3}f^2 T^3$. Result (3.17) becomes

$$G(x_f, t_f, x_i, t_i) = \sqrt{\frac{m}{2\pi i\hbar T}} \exp\left\{\frac{i}{\hbar}\left(\frac{1}{2}m\frac{(x_f - x_i)^2}{T} + \frac{1}{2}(x_f + x_i)fT - \frac{f^2 T^3}{24m}\right)\right\}. \tag{3.18}$$

Result (3.18) also follows from the operator result (1.57) for the linear potential transition element between a momentum eigenstate and a position eigenstate. Setting $\tau = -iT/\hbar$ in result (1.57) and using equation (1.37) gives

$$
\begin{aligned}
G_J(x_f, t_f, x_i, t_i) &= \int_{-\infty}^{\infty} \mathrm{d}p \, \langle x_f | e^{-iT(-f\hat{X}+\hat{P}^2/2m)/\hbar} | p \rangle \langle p | x_i \rangle \\
&= \int_{-\infty}^{\infty} \frac{\mathrm{d}p}{2\pi\hbar} \exp\left\{ -\frac{iT}{\hbar}\left(\frac{p^2}{2m} - p\left(\frac{x_f - x_i}{T} - \frac{fT}{2m} \right) + \frac{f^2T^2}{6m} - fx_f \right) \right\}.
\end{aligned}
\tag{3.19}
$$

The Gaussian integral of equation (3.19) immediately yields the path integral result (3.18).

3.2 Classical paths and continuum methods

The presence of the classical action in the path integral allows classical trajectories and the fluctuations around them to be used to analyze quantum processes. Although this technique can be applied to either the phase space or configuration form of the path integral, its implementation is easier when the Lagrangian appearing in the path integral takes the form $\mathcal{L}(x, \dot{x})$. This approach will therefore be developed using the configuration space path integral (2.13). The starting point is the simple observation that the variables of integration can be translated by a function of time $f(t)$, so that $x_j \to y_j + f(t_j)$. This leaves the path integral invariant as long as the range of integration for these variables is infinite. Unlike the variables of integration, the function $f(t)$ can be assumed to obey the standard derivative definition $\lim_{\epsilon\to 0}(f(t_j + \epsilon) - f(t_j))/\epsilon = \dot{f}(t_j)$, where $\dot{f}(t_j)$ is the time derivative of the function $f(t)$ evaluated at the time t_j. As a result, the formal time derivatives in the action of the path integral change under the translation according to

$$
\dot{x}_j = \lim_{\epsilon\to 0} \frac{x_{j+1} - x_j}{\epsilon} \to \lim_{\epsilon\to 0} \frac{y_{j+1} - y_j}{\epsilon} + \dot{f}(t_j). \tag{3.20}
$$

The action appearing in the path integral has the associated classical Euler–Lagrange equation (1.7), which typically takes the form $m\ddot{x} + \partial V(x)/\partial x = 0$ for a point mass moving in one dimension. It is assumed there is a classical trajectory, denoted $x_c(t)$, that satisfies the Euler–Lagrange equation *and* matches the boundary conditions defined by the path integral, $x_c(t_i) = x_i$ and $x_c(t_f) = x_f$. The variables of integration appearing in the action sum of equation (2.13) are then translated by the classical trajectory, so that $x_j \to y_j + x_c(t_j)$ and equation (3.20) results in $\dot{x}_j \to \dot{y}_j + \dot{x}_c(t_j)$. In this expression $\dot{y}_j = (y_{j+1} - y_j)/\epsilon$ is still only a notational convenience, but $\dot{x}_c(t_j)$ is the time derivative of the classical trajectory at t_j. By virtue of the translation, the new variables have the endpoint identifications $y_N = x_N - x_c(t_N) = x_f - x_c(t_f) = 0$ and $y_0 = x_0 - x_c(t_i) = x_i - x_c(0) = 0$. As a result, the kinetic energy term in the action sum becomes

$$
\frac{1}{2}m\dot{x}_j^2 \to \frac{1}{2}m\dot{x}_c^2(t_j) + m\dot{y}_j\,\dot{x}_c(t_j) + \frac{1}{2}m\dot{y}_j^2. \tag{3.21}
$$

The sum over the second term appearing in the path integral action can undergo the integration by parts of equation (3.11), with the result

$$\begin{aligned}\sum_{j=0}^{N-1} \epsilon\, m\dot{y}_j\, \dot{x}_c(t_j) &= my_n\, \dot{x}_c(t_{n-1}) - my_0\, \dot{x}_c(t_i) - \sum_{j=1}^{N-1} \epsilon\, m\, y_j \frac{(\dot{x}_c(t_j) - \dot{x}_c(t_{j-1}))}{\epsilon} \\ &= -\sum_{j=1}^{N-1} \epsilon\, y_j\, m\ddot{x}_c(t_j),\end{aligned} \tag{3.22}$$

where $y_N = y_0 = 0$ and $\dot{x}_c(t_{j-1}) = \dot{x}_c(t_j - \epsilon) = -\epsilon\, \ddot{x}_c(t_j)$ were used. Finally, the translated potential $V(x_j, t_j) \to V(y_j + x_c(t_j), t_j)$ appearing in the action sum of equation (2.13) is expanded in a Taylor series around x_c,

$$V(y_j + x_c, t_j) = V(x_c, t_j) + \frac{\partial V(x_c, t_j)}{\partial x_c} y_j + \frac{1}{2}\frac{\partial^2 V(x_c, t_j)}{\partial x_c{}^2} y_j^2 + \cdots, \tag{3.23}$$

where it is understood the classical trajectory is evaluated at $t = t_j$ and the ellipsis represents the higher order terms in the Taylor expansion. If the original action was quadratic, then there will be no higher order terms. Obviously, if they exist, the higher order terms will be cubic and higher in the new integration variable y_j. Combining the sum over the second term in equation (3.23) with the sum in equation (3.22) shows that the terms linear in y_j vanish since

$$-\sum_{j=1}^{N-1} \epsilon\, y_j \left(m\ddot{x}_c(t_j) + \frac{\partial}{\partial x_c(t_j)} V(x_c(t_j), t_j) \right) = 0 \tag{3.24}$$

by virtue of x_c satisfying the Euler–Lagrange equation. The sum over the first terms in equations (3.21) and (3.23) becomes an exact integral in the limit $\epsilon \to 0$, combining to give

$$\int_{t_i}^{t_f} \mathrm{d}t \left(\frac{1}{2} m\dot{x}_c^2(t) - V(x_c(t), t) \right) = S_c(x_f, x_i, t_f, t_i), \tag{3.25}$$

which is the value of the classical action along the classical trajectory. The value of the classical action will depend on the initial and final positions x_i and x_f for the transition element, as well as the times t_f and t_i, typically in the form $T = t_f - t_i$.

Subsequent to this translation the path integral becomes

$$G(x_f, x_i, t_f, t_i) = P(x_f, x_i, t_f, t_i)\exp\left(\frac{i}{\hbar} S_c(x_f, x_i, t_f, t_i) \right), \tag{3.26}$$

where $P(x_f, x_i, t_f, t_i)$, known as the *prefactor*, is given by the modified path integral

$$P(x_f, x_i, t_f, t_i) = \int [\tilde{\mathcal{D}}y] \exp\left\{ \frac{i}{\hbar} \sum_{j=0}^{N-1} \epsilon \left(\frac{1}{2} m\dot{y}_j^2 - \frac{1}{2} V''(x_c, t_j)\, y_j^2 + \ldots \right) \right\}. \tag{3.27}$$

The modified path integral of equation (3.27) has the familiar formal time derivative $\dot{y}_j = (y_{j+1} - y_j)/\epsilon$, while V'' is shorthand for the second derivative of the original potential present in the Taylor series expansion (3.23). The measure appearing in equation (3.27) is the same as equation (2.14), but the prefactor has the boundary conditions $y_N = y_0 = 0$. The path integral of equation (3.27) is often referred to as the *fluctuation integral* since it integrates over the deviations y_j from the classical trajectory. In this sense, it encapsulates the quantum fluctuations around the classical motion of the particle. The combination of the classical action factor and the prefactor in equation (3.26) is identical to the original path integral representation of the transition element.

It is instructive to apply this result to the free particle, where $V = 0$. The classical trajectory satisfying the original boundary conditions is given by $x_c(t) = x_i + (x_f - x_i)(t - t_i)/T$, and the associated classical action is given by

$$S_c(x_f, x_i, t_f, t_i) = \int_0^T \mathrm{d}t\, \frac{1}{2} m\dot{x}_c^2 = \frac{1}{2} m \frac{(x_f - x_i)^2}{T}. \tag{3.28}$$

Comparison to the previously computed transition element (3.6) shows that the classical action factor $\exp\{iS_c(x_f, x_i, t_f, t_i)/\hbar\}$ is indeed present. For the free case, comparison to equation (3.6) also shows that the prefactor must yield

$$P(x_f, x_i, t_f, t_i) = \int [\tilde{\mathcal{D}}y] \exp\left\{\frac{i}{\hbar} \sum_{j=0}^{N-1} \epsilon\left(\frac{1}{2} m\dot{y}_j^2\right)\right\} = \sqrt{\frac{m}{2\pi i\hbar T}}, \tag{3.29}$$

where the measure is given by equation (3.2). While result (3.29) can be verified by performing the intermediate Gaussian integrations over the variables y_j, it will be useful and instructive to develop another method to obtain result (3.29).

The boundary conditions $y_0 = y_N = 0$ suggest the use of Fourier sine series methods. The y_j variables in equation (3.27) are rewritten as a Fourier sine series, so that

$$y_j = \sqrt{\frac{2}{T}} \sum_{k=1}^{N-1} a_k \sin\left(\frac{\pi k(t_j - t_i)}{T}\right) = \sqrt{\frac{2}{T}} \sum_{k=1}^{N-1} a_k \sin\left(\frac{\pi k j}{N}\right), \tag{3.30}$$

where $t_j = t_i + j\epsilon = t_i + jT/N$ was used. The factor $\sqrt{2/T}$ normalizes the sine functions over the interval $T = t_f - t_i$. Expansion (3.30) obviously satisfies the required boundary conditions at $j = 0$ and $j = N$. It is important to note that this is a change of integration variables from y_j to the Fourier coefficients a_k. As a result, since there are $N - 1$ variables y_j, there must be $N - 1$ new variables a_k, and so the sum is finite. Of course, it is understood that the limit $N \to \infty$ will be taken, but that will occur at the end of the calculation. In addition, all changes of variables in a multivariable integral must include the Jacobian J, given by equation (1.103) for the case of c-number variables. The Jacobian is given by $J = \det \mathbf{A}$, where the $(N-1) \times (N-1)$ matrix $\mathbf{A}$ in this case has the elements

$$A_{jk} = \frac{\partial y_j}{\partial a_k} = \sqrt{\frac{2}{T}} \sin\left(\frac{k\pi j\epsilon}{T}\right) = \sqrt{\frac{2}{T}} \sin\left(\frac{jk\pi}{N}\right). \tag{3.31}$$

The Jacobian can be evaluated exactly by using the fact that for large N the orthogonality of the sine functions results in $\mathbf{A}^2 = (N/T)\,\mathbf{I}$, where $\mathbf{I}$ is the unit matrix with dimension $(N-1) \times (N-1)$. This is demonstrated by converting the elements of the matrix product into an integral using the variable $x = j/N$, consistent with $\mathrm{d}j = N\mathrm{d}x = 1$. For large N the matrix product elements are

$$\begin{aligned}\lim_{N\to\infty} \sum_{j=1}^{N-1} A_{kj} A_{jk'} &= \lim_{N\to\infty} \sum_{j=1}^{N-1} \frac{2}{T} \sin\left(\frac{\pi kj}{N}\right) \sin\left(\frac{\pi jk'}{N}\right) \\ &= \frac{2N}{T} \int_0^1 dx\, \sin(k\pi x) \sin(k'\pi x) = \frac{N}{T}\delta_{kk'},\end{aligned} \tag{3.32}$$

which are the matrix elements of the $(N-1) \times (N-1)$ matrix $(N/T)\,\mathbf{I}$. The determinant of $\mathbf{A}$ is then computed by using $\det(\mathbf{A}^2) = (\det \mathbf{A})^2$. This gives

$$(\det \mathbf{A})^2 = \det\left(\frac{N}{T}\mathbf{I}\right) = \left(\frac{N}{T}\right)^{N-1} \implies J = \det \mathbf{A} = \left(\frac{N}{T}\right)^{\frac{1}{2}(N-1)}. \tag{3.33}$$

Using equation (3.30) and $t_{j+1} = t_j + \epsilon = (j+1)T/N$ shows that the difference $y_{j+1} - y_j$ present in the fluctuation integral (3.29) is given by

$$\begin{aligned}y_{j+1} - y_j &= \sum_{k=1}^{N-1} a_k \sqrt{\frac{2}{T}} \left(\sin\left(\frac{\pi k(t_j + \epsilon)}{T}\right) - \sin\left(\frac{\pi k t_j}{T}\right)\right) \\ &= \sum_{k=1}^{N-1} a_k \sqrt{\frac{2}{T}} \left(\sin\left(\frac{\pi jk}{N} + \frac{\pi k}{N}\right) - \sin\left(\frac{\pi jk}{N}\right)\right).\end{aligned} \tag{3.34}$$

While it is tempting to expand the first sum in equation (3.34) in a Taylor series and retain only the first order in ϵ, it should be clear from the second line that the term $\pi k/N$ is *not infinitesimal* for values of k comparable to N. Treating $\epsilon\pi k/T$ as infinitesimal, when it is actually $\pi k/N$, is an error which has its roots in the fact that $(y_{j+1} - y_j)/\epsilon$ is *not* a time derivative. Instead, the difference in the two sine functions will be rewritten using basic trigonometric identities as

$$\sin\left(\frac{\pi jk}{N} + \frac{\pi k}{N}\right) - \sin\left(\frac{\pi jk}{N}\right) = \alpha_k \cos\left(\frac{\pi jk}{N}\right) + \beta_k \sin\left(\frac{\pi jk}{N}\right), \tag{3.35}$$

where $\alpha_k = \sin(\pi k/N)$ and $\beta_k = \cos(\pi k/N) - 1$. The action sum can be computed using the orthogonality relation (3.32) and the related large N identities,

$$\sum_{j=1}^{N-1} \cos\left(\frac{\pi jk}{N}\right)\cos\left(\frac{\pi jk'}{N}\right) = \frac{N}{2}\delta_{kk'},$$
$$\sum_{j=1}^{N-1} \cos\left(\frac{\pi jk}{N}\right)\sin\left(\frac{\pi jk'}{N}\right) = 0. \tag{3.36}$$

Combining equations (3.32), (3.35), (3.36), α_k, and β_k with some trigonometry gives

$$\begin{aligned}\sum_{j=0}^{N-1}\left[\frac{1}{2}m\frac{(y_{j+1}-y_j)^2}{\epsilon}\right] &= \frac{mN^2}{2T^2}\sum_{k=1}^{N-1} a_k{}^2\left(\alpha_k^2+\beta_k^2\right)\\ &= \frac{2mN^2}{T^2}\sum_{k=1}^{N-1} a_k^2 \sin^2\left(\frac{k\pi}{2N}\right),\end{aligned} \tag{3.37}$$

where the $N \to \infty$ limit is understood.

After the change of variables the fluctuation integral becomes

$$P(x_f, x_i, t_f, t_i) = \int [\tilde{\mathcal{D}}a]\exp\left\{\frac{i}{\hbar}\left(\frac{2mN^2}{T^2}\right)\sum_{k=1}^{N-1} a_k^2\left(\sin^2\left(\frac{\pi k}{2N}\right)\right)\right\}, \tag{3.38}$$

while the measure of equation (2.14) is now given by

$$[\tilde{\mathcal{D}}a] = \left(\frac{m}{2\pi i\epsilon\hbar}\right)^{N/2} J\, \mathrm{d}a_1\cdots \mathrm{d}a_{N-1}, \tag{3.39}$$

where J is the Jacobian of equation (3.33). The range of integrations for each of the Fourier coefficients is $(-\infty, \infty)$ in order to match the range of integrations for the original y_j variables. The $N-1$ Gaussian integrations of equation (3.38) give

$$\begin{aligned}P(x_f, x_i, t_f, t_i) &= \left(\frac{\pi i\hbar T^2}{2\,mN^2}\right)^{(N-1)/2}\left(\frac{mN}{2\pi i\hbar T}\right)^{N/2}\left(\prod_{k=1}^{N-1}\sin\left(\frac{\pi k}{2N}\right)\right)^{-1} J\\ &= \frac{1}{2^N}\sqrt{\frac{2\,mN}{\pi i\hbar T}}\left(\prod_{k=1}^{N-1}\sin\left(\frac{\pi k}{2N}\right)\right)^{-1}.\end{aligned} \tag{3.40}$$

The product of sine functions appearing in equation (3.40) obeys the remarkable identity

$$\left(\prod_{k=1}^{N-1}\sin\left(\frac{k\pi}{2N}\right)\right)^{-1} = \frac{2^N}{2\sqrt{N}}, \tag{3.41}$$

and using equation (3.41) in equation (3.40) yields the prefactor $\sqrt{m/2\pi i\hbar T}$ anticipated by equation (3.29).

Given the number and complexity of steps from equations (3.30) to (3.41), the Fourier method could scarcely be considered a calculational improvement for analyzing the free particle path integral. Nevertheless, it reveals an important aspect of the contribution of various types of fluctuation paths. There is a nondenumerable infinity of paths that leave the origin and return back. However, they can be broken into two generic classes. The first are paths that are differentiable or *smooth*, and these are the paths typically associated with classical mechanics. The second are paths that are not differentiable, referred to as *unruly paths*. These are not present in the solutions to the classical equations of motion since the classical action is formulated by assuming the classical trajectory is differentiable. However, it is possible to isolate the contribution of the smooth paths to the fluctuation integral using the difference equation (3.34). A smooth path will be such that $(y_{j+1} - y_j)/\epsilon$ is given by a time derivative. Using this definition for a smooth path, expression (3.34) becomes

$$
\begin{aligned}
y_{j+1} - y_j &= \sum_{k=1}^{N-1} a_k \sqrt{\frac{2}{T}} \left(\sin\left(\frac{\pi k(t_j + \epsilon)}{T}\right) - \sin\left(\frac{\pi k t_j}{T}\right)\right) \\
&= \sum_{k=1}^{N-1} \epsilon\, a_k \sqrt{\frac{2}{T}} \left(\frac{\pi k}{T}\right) \cos\left(\frac{\pi k t_j}{T}\right).
\end{aligned}
\tag{3.42}
$$

Converting the action into a time integral and using the orthogonality of the cosine functions gives the free action for smooth fluctuations,

$$
\begin{aligned}
\sum_{j=0}^{N-1} \frac{1}{2} m \frac{(y_{j+1} - y_j)^2}{\epsilon} &= \sum_{k,k'=1}^{N-1} a_k a_{k'} \left(\frac{m\pi^2 k k'}{T^3}\right) \sum_{j=0}^{N-1} \epsilon \cos\left(\frac{\pi k t_j}{T}\right) \cos\left(\frac{\pi k' t_j}{T}\right) \\
&= \sum_{k=1}^{N-1} \frac{1}{2} m \left(\frac{\pi^2 k^2}{T^2}\right) a_k^2.
\end{aligned}
\tag{3.43}
$$

It is important to note that $\pi^2 k^2/T^2$ is the eigenvalue of the differential operator $-\mathrm{d}^2/\mathrm{d}t^2$ associated with the eigenfunction $\sin(\pi k t/T)$. Restricting to smooth paths allows the fluctuation variables to become differentiable functions $y_j = y(t_j)$. The fluctuation action then becomes an integral that can be integrated by parts,

$$
\sum_{j=0}^{N-1} \frac{1}{2} m \frac{(y_{j+1} - y_j)^2}{\epsilon} = \int_{t_i}^{t_f} \mathrm{d}t\, \frac{1}{2} m \dot{y}^2 = \int_{t_i}^{t_f} \mathrm{d}t\, \frac{1}{2} m y \left(-\frac{\mathrm{d}^2}{\mathrm{d}t^2}\right) y. \tag{3.44}
$$

Expression (3.30) can then be understood as an expansion of the fluctuation variables in terms of the orthonormal eigenfunctions $f_k(t)$ of the differential operator $-\mathrm{d}^2/\mathrm{d}t^2$ that satisfy the fluctuation boundary conditions $f_k(t_i) = f_k(t_f) = 0$.

The technique of assuming differentiable or *smooth fluctuations*, as in equation (3.44), is almost universally employed in the analysis of path integrals. However, there is a subtle but nonfatal error in doing so. Inserting equation (3.43) into the fluctuation integral (3.29), performing the $N - 1$ Gaussian integrals, and using J from equation (3.33) gives the prefactor

$$P(x_f, x_i, t_f, t_i) = \frac{N^{N-\frac{1}{2}}}{\pi^{N-1}(N-1)!}\sqrt{\frac{m}{2\pi i\hbar T}}. \tag{3.45}$$

Expression (3.45) can be understood by using Stirling's formula, which states that as N becomes large $(N-1)!$ is given by

$$(N-1)! \approx N^N e^{-N}\sqrt{\frac{2\pi}{N}}. \tag{3.46}$$

Using equation (3.46) in equation (3.45) shows that the prefactor obtained using smooth fluctuations is given by

$$P = \left(\frac{e}{\pi}\right)^{N-1}\sqrt{\frac{e^2}{2\pi}}\sqrt{\frac{m}{2\pi i\hbar T}}. \tag{3.47}$$

The prefactor computed using only smooth fluctuation paths *vanishes* in the $N \to \infty$ limit since $e/\pi < 1$. The smooth fluctuation paths are therefore a set of *measure zero*, and it is the *unruly paths* that dominate the quantum fluctuations. Treating the fluctuation variables as smooth does not weight the high frequencies correctly in equation (3.34), and it is these high frequency contributions that are responsible for generating the unruly paths and giving the correct answer (3.29). This problem also occurs in the theory of Fourier series, where the series representation of discontinuous functions does not converge properly in the neighborhood of the discontinuity. There it is remedied by inserting *Lanczos factors* to improve convergence.

This problem is related to how the path integral is *regularized* or made finite. Examining the measure (3.2) shows that it has a factor that is divergent in the limit $N \to \infty$ since it is proportional to $N^{N/2}$. This *regularizing factor* was obtained by performing the Gaussian momentum integrals originating from the Hamiltonian and its presence exactly cancels the factors generated by the intermediate integrations *when both smooth and unruly paths are included.* In dealing with regularization it is very useful to note that the correct prefactor for the free particle is present in the smooth path contribution of equation (3.45). As a result, this problem is often remedied in the following way. The fluctuation integral is restricted to smooth paths and the result is *normalized* to the *free case* of equation (3.29). This simply means that the result is multiplied by the factor found from equation (3.45),

$$\mathcal{N}_S = \frac{\pi^{N-1}(N-1)!}{N^{N-\frac{1}{2}}}. \tag{3.48}$$

Doing so cancels the incorrect factor in equation (3.45) and *normalizes* the smooth fluctuation integral to yield the correct prefactor.

This procedure will be shown to yield the correct prefactor for the simple harmonic oscillator in the next section. Because $\mathcal{N}_S \to \infty$ in the large N limit, multiplying by $\mathcal{N}_S$ adjusts the measure of the smooth paths to give the correct prefactor. In path integral literature this factor is often combined with the correct Jacobian J to give a *normalization factor* $\mathcal{N}$ to use in conjunction with smooth fluctuations,

$$\mathcal{N} = \mathcal{N}_S J = \frac{\pi^{N-1}(N-1)!}{N^{N-\frac{1}{2}}}\left(\frac{N}{T}\right)^{\frac{1}{2}(N-1)} = \frac{(N-1)!}{N^{N/2}}\left(\frac{\pi^2}{T}\right)^{\frac{1}{2}(N-1)}. \tag{3.49}$$

This factor can be used whenever the Fourier expansion (3.30) is employed to analyze a configuration space fluctuation integral *that can be normalized to the free particle case*. However, there are path integrals, such as the coherent state path integral, for which this approach fails.

The value of this approach is that it can be applied whenever a classical trajectory is available that reduces the path integral to evaluating the fluctuation path integral of equation (3.26). Treating the fluctuation variables as differentiable functions will give rise to a general differential operator $\hat{D}(t) = -\mathrm{d}^2/\mathrm{d}t^2 + V''(x_c, t)$ in the fluctuation integral after an integration by parts. Obviously the free particle case gives $\hat{D}_0(t) = -\mathrm{d}^2/\mathrm{d}t^2$. The fluctuation integral then takes the form

$$P(x_f, x_i, t_f, t_i) = \int [\tilde{\mathcal{D}}_S y] \exp\left\{\frac{i}{\hbar}\sum_{j=0}^{N-1} \epsilon \frac{1}{2} m y(t_j)\left(\hat{D}(t_j) y(t_j)\right)\right\}, \tag{3.50}$$

where the smooth measure $[\tilde{\mathcal{D}}_S y]$ is obtained from the previous fluctuation measure (3.2) by replacing each $\mathrm{d}y_j$ with $\mathrm{d}y(t_j)$ and retaining the overall factor $(m/2\pi i\epsilon\hbar)^{N/2}$.

The differential operator in equation (3.50) will typically include the term $V''(x_c, t_j)$ from the Taylor series expansion of the potential given by equation (3.23) around the classical trajectory $x_c(t)$. It is assumed that the differential operator is of the Sturm–Liouville type and possesses a *complete set of orthonormal eigenfunctions*, denoted $f_k(t)$, which satisfy $\hat{D}(t) f_k(t) = \lambda_k f_k(t)$ as well as the boundary conditions $f_k(t_f) = f_k(t_i) = 0$. Since the set is complete, the fluctuation variables can be expanded in them as in equation (3.30), so that $y_j = \sum_{k=1}^{N-1} a_k f_k(t_j)$. Since these eigenfunctions are orthonormal, the action in the smooth fluctuation integral takes the general form

$$\begin{aligned}\frac{i}{\hbar}\sum_{j=0}^{N-1} \epsilon \frac{1}{2} m y(t_j)\left(\hat{D}(t_j) y(t_j)\right) &= \frac{i}{\hbar}\sum_{k,k'=1}^{N-1} \frac{1}{2} m a_k a_{k'} \lambda_k \int_{t_i}^{t_f} \mathrm{d}t\, f_{k'}(t) f_k(t) \\ &= \frac{i}{\hbar}\sum_{k,k'=1}^{N-1} \frac{1}{2} m a_k a_{k'} \lambda_k\, \delta_{kk'} = \frac{i}{\hbar}\sum_{k=1}^{N-1} \frac{1}{2} m \lambda_k a_k^2.\end{aligned} \tag{3.51}$$

The a_k integrals are simple Fresnel Gaussians and combining them with the measure $[\tilde{\mathcal{D}}a]$ of equation (3.39) gives the fluctuation integral

$$\begin{aligned}\int [\tilde{\mathcal{D}}a]\, \mathcal{N} \exp\left\{\frac{i}{\hbar}\sum_{k=1}^{N-1}\left(\frac{1}{2} m \lambda_k a_k^2\right)\right\} &= \mathcal{N}\sqrt{\frac{m}{2\pi i\hbar}}\left(\frac{N}{T}\right)^{N/2}\left(\prod_{k=1}^{N-1} \lambda_k\right)^{-\frac{1}{2}} \\ &= (N-1)!\left(\frac{\pi}{T}\right)^{N-1}\left(\prod_{k=1}^{N-1} \lambda_k\right)^{-\frac{1}{2}} \sqrt{\frac{m}{2\pi i\hbar T}}.\end{aligned} \tag{3.52}$$

The product of the eigenvalues of the differential operator $\hat{D}(t)$ appearing in equation (3.52) is reminiscent of the determinant of a matrix (1.107), which is given by the product of its eigenvalues. The product of eigenvalues in equation (3.52) is therefore written

$$\lim_{N\to\infty}\prod_{k=1}^{N-1}\lambda_k = \det \hat{D}(t), \tag{3.53}$$

and is referred to as the *functional determinant* of the differential operator $\hat{D}(t)$.

The result of the Gaussian integral (3.52) is then formally the same as the Gaussian integral (1.110), with the inverse square root of the functional determinant appearing in place of the matrix determinant. Comparing equation (3.52) with equation (3.43) or applying $-\mathrm{d}^2/\mathrm{d}t^2$ to the eigenfunctions used in the Fourier expansion (3.30) shows that $\lambda_k = \pi^2k^2/T^2$ for the free particle. As a result, the factor $(N-1)!\,(\pi/T)^{N-1}$ in equation (3.52) is identified as $\prod_{k=1}^{N-1}\lambda_k^{1/2} = (\det \hat{D}_0(t))^{1/2}$, where $\hat{D}_0(t) = -\mathrm{d}^2/\mathrm{d}t^2$ is the free particle differential operator. Using this notation the quadratic fluctuation integral can be written very compactly as

$$P(x_f, x_i, t_f, t_i) = \left(\frac{\det \hat{D}_0(t)}{\det \hat{D}(t)}\right)^{1/2}\sqrt{\frac{m}{2\pi i\hbar T}} = \frac{\mathcal{N}}{\sqrt{\det \hat{D}(t)}}, \tag{3.54}$$

where $\mathcal{N} = \sqrt{m \det \hat{D}_0(t)/2\pi i\hbar T}$ normalizes the prefactor. In this case $\mathcal{N}$ has been chosen to normalize the prefactor to the free particle case. Expression (3.54) correctly reduces to the free particle prefactor since $\hat{D}(t) = \hat{D}_0(t)$ for $V = 0$. It is important to note that expression (3.54) is valid *only if* $\hat{D}(t)$ can be reduced to $\hat{D}_0(t)$ by setting some parameters to zero. For example, $\hat{D}(t) = \mathrm{d}^2/\mathrm{d}t^2 + \omega^2$ becomes $\hat{D}_0(t)$ for $\omega = 0$, while $\hat{D}(t) = \mathrm{d}/\mathrm{d}t + i\omega$ *does not*. The latter differential operator therefore requires a different normalization procedure. The general method for normalizing path integrals is revisited when the partition function is analyzed later. The Gel'fand–Yaglom theorem and the van Vleck–Pauli–Morette determinant provide methods to find the regularized functional determinant even if the explicit eigenvalues are unknown. Both approaches are presented in the section on the semiclassical approximation.

3.3 The harmonic oscillator

The harmonic oscillator path integral, where the potential energy is $V(x) = \frac{1}{2}m\omega^2x^2$, can be treated using the smooth fluctuation approach of the previous section. The first step is to solve the classical equation of motion $m\ddot{x} + m\omega^2x = 0$ subject to the boundary conditions $x_c(t_i) = x_i$ and $x_c(t_f) = x_f$. The solution is given by

$$x_c(t) = \frac{x_i}{\sin\omega T}\sin(\omega(t_f - t)) + \frac{x_f}{\sin\omega T}\sin(\omega(t - t_i)). \tag{3.55}$$

Integrating the classical action by parts and using equation (3.55) gives

$$\begin{aligned} S_c &= \int_{t_i}^{t_f} \mathrm{d}t \left(\frac{1}{2} m \dot{x}_c^2 - \frac{1}{2} m\omega^2 x_c^2 \right) \\ &= \int_{t_i}^{t_f} \mathrm{d}t \left\{ \frac{1}{2} m \frac{\mathrm{d}}{\mathrm{d}t}(x_c \dot{x}_c) - \frac{1}{2} m x_c (\ddot{x}_c + \omega^2 x_c) \right\} \\ &= \frac{1}{2} m x_c(t) \dot{x}_c(t) \Big|_{t_i}^{t_f} = \frac{m\omega}{2 \sin \omega T} \left((x_f^2 + x_i^2) \cos \omega T - 2 x_i x_f \right). \end{aligned} \tag{3.56}$$

Result (3.56) reduces to the free particle action (3.28) for $\omega \to 0$.

The second step is to calculate the fluctuation integral. In the case of the harmonic oscillator potential, the Taylor series expansion (3.23) generates a quadratic term $V'' = m\omega^2$, but no higher order terms. Using the smooth path approach of the previous section, an integration by parts gives the fluctuation integral

$$P(x_i, x_f, t_f, t_i) = \int [\tilde{\mathcal{D}}_S y] \exp \left\{ \frac{i}{\hbar} \sum_{j=0}^{N-1} \epsilon \frac{1}{2} m y(t_j) \left(-\frac{\mathrm{d}^2}{\mathrm{d}t_j^2} - \omega^2 \right) y(t_j) \right\}. \tag{3.57}$$

The differential operator $\hat{D}(t) = -\mathrm{d}^2/\mathrm{d}t^2 - \omega^2$ has the same orthonormal eigenfunctions, $\sqrt{2/T}\,\sin(\pi k(t - t_i)/T)$, that were used with the free particle in equation (3.30). The eigenvalues are given by $\lambda_k = (\pi^2 k^2/T^2) - \omega^2$. This eigenvalue can be zero if $\pi^2 k^2 - \omega^2 T^2 = 0$, and this would render the inverse functional determinant undefined. The Wick rotation, $T \to -iT$, is often performed to avoid this problem, since the eigenvalues then take the negative definite form $\lambda_k = -\pi^2 k^2/T^2 - \omega^2$. Using the Wick rotation creates well-defined Gaussian integrals of the form (1.99). After performing the $N - 1$ Gaussian integrals, followed by the inverse Wick rotation $T \to iT$, the last line in equation (3.52) is given in the large N limit by the expression

$$\begin{aligned} P(x_f, x_{t_f}, t_i) &= (N-1)! \left(\frac{\pi}{T} \right)^{N-1} \left\{ \prod_{k=1}^{N-1} \left(\omega^2 - \frac{\pi^2 k^2}{T^2} \right)^{-\frac{1}{2}} \right\} \sqrt{\frac{m}{2\pi i \hbar T}} \\ &= \sqrt{\frac{m}{2\pi i \hbar T}} \prod_{k=1}^{N-1} \left(1 - \frac{\omega^2 T^2}{\pi^2 k^2} \right)^{-\frac{1}{2}} = \sqrt{\frac{m\omega}{2\pi i \hbar \sin \omega T}}, \end{aligned} \tag{3.58}$$

where the product identity was first proved by Euler using factorization arguments.

Because the inverse square root of $(1 - \omega^2 T^2/\pi^2 k^2)$ appears in the prefactor (3.58), there is a subtlety that requires attention. For large values of T there will be ν terms in the product of equation (3.58) that become negative, where ν is the largest integer bounded by $\omega T/\pi$. If the term is negative but nonzero the Fresnel integral in equation (3.52) is still valid. However, the inverse square root of each of these negative factors is given by $1/i\sqrt{|1 - \omega^2 T^2/\pi^2 k^2|}$, so that an overall factor of $(-i)^\nu = \exp(-i\nu\pi/2)$ appears in the prefactor. The integer ν is known as the

Maslov index of the differential operator. The harmonic oscillator propagator is therefore given by

$$G(x_f, x_i, t_f, t_i) = e^{-i\nu\pi/2}\sqrt{\frac{m\omega}{2\pi i\hbar|\sin\omega T|}}\exp\left\{\frac{i}{\hbar}S_c\right\}, \tag{3.59}$$

where S_c is given by equation (3.56). Result (3.59) becomes the free particle propagator (3.6) in the limit $\omega \to 0$, *as it must*, and this justifies the free particle normalization method of equation (3.54). Result (3.59) is valid for all times T if the correct Maslov index is used. In the case of the harmonic oscillator the Maslov index ν coincides with the number of *classical turning points* in the time interval T. A turning point corresponds to the classical trajectory giving $\dot{x}_c(t) = 0$. This observation can be generalized to other systems and is discussed again in the section on the semiclassical approximation. Despite this subtlety, the method of functional determinants used to find (3.59) is often simpler than performing the intermediate integrations.

It is useful to add an external source term to the oscillator action, so that the classical action is given by $\mathcal{L} = \frac{1}{2}m\dot{x}^2 - \frac{1}{2}m\omega^2x^2 + xJ(t)$, where $J(t)$ is an arbitrary integrable function of time. Using the methods of the previous section this is easily accommodated. The only change in the path integral occurs in the factor involving the classical action since the term being added to the action is linear in x and does not contribute to the prefactor fluctuation integral. As a result, the problem reduces to solving the altered classical equation of motion, $m\ddot{x}_c + m\omega^2x_c - J(t) = 0$, and then evaluating the classical action using the result. Translating the path integral variables by this classical solution will remove the linear term $x_jJ(t_j)$ in the path integral action. This technique can be used in a number of path integral applications, both in quantum mechanics and in quantum field theory.

Since the source function is arbitrary, solving the equation of motion requires a Green's function approach. The Green's function $\Delta(t, t')$ satisfies

$$\left(m\frac{\mathrm{d}^2}{\mathrm{d}t^2} + m\omega^2\right)\Delta(t, t') = \delta(t - t'), \tag{3.60}$$

along with the boundary conditions $\Delta(t_f, t') = \Delta(t_i, t') = 0$ for the case that both t and t' are in the interval (t_i, t_f). The *harmonic oscillator impulse function* $I(t)$ is then defined by the integral

$$I(t) = \int_{t_i}^{t_f}\mathrm{d}t'\,\Delta(t, t')J(t') \implies \left(m\frac{\mathrm{d}^2}{\mathrm{d}t^2} + m\omega^2\right)I(t) = J(t). \tag{3.61}$$

The Green's function satisfying equation (3.60) and the boundary conditions is given by

$$\Delta(t, t') = \frac{1}{m\omega}\left\{\theta(t - t')\sin\omega(t - t') - \frac{\sin\omega(t - t_i)\sin\omega(t_f - t')}{\sin\omega T}\right\}, \tag{3.62}$$

which uses equation (1.17) to satisfy equation (3.60). The boundary conditions on $\Delta(t, t')$ give $I(t_f) = I(t_i) = 0$, and so the classical trajectory is given by

$$x_c(t) = \left(\frac{x_i}{\sin \omega T}\right)\sin(\omega(t_f - t)) + \left(\frac{x_f}{\sin \omega T}\right)\sin(\omega(t - t_i)) + I(t). \tag{3.63}$$

The classical action in the presence of the source term is then given by

$$\begin{aligned} S_c &= \int_{t_i}^{t_f} \mathrm{d}t \left(\frac{1}{2}m\frac{\mathrm{d}}{\mathrm{d}t}(x_c\dot{x}_c) - \frac{1}{2}mx_c\ddot{x}_c - \frac{1}{2}m\omega^2 x_c^2 + x_c J(t)\right) \\ &= \frac{1}{2}m\big(x_c(t_f)\dot{x}_c(t_f) - x_c(t_i)\dot{x}_c(t_i)\big) + \int_{t_i}^{t_f} \mathrm{d}t\, \frac{1}{2}x_c(t)J(t). \end{aligned} \tag{3.64}$$

Using equations (3.63) and (3.62) in equation (3.64) gives, after considerable algebraic manipulation, the classical action,

$$\begin{aligned} S_c[J] = \frac{m\omega}{2\sin\omega T}\Big\{&(x_f^2 + x_i^2)\cos\omega T - 2x_f x_i \\ &+ \frac{2x_i}{m\omega}\int_{t_i}^{t_f} dt \sin\omega(t_f - t)\, J(t) + \frac{2x_f}{m\omega}\int_{t_i}^{t_f} dt \sin\omega(t - t_i)\, J(t) \\ &- \frac{2}{m^2\omega^2}\int_{t_i}^{t_f} dt \sin\omega(t_f - t)\, J(t)\int_{t_i}^{t} dt' \sin\omega(t' - t_i)\, J(t')\Big\}. \end{aligned} \tag{3.65}$$

The $J = 0$ form of equation (3.65) clearly reduces to equation (3.56). The final form of the harmonic oscillator propagator in the presence of an external source $J(t)$ is given by

$$Z_{if}[J] = e^{-i\nu\pi/2}\sqrt{\frac{m\omega}{2\pi i\hbar|\sin\omega T|}}\exp\left\{\frac{i}{\hbar}S_c[J]\right\}. \tag{3.66}$$

The functionals (3.17) and (3.66) are revisited in the section on perturbation theory.

3.4 The semiclassical approximation

The two most complicated path integral results obtained so far, equations (3.17) and (3.66), are still associated with quadratic Lagrangians, and thus were exactly solvable path integrals. The method of using smooth or differentiable paths required normalizing the fluctuation integral with the free particle functional determinant (3.54). This method was applied to the harmonic oscillator and shown to yield the correct result. This method can be applied to any quadratic action that reduces to the free action for some choice of parameters to obtain the exact result.

Because the kinetic energy is assumed to be quadratic, it is the higher order terms beyond quadratic in the potential that prevent an exact evaluation by performing Gaussian integrals. A useful approach to analyzing such nonquadratic theories is to evaluate the path integral using the *semiclassical approximation*. This treats the path integral in terms of an expansion in $\hbar$ and retains only the lowest order contribution. The semiclassical approximation is relatively easy to formulate by using the fluctuation variable Taylor series around the classical solution given by equation (3.23). Because the terms linear in the fluctuation variables y_j vanish, the remaining terms are quadratic and higher in the y_j. Scaling the fluctuation variables according to $y_j \to \sqrt{\hbar}\, y_j$ therefore cancels the inverse power of $\hbar$ in the action, and all but one of the inverse factors of $\sqrt{\hbar}$ in the measure $[\mathcal{D}y]$ given by equation (3.2). However, all terms cubic and higher in y_j appearing in equation (3.23) retain coefficients involving powers of $\sqrt{\hbar}$. The *semiclassical approximation* suppresses these higher order terms in equation (3.27) in the limit $\hbar \to 0$. The semiclassical approximation therefore takes the form (3.26), consisting of the classical action factor and the prefactor, in this case given by equation (3.54). In the semiclassical approximation the differential operator appearing in equation (3.54) is given by

$$\hat{D}(t) = -\frac{d^2}{dt^2} - \frac{1}{m}V''(x_c(t), t) \equiv -\frac{d^2}{dt^2} - \tilde{\omega}^2(t). \tag{3.67}$$

If there are terms cubic and higher in the potential, these will give rise to a time-dependent frequency $\tilde{\omega}(t)$ that depends on the classical trajectory *even if* V itself has no explicit time-dependence. The semiclassical approximation is often referred to as the *WKB (Wentzel–Kramers–Brillouin) approximation*, although the original WKB approximation was developed to find solutions to the Schrödinger equation. A similar method is also used to approximate certain types of integrals, where it is referred to as the *stationary phase approximation* or the *method of steepest descent*. This is discussed again in the next section.

Rather than attempting a direct calculation of the eigenvalues of equation (3.67) and then finding their product, the determinant of the differential operator (3.67) can often be found by easier methods. The first method is provided by the Gel'fand–Yaglom (GY) theorem, a variant of which states that the functional determinant ratio in equation (3.54) is given by a remarkably simple expression,

$$\frac{\det \hat{D}_0(t)}{\det \hat{D}(t)} = \frac{T}{f_0(t_f)}, \tag{3.68}$$

where the function $f_0(t)$ satisfies the *zero mode* equation for $\hat{D}(t)$,

$$\hat{D}(t)f_0(t) = -\left(\frac{d^2}{dt^2} + \tilde{\omega}^2(t)\right)f_0(t) = 0, \tag{3.69}$$

along with the boundary conditions $f_0(t_i) = 0$ and $\dot{f}_0(t_i) = 1$. For brevity, the lengthy proof of the GY theorem will not be presented here, but is available in Kleinert. It is useful to note that a solution to equation (3.69) is always available from the solutions x_c of the Euler–Lagrange equation. This follows from

$$\frac{d}{dt}\left(m\ddot{x}_c + \frac{\partial V}{\partial x_c}\right) = m\left(\frac{d^2}{dt^2} + \frac{1}{m}V''(x_c)\right)\dot{x}_c = 0, \tag{3.70}$$

so that f_0 can be found from $\dot{x}_c$ *if the boundary conditions for* f_0 *can be satisfied.*

For example, a general free particle trajectory $x_c(t) = v(t - t_i) + x_i$ satisfies equation (3.69) since $\tilde{\omega}^2(t) = 0$. Using x_c for f_0 requires choosing $x_i = 0$ and $v = 1$, so that $f_0(t_i) = 0$ and $\dot{f}_0(t_i) = 1$. The resulting function $f_0(t) = t - t_i$ gives $f_0(t_f) = T$ and unity for the right-hand side of equation (3.68), consistent with a free particle. In this case $\dot{x}_c$ satisfies equation (3.69), but it cannot satisfy the boundary condition $\dot{f}_0 = 1$ since $\ddot{x}_c = 0$. For the simple harmonic oscillator, where $\tilde{\omega}(t) = \omega$, choosing $f_0(t) = \sin(\omega(t - t_i))/\omega$ satisfies both equation (3.69) and the boundary conditions. Using $f_0(t_f)$ in equation (3.68) and returning the result to equation (3.54) immediately gives the prefactor present in equation (3.59). In this case, choosing $x_c(t) = -\cos(\omega(t - t_i))/\omega^2$ gives $f_0 = \dot{x}_c$.

Another method for calculating the prefactor can be inferred using the definition of the propagator, which gives

$$\int_{-\infty}^{\infty} \mathrm{d}x_i\, G(x_f, x_i, T) G^*(x'_f, x_i, T) = \int_{-\infty}^{\infty} \mathrm{d}x_i \langle x_f | e^{-i\hat{H}T/\hbar} | x_i \rangle \langle x_i | e^{i\hat{H}T/\hbar} | x'_f \rangle$$
$$= \langle x_f | x'_f \rangle = \delta(x_f - x'_f) = \int_{-\infty}^{\infty} \frac{\mathrm{d}p}{2\pi\hbar} \exp\left(\pm \frac{i}{\hbar} p\,(x_f - x'_f)\right), \tag{3.71}$$

where the Fourier transform version of the Dirac delta (1.38) has been chosen for consistency with the range of integration on x_i. It is important to note that either choice of sign in the exponential for the Dirac delta can be made. Based on previous results, the prefactor in the general form (3.26) for the propagator is written $P(x_f, x_i, T) = \sqrt{-i}\, R(x_f, x_i, T)$. For a sufficiently small value of T it is *assumed* that $R(x_f, x_i, T)$ is a positive real valued function. Using this form of the prefactor in equation (3.71) gives the relation

$$\int_{-\infty}^{\infty} \mathrm{d}x_i\, R(x_f, x_i, T) R(x'_f, x_i, T) \exp\left\{\frac{i}{\hbar}\left(S_c(x_f, x_i, T) - S_c(x'_f, x_i, T)\right)\right\}$$
$$= \int_{-\infty}^{\infty} \frac{\mathrm{d}p}{2\pi\hbar} \exp\left(\pm \frac{i}{\hbar} p\,(x_f - x'_f)\right). \tag{3.72}$$

Result (3.72) holds for arbitrarily small values of $x_f - x'_f$, and this allows both $R(x'_f, x_i, T)$ and $S_c(x'_f, x_i, T)$ to be expanded around a value of $x_f - x'_f$ such that $(x_f - x'_f)^2 \approx 0$. This gives $R(x'_f, x_i, T) \approx R(x_f, x_i, T) + (x'_f - x_f)\partial R(x_f, x_i, T)/\partial x_f$ and $S_c(x'_f, x_i, T) \approx S_c(x_f, x_i, T) + (x'_f - x_f)\partial \tilde{S}_c(x_f, x_i, T)/\partial x_f$. To leading order in $(x_f - x'_f)$ expression (3.72) is then given by

$$\int_{-\infty}^{\infty} \mathrm{d}x_i\, R^2(x_f, x_i, T) \exp\left\{\frac{i}{\hbar} \frac{\partial S_c(x_f, x_i, T)}{\partial x_f}(x_f - x'_f)\right\}$$
$$= \int_{-\infty}^{\infty} \frac{\mathrm{d}p}{2\pi\hbar} \exp\left(\pm \frac{i}{\hbar} p\,(x_f - x'_f)\right). \tag{3.73}$$

The two sides of equation (3.73) are matched by changing the variable of integration x_i to $q = \pm\partial S_c(x_f, x_i, T)/\partial x_f$. In the one-dimensional case the Jacobian of this change is $J = \pm\partial x_i/\partial q = (\pm\partial q/\partial x_i)^{-1} = (\pm\partial^2 S_c(x_f, x_i, T)/\partial x_i\,\partial x_f)^{-1}$, which gives

$$\int_{-\infty}^{\infty} \mathrm{d}q\; R^2(x_f, x_i, T)\left(\pm\frac{\partial^2 S_c(x_f, x_i, T)}{\partial x_i\,\partial x_f}\right)^{-1} \exp\left\{\pm\frac{i}{\hbar}q\,(x_f - x_f')\right\} = \int_{-\infty}^{\infty} \frac{\mathrm{d}p}{2\pi\hbar}\exp\left(\pm\frac{i}{\hbar}p\,(x_f - x_f')\right). \tag{3.74}$$

The product of the two factors in equation (3.74) must therefore be $1/2\pi\hbar$, and taking the square root to find R creates an additional pair of sign options, resulting in

$$\begin{aligned} P(x_f, x_i, T) &= \sqrt{-i}\; R(x_f, x_i, T) = \pm\sqrt{-i}\left(\pm\frac{1}{2\pi\hbar}\frac{\partial^2 S_c(x_f, x_i, T)}{\partial x_i\,\partial x_f}\right)^{1/2} \\ &= e^{-i\nu\pi/2}\left(\frac{1}{2\pi i\hbar}\left|\frac{\partial^2 S_c(x_f, x_i, T)}{\partial x_i\,\partial x_f}\right|\right)^{1/2}, \end{aligned} \tag{3.75}$$

where ν can be 0, 1, 2, or 3 to reflect all four possible sign combinations. The right-hand side of (3.75) is known as the *van Vleck–Pauli–Morette determinant*. By assumption $R(x_f, x_i, T)$ is a real valued function, and so a choice for ν is made that creates a positive value in the argument of the square root. This is the origin of the absolute value sign appearing in the second line of equation (3.75). In what follows ν will be identified as the Maslov index *modulo* 4.

The interpretation of the sign ambiguities in this derivation is ad hoc, but it can be clarified by a more rigorous but considerably lengthier proof involving variants of the GY theorem. The more detailed derivation relates the Maslov index ν to the number of turning points, where $\dot{x}_c = 0$, in the classical trajectory during the time interval T. The Maslov index is given by the number of smooth turning points, where the momentum is continuous, plus twice the number of hard reflections off a barrier, where the momentum is discontinuous. This is consistent with the results already obtained. The free particle and the harmonic oscillator actions, equations (3.28) and (3.56), yield the second derivatives $-m/T$ and $-m\omega/\sin\omega T$ respectively. Using these in equation (3.75) gives equation (3.6), where there are no classical turning points, and equation (3.59), where there are smooth turning points whenever $\cos\omega t = 0$.

Result (3.75) is easily generalized to the case of higher dimensions, where the components of the initial and final positions $\boldsymbol{x}_i$ and $\boldsymbol{x}_f$ are denoted x_f^a and x_i^b. In n spatial dimensions $P(\boldsymbol{x}_f, \boldsymbol{x}_i, T) = R(\boldsymbol{x}_f, \boldsymbol{x}_i, T)/\sqrt{i^n}$ and the generalized version of equation (3.73) is given by

$$\int_{-\infty}^{\infty} \mathrm{d}^n\boldsymbol{x}_i\, R^2(\boldsymbol{x}_f, \boldsymbol{x}_i, T)\exp\left\{\frac{i}{\hbar}\sum_{a=1}^{n} \frac{\partial S_c(\boldsymbol{x}_f, \boldsymbol{x}_i, T)}{\partial x_f^a}\,(x_f^a - x_f^{'a})\right\}$$
$$= \int_{-\infty}^{\infty} \frac{\mathrm{d}^n\boldsymbol{p}}{(2\pi\hbar)^n} \exp\left(\pm\frac{i}{\hbar}\sum_{a=1}^{n} p^a\,(x_f^a - x_f^{'a})\right). \tag{3.76}$$

A similar change of integration variables in the first integral is given by $q^a = \pm\partial S_c/\partial x_i^a$. This is accompanied by the Jacobian, the determinant of the matrix $\pm\partial\boldsymbol{x}_i/\partial\boldsymbol{q} = \pm(\partial\boldsymbol{q}/\partial\boldsymbol{x}_i)^{-1}$. The elements of the second matrix are given by $\pm\partial q^a/\partial x_i^b = \pm\partial^2 S_c/\partial x_i^b\,\partial x_f^a$. The prefactor for n-dimensional systems in the semiclassical approximation is then given by

$$P(\boldsymbol{x}_f, \boldsymbol{x}_i, T) = e^{-i\nu\pi/2}\left(\frac{1}{(2\pi i\hbar)^n}\left|\det\left(\frac{\partial^2 S_c(\boldsymbol{x}_f, \boldsymbol{x}_i, T)}{\partial\boldsymbol{x}_f\,\partial\boldsymbol{x}_i}\right)\right|\right)^{1/2}. \tag{3.77}$$

This result has been obtained by working in Cartesian coordinates. It can be transformed into another coordinate system $\boldsymbol{y} = \boldsymbol{y}(\boldsymbol{x})$ using the chain rule $\partial/\partial x^a = \sum_b(\partial y^b/\partial x^a)(\partial/\partial y^b)$. This transformation results in the two associated Jacobians appearing in the prefactor since $\det(\mathbf{A}\cdot\mathbf{B}\cdot\mathbf{C}) = \det\mathbf{A}\det\mathbf{B}\det\mathbf{C}$, so that

$$\det\left(\frac{\partial^2 S_c(\boldsymbol{x}_f, \boldsymbol{x}_i, T)}{\partial\boldsymbol{x}_f\,\partial\boldsymbol{x}_i}\right) = \det\left(\frac{\partial\boldsymbol{y}_i}{\partial\boldsymbol{x}_i}\right)\det\left(\frac{\partial\boldsymbol{y}_f}{\partial\boldsymbol{x}_f}\right)\det\left(\frac{\partial^2 S_c(\boldsymbol{y}_f, \boldsymbol{y}_i, T)}{\partial\boldsymbol{y}_f\,\partial\boldsymbol{y}_i}\right). \tag{3.78}$$

The relation (3.71) that led to equations (3.75) and (3.77) assumed that the integration over the initial position had an infinite range. It also assumed that there was a unique classical solution between the endpoints that led to a unique form for the classical action. Neither of these assumptions may be valid. For example, in the case of the particle constrained to lie on a circle, the associated configuration space path integral (2.44) requires an integration over the initial angle with a finite range. In addition, there is a sum over an infinite number of initial angles that are equivalent to θ_i, given by $\theta_i + 2\pi k$, where k is the winding number. In the semiclassical approximation for the free particle on a circle the winding number appears because there are an infinite number of classical paths going from θ_i to θ_f, written $\theta_c(k, t) = \theta_i + (\theta_f - \theta_i + 2\pi k)(t - t_i)/T$. In the absence of a potential the classical action for $\theta_c(k, t)$ depends only on $\dot{\theta}_c$ and is given by

$$S_c^{(k)}(\theta_f, \theta_i, T) = \int_{t_i}^{t_f} \mathrm{d}t\, \frac{1}{2}mR^2\dot{\theta}_c^2(k, t) = \frac{1}{2}m\frac{R^2}{T}(\theta_f - \theta_i + 2\pi k)^2. \tag{3.79}$$

It is important to understand how (3.79) appears in the semiclassical approximation to the propagator. This is facilitated by noting that the path integral (2.44) can be evaluated exactly in the absence of a potential. Apart from the sum over the initial angles, the path integral (2.44) is identical to the free particle path integral (3.1), and so the Gaussian integral (3.3) can be used to evaluate it. Alternatively, the transition element can be evaluated exactly using the operator techniques of equation (3.7) followed by applying the Poisson resummation. Both methods yield the result

$$
\begin{aligned}
G(\theta_f, \theta_i, T) &= \sqrt{\frac{mR^2}{2\pi i\hbar T}} \sum_{k=-\infty}^{\infty} \exp\left\{\frac{i}{\hbar}\left(\frac{1}{2}m\frac{R^2}{T}(\theta_f - \theta_i + 2\pi k)^2\right)\right\} \\
&= \sum_{k=-\infty}^{\infty} \left(\frac{1}{2\pi i\hbar}\left|\frac{\partial^2 S_c^{(k)}(\theta_f, \theta_i, T)}{\partial\theta_i\, \partial\theta_f}\right|\right)^{1/2} \exp\left(\frac{i}{\hbar}S_c^{(k)}(\theta_f, \theta_i, T)\right),
\end{aligned}
\tag{3.80}
$$

where the second line follows from the classical action (3.79).

Since the exact propagator and the semiclassical approximation must match in this case, result (3.80) shows how to define the semiclassical approximation for the case that there is a family of classical trajectories that match the boundary conditions. For this particular case there are no turning points in the classical trajectory. Of course, the number of unique solutions depends on the system being analyzed, and so the final form of the general semiclassical approximation in an n-dimensional system is inferred to be given by

$$
\begin{aligned}
G(\boldsymbol{x}_f, \boldsymbol{x}_i, t_f, t_i) = \sum_k e^{-i\nu_k\pi/2}&\left(\frac{1}{(2\pi i\hbar)^n}\left|\det\frac{\partial^2 S_c^{(k)}(\boldsymbol{x}_f, \boldsymbol{x}_i, t_f, t_i)}{\partial\boldsymbol{x}_i\, \partial\boldsymbol{x}_f}\right|\right)^{1/2} \\
&\times \exp\left\{\frac{i}{\hbar}S_c^{(k)}(\boldsymbol{x}_f, \boldsymbol{x}_i, t_f, t_i)\right\},
\end{aligned}
\tag{3.81}
$$

where the sum is over all possible classical trajectories $\boldsymbol{x}_c^{(k)}$ matching the boundary conditions $\boldsymbol{x}_c^{(k)}(t_i) = \boldsymbol{x}_i$ and $\boldsymbol{x}_c^{(k)}(t_f) = \boldsymbol{x}_f$.

Using equation (3.81) allows the path integral (2.36) for a particle in a box to be evaluated by the semiclassical method. The classical trajectories correspond to a point particle bouncing a total of n times off the box walls while conserving energy. The total distance traveled by a point particle bouncing between the two walls at a constant speed is given by $D = |x_f - (-1)^r x_i + 2kL|$, where $r = 0$ corresponds to an even number n of bounces and $r = 1$ corresponds to an odd number n of bounces. If n is even, the integer k can be either $\frac{1}{2}n$ or $-\frac{1}{2}n$, while if n is odd the integer k can be either $\frac{1}{2}(n - 1)$ or $-\frac{1}{2}(n + 1)$. For n either even or odd, the possible values for k run from $-\infty$ to ∞ with no repetition in value. The constant speed of the particle is $v = D/T$, and so the action is $S_c(x_f, x_i, k, r) = \frac{1}{2}mv^2T = \frac{1}{2}mD^2/T = \frac{1}{2}m(x_f - (-1)^r x_i + 2kL)^2/T$. All the bounces are classified as hard since the momentum changes discontinuously. Since hard reflections count twice in the Maslov index, the Maslov index for an even number of bounces $2n$ gives $\nu = 2 \cdot 2n = 4n$, while an odd number of bounces $2n + 1$ corresponds to the Maslov index $\nu = 2 \cdot (2n + 1) = 4n + 2$. Therefore, the Maslov index gives a factor of $e^{i\nu\pi/2} = e^{i2\pi n} = 1$ for an even number of bounces, and $e^{i\nu\pi/2} = e^{i2\pi n + i\pi} = -1$ for an odd number of bounces. Since $|\partial^2 S_c/\partial x_i\, \partial x_f| = m/T$ for the action, the general form (3.81) gives the semiclassical approximation for the propagator of a particle in a box as

$$G(x_f, x_i, T) = \sum_{k=-\infty}^{\infty} \sqrt{\frac{m}{2\pi i\hbar T}} \left(\exp\left\{ \frac{i}{\hbar}\frac{m}{2T}(x_f - x_i + 2kL)^2 \right\} \right. \\ \left. - \exp\left\{ \frac{i}{\hbar}\frac{m}{2T}(x_f + x_i + 2kL)^2 \right\} \right). \tag{3.82}$$

The path integral (2.36) for a particle confined to a box gives *exactly the same result* as (3.82) since it is simply a sum of free particle path integrals with different initial positions. The negative Maslov index factors in equation (3.82), arising from an odd number of bounces, have their counterpart in the negative term in the sine functions occurring in the derivation of the path integral (2.36). Although the same semiclassical approximation (3.81) can be derived by other means, the correctly formulated path integral reveals the relation to classical trajectories in the most transparent manner.

3.5 Energy eigenfunctions and the path integral

So far the path integral and its applications have centered primarily around the propagator between configuration states. However, there are many important quantum mechanical results that are revealed by the nature of the energy eigenfunctions and their eigenvalues. For the case that the Hamiltonian has no manifest time-dependence the standard method for finding these is solving the Schrödinger equation. An important quantum mechanical result is the presence of energy eigenfunctions that correspond to *tunneling* through classically forbidden regions. In its simplest manifestation, a particle of mass m in a one-dimensional system with a constant potential barrier of positive height V_0 and width L possesses energy eigenfunctions that correspond to the superposition of an incident wave $\psi_E^I(x)$ and a reflected wave $\psi_E^R(x)$ in the form $A\psi_E^I(x) + B\psi_E^R(x)$ for x on one side of the potential barrier. In addition, there is a transmitted wave $C\psi_E^T(x)$ for x on the other side of the barrier. It is the presence of the nonzero transmitted wave that is the manifestation of tunneling. In the case of the Schrödinger equation, the details of the tunneling process are revealed in the coefficients of the normalized eigenfunction that result from demanding continuity of the wave function and its derivatives across the two interfaces with the barrier.

In the case of more complicated barriers tunneling is often analyzed using the *WKB approximation* to the wave function. In order to use path integrals to investigate tunneling, the product $\psi_E^T(x_f)(A\psi_E^{I*}(x_i) + B\psi_E^{R*}(x_i))$ will be found, where x_i and x_f are on opposite sides of the barrier. The path integral provides a straightforward way to find this product of eigenfunctions using classical trajectories, and from it the details of tunneling can be extracted. This product of eigenfunctions will be shown to be nonzero *even if* the energy eigenvalue E is such that $E < V_0$, in which case $\psi_E^T(x_f)$ is found to be proportional to $\exp(-L\sqrt{2m(V_0 - E)}/\hbar)$.

Demonstrating energy dependent quantum processes such as tunneling in the context of a path integral requires projecting the path integral into an energy representation. The goal is to extract energy related quantum processes from the path integral version of the configuration space propagator. This is accomplished by using the path integral representation of the *retarded propagator*, which is obtained from the usual propagator by the definition $G_R(x_f, x_i, t_f - t_i) = \theta(t_f - t_i)$ $G(x_f, x_i, t_f - t_i)$. Combining (1.17), $G(x_f, x_i, 0) = \delta(x_f - x_i)$, and $(\hat{H} - i\hbar\, \partial/\partial t)$ $G(x_f, x_i, t_f - t_i) = 0$, where $\hat{H}$ is in configuration representation, yields

$$\left(\hat{H}\left(-i\hbar\frac{\partial}{\partial x_f}, x_f\right) - i\hbar\frac{\partial}{\partial t_f}\right)G_R(x_f, x_i, t_f - t_i) = i\hbar\, \delta(t_f - t_i)\, \delta(x_f - x_i), \tag{3.83}$$

so that the retarded propagator is a Green's function for the Schrödinger operator. Contact with the energy eigenfunctions follows from the Fourier transform of the retarded propagator, defined by

$$\begin{aligned} \tilde{G}_R(x_f, x_i, E) &= \int_{-\infty}^{\infty} \mathrm{d}T\; G_R(x_f, x_i, T)\, e^{iET/\hbar} \\ &= \int_0^{\infty} \mathrm{d}T\; G(x_f, x_i, T)\, e^{iET/\hbar}, \end{aligned} \tag{3.84}$$

where the step function has been absorbed into the integration limits. The path integral representation (2.11) for the propagator can now be used in equation (3.84). The relationship of equation (3.84) to the energy eigenstates is revealed from the definition (2.8) of the propagator. Because the exponential in equation (3.84) is oscillatory, the integral is defined by taking the limit of a damping factor $i\eta$ added to E. This gives

$$\begin{aligned} \tilde{G}_R(x_f, x_i, E) &= \lim_{\eta\to 0^+} \int_0^{\infty} \mathrm{d}T\; \langle x_f|e^{-i(\hat{H}-E-i\eta)T/\hbar}|x_i\rangle \\ &= \lim_{\eta\to 0^+} \langle x_f|\frac{i\hbar}{E - \hat{H} + i\eta}|x_i\rangle, \end{aligned} \tag{3.85}$$

where $i\hbar/(E - \hat{H} + i\eta)$ is referred to as the *resolvent*. After inserting a complete set of discrete energy eigenstates, expression (3.85) becomes

$$\begin{aligned} \tilde{G}_R(x_f, x_i, E) &= \lim_{\eta\to 0^+} \sum_k \langle x_f|\frac{i\hbar}{E - \hat{H} + i\eta}|E_k\rangle\langle E_k|x_i\rangle \\ &= \lim_{\eta\to 0^+} \sum_k \psi_k(x_f)\frac{i\hbar}{E - E_k + i\eta}\psi_k^*(x_i), \end{aligned} \tag{3.86}$$

showing the appearance of the energy eigenfunctions ψ_k in the useful product form discussed earlier in regard to tunneling. Result (3.86) possesses a sequence of poles at $E = E_k - i\eta$. If E_k is a continuous energy, the sum becomes an integral and the sequence of poles becomes a cut. For the case that E is an eigenvalue, so that $E \approx E_j$, the sum is dominated by the jth contribution and $\tilde{G}_R(x_f, x_i, E_j) \propto \psi_j(x_f)\psi_j^*(x_i)$.

The simple example of the free particle serves to demonstrate the results obtained so far. Using result (3.6) for the free particle propagator in equation (3.84) gives a variant of the Fresnel integral (1.101),

$$\begin{aligned}\tilde{G}_R(x_f, x_i, E) &= \int_0^\infty \mathrm{d}T \sqrt{\frac{m}{2\pi i\hbar T}} \exp\left\{\frac{i}{\hbar}\left(\frac{m(x_f - x_i)^2}{2T} + ET\right)\right\} \\ &= \sqrt{\frac{2m}{\pi i\hbar}} \int_0^\infty \mathrm{d}y \exp\left\{\frac{i}{\hbar}\left(\frac{m(x_f - x_i)^2}{2y^2} + Ey^2\right)\right\} \\ &= \sqrt{\frac{m}{2E}} \exp\left\{\frac{i}{\hbar}\sqrt{2\,mE(x_f - x_i)^2}\right\} = \frac{m}{p} \exp\left\{\frac{i}{\hbar}p(x_f - x_i)\right\},\end{aligned} \tag{3.87}$$

where $p = \sqrt{2\,mE}$. Result (3.87) shows the product of free particle energy eigenfunctions associated with $E = p^2/2m$ predicted by equation (3.86). Similarly, setting $E = p^2/2m$ and $E_n = k^2/2m$ in equation (3.86) while using the energy eigenfunction $\psi_k(x) = e^{ikx/\hbar}/\sqrt{2\pi\hbar}$ yields

$$\tilde{G}_R\left(x_f, x_i, \frac{p^2}{2m}\right) = \lim_{\eta\to 0^+} \int_{-\infty}^{\infty} \frac{\mathrm{d}k}{2\pi\hbar} \frac{2\,mi\hbar}{p^2 - k^2 + i2\,m\eta} e^{ik(x_f - x_i)/\hbar}. \tag{3.88}$$

The integral (3.88) can be evaluated using the Cauchy residue theorem. The parameter η serves to move the poles off the real axis so that, to $O(\eta)$, the denominator in equation (3.88) can be written $p^2 - k^2 + i2\,m\eta = -(k + p + im\eta/p)(k - p - im\eta/p)$. For $x_f - x_i > 0$ and $p > 0$, the contour is closed in the upper half of the complex k plane and the enclosed simple pole occurs at $k = p + im\eta/p$. Evaluating $2\pi i$ times the residue in the limit $\eta \to 0$ immediately reproduces result (3.87).

Using expression (2.11) or (2.13) for the propagator in equation (3.84) enables a path integral derivation of the product of energy eigenfunctions. It will now be shown that the semiclassical approximation (3.81) reproduces the WKB approximation for the energy eigenfunctions. For a one-dimensional system this starts by noting that

$$\tilde{G}_R(x_f, x_i, E) = \int_0^\infty \mathrm{d}T \sqrt{\frac{1}{2\pi i\hbar}\left|\frac{\partial^2 S_c}{\partial x_f\,\partial x_i}\right|} \exp\left\{\frac{i}{\hbar}(S_c + ET)\right\}, \tag{3.89}$$

where the classical action $S_c(x_f, x_i, T)$ is obtained from the classical path connecting x_i to x_f in the time T. Evaluating the integral (3.89) uses the *method of steepest descent*, which begins by expanding both the prefactor and the argument of the exponential in a Taylor series around the value of T, denoted T_c, that satisfies

$$\frac{\partial}{\partial T}(S_c + ET)\bigg|_{T=T_c} = E + \frac{\partial S_c}{\partial T}\bigg|_{T=T_c} = 0, \tag{3.90}$$

where E was treated as a parameter with no dependence on T. Since ET is a monotonically increasing function of T for $E > 0$, it follows that T_c is a *minimum* of $S_c + ET$ and is assumed to be unique. As a result, $\partial^2 S_c/\partial T_c^2$ must be a positive quantity. Using equation (1.13), equation (3.90) is satisfied by $E_c(x_i, x_f, T_c) = E$, so that T_c is determined as a function of x_i, x_f, and E. Using condition (3.90) shows that the Taylor series expansions for the factors in equation (3.89) are given by

$$\begin{aligned} S_c(x_i, x_f, T) + ET &= S_c(x_i, x_f, T_c) + ET_c + \frac{1}{2}\frac{\partial^2 S_c(x_i, x_f, T_c)}{\partial T_c^2}(T - T_c)^2 + \cdots \\ \frac{\partial^2 S_c(x_i, x_f, T)}{\partial x_f\, \partial x_i} &= \frac{\partial^2 S_c(x_i, x_f, T_c)}{\partial x_f\, \partial x_i} + \frac{\partial^3 S_c(x_i, x_f, T_c)}{\partial x_f\, \partial x_i\, \partial T_c}(T - T_c) + \cdots \end{aligned} \tag{3.91}$$

The next step is to change the variable of integration in (3.89) to $T - T_c = \sqrt{\hbar}\, y$ to cancel the inverse factor of $\sqrt{\hbar}$ in the prefactor of equation (3.89). In the semiclassical limit, $\hbar \to 0$, this suppresses all terms higher than quadratic in the exponential and all dependence on $T - T_c$ in the prefactor. For $\hbar \to 0$ the limits on the integral (3.89) over y become $(-T_c/\sqrt{\hbar}, \infty) \to (-\infty, \infty)$, so that

$$\begin{aligned} \tilde{G}_R(x_i, x_f, E) = &\sqrt{\frac{1}{2\pi i}\left|\frac{\partial^2 S_c(x_i, x_f, T_c)}{\partial x_f\, \partial x_i}\right|} \\ &\exp\left\{\frac{i}{\hbar}(S_c(x_i, x_f, T_c) + ET_c)\right\} \\ &\times \int_{-\infty}^{\infty} \mathrm{d}y \exp\left\{\frac{i}{2}\frac{\partial^2 S_c(x_i, x_f, T_c)}{\partial T_c^2}\, y^2\right\}. \end{aligned} \tag{3.92}$$

Result (3.92) and the Fresnel theorem yield the semiclassical form for $\tilde{G}_R$,

$$\tilde{G}_R(x_i, x_f, E) = \sqrt{\left(\frac{\partial^2 S_c}{\partial T_c^2}\right)^{-1}\left|\frac{\partial^2 S_c}{\partial x_f\, \partial x_i}\right|}\ \exp\left\{\frac{i}{\hbar}(S_c + ET_c)\right\}, \tag{3.93}$$

where $S_c = S_c(x_i, x_f, T_c)$. The semiclassical approximation (3.93) is exact for a quadratic Hamiltonian. This is easily demonstrated for the free particle, where $S_c(x_i, x_f, T) = m(x_f - x_i)^2/2T$. Solving equation (3.90) for T_c gives $T_c^2 = m(x_f - x_i)^2/2E$. It follows that $\partial^2 S_c(T_c)/\partial T_c^2 = m(x_f - x_i)^2/T_c^3$ and $|\partial^2 S_c/\partial x_i\, \partial x_f| = m/T_c$. This gives the prefactor $\sqrt{T_c^2/(x_f - x_i)^2} = \sqrt{m/2E} = m/p$, identical to result (3.87).

For the case that the Hamiltonian takes the form $H = p^2/2m + V(x)$, the relation $p_c(x) = \sqrt{2m(E_c(x_i, x_f, T_c) - V(x))}$ can be used. Relation (1.12) and the requirement that the trajectory has the energy $E_c(x_i, x_f, T_c) = E$ allows the argument of the exponential to be written

$$S_c(x_i, x_f, T_c) + ET_c = \int_{x_i}^{x_f} \mathrm{d}x\, p_c(x) = \int_{x_i}^{x_f} \mathrm{d}x\, \sqrt{2m(E_c(x_i, x_f, T_c) - V(x))}. \tag{3.94}$$

Using the relations appearing in equation (1.13) gives

$$\begin{aligned}\frac{\partial^2 S_c(x_i, x_f, T_c)}{\partial x_f\, \partial x_i} &= -\frac{\partial p_c(x_i)}{\partial x_f} = -\frac{m}{\sqrt{2\, m(E_c(x_i, x_f, T_c) - V(x_i))}}\frac{\partial E_c}{\partial x_f} \\ &= \frac{m}{p_c(x_i)}\frac{\partial}{\partial x_f}\frac{\partial S_c}{\partial T_c} = \frac{m}{p_c(x_i)}\frac{\partial}{\partial T_c}\frac{\partial S_c}{\partial x_f} = \frac{m}{p_c(x_i)}\frac{\partial p_c(x_f)}{\partial T_c} \\ &= \frac{m^2}{p_c(x_i)\, p_c(x_f)}\frac{\partial E_c}{\partial T_c} = -\frac{m^2}{p_c(x_i)\, p_c(x_f)}\frac{\partial^2 S_c}{\partial T_c^2}.\end{aligned} \tag{3.95}$$

Because $\partial^2 S_c/\partial T_c^2$ is positive, substituting equation (3.95) into equation (3.93) gives

$$\tilde{G}_R(x_i, x_f, E) = \sqrt{\frac{m^2}{|p_c(x_i)|\, |p_c(x_f)|}}\exp\left\{\frac{i}{\hbar}\int_{x_i}^{x_f} \mathrm{d}x\, \sqrt{2\, m(E - V(x))}\right\}, \tag{3.96}$$

which is the product of energy eigenfunctions in the standard WKB approximation, where $m/p_c(x) = 1/v_c(x)$ is the inverse of the classical velocity v_c at the point x. Result (3.96) shows the singularity at the classical trajectory's turning points, which occur where $v_c(x) = 0$, signaling a breakdown of the WKB approximation.

For the case of barrier penetration (3.96) reproduces the usual WKB result. For x_i and x_f on opposite sides of the barrier of width L and height $V_0 > E = p^2/2m$, the exponential factor in equation (3.96) yields $\exp\{ip(x_f - x_i - L)/\hbar\}\exp\{-L\sqrt{2\, mV_0 - p^2}\,/\hbar\}$, demonstrating the damping of energy eigenfunctions across the barrier. The modulus squared is proportional to $\exp\{-2L\sqrt{2\, mV_0 - p^2}\,/\hbar\}$, which gives the relative probability of tunneling for the specific case being considered. However, it must be stressed that the definition (3.84) no longer has any reference to the time involved in the original transition element since it has been integrated to find the projection onto energy eigenstates. As a result, calculating the tunneling rate requires the incident flux of particles at x_i, which is obtained from the spatial part of the probability current at x_i, $J(x_i, t) = (i\hbar/2m)(\nabla\Psi^*\Psi - \Psi^*\nabla\Psi)$, where Ψ is the wave function at x_i at the time t. For consistency with equation (3.96) the system of a single particle is prepared in a momentum eigenstate, $\Psi(x_i, t) = V^{-1/2}\exp\{i/\hbar(px_i - E_pt)\}$, where V is the volume of space to the left of the barrier at x_i. For such a state $J(x_i, t) = p/mV$, which has the units of inverse time. The probability per unit time $\mathrm{d}P/\mathrm{d}t$ for observing the particle on the other side of the barrier is then $\mathrm{d}P/\mathrm{d}t = (p/mV)\exp\{-2L\sqrt{2\, mV_0 - p^2}\,/\hbar\}$ for $p^2 < 2\, mV_0$. As a final note, the path integral representation of the propagator appearing in expression (3.84) for $\tilde{G}_R$ need not be evaluated by using the semiclassical approximation as it was in equation (3.93). Using the full path integral (2.11) or (2.13) is equivalent to the Schrödinger equation for finding the product of the exact energy eigenfunctions.

Another commonly employed method for the path integral analysis of tunneling starts with the definition of $\tilde{G}_R$ made in equation (3.84). However, instead of the method of steepest descent, the previously developed classical trajectory expansion of equation (3.26) is adapted to *imaginary time*. This is possible because the Wick rotation $t \to -i\tau$ effectively inverts the potential in the exponential,

$$\frac{i}{\hbar}\int_0^T \mathrm{d}t\left(\frac{1}{2}m\left(\frac{\mathrm{d}x}{\mathrm{d}t}\right)^2 - V(x)\right) \to -\frac{1}{\hbar}\int_0^\beta \mathrm{d}\tau\left(\frac{1}{2}m\left(\frac{\mathrm{d}x}{\mathrm{d}\tau}\right)^2 + V(x)\right), \tag{3.97}$$

where $\beta = iT$. The path integral after the Wick rotation is often referred to as the *Euclidean path integral* for reasons that are clearer in the context of the field theory path integral. The action of equation (3.97) is associated with the *Euclidean equation of motion* for $x_w = x(\tau)$,

$$m\frac{\mathrm{d}^2x_w}{\mathrm{d}\tau^2} - \frac{\partial V(x_w)}{\partial x_w} = 0, \tag{3.98}$$

Like equation (3.26), the Wick rotated action of equation (3.97) can be expanded around a solution to equation (3.98), denoted $x_{wc}(\tau)$, which satisfies the boundary conditions $x_{wc}(0) = x_i$ and $x_{wc}(\beta) = x_f$. In the Euclidean action of equation (3.97) the potential barrier becomes a potential well, and this will allow a well-defined solution to the equation of motion (3.98) parameterized by the imaginary time τ. In effect, this represents the *analytic continuation* of the original path integral.

The Wick rotated velocity is $v_w = \mathrm{d}x_w/\mathrm{d}\tau = i\,\mathrm{d}x/\mathrm{d}t = iv$, and so the Wick rotation of the real time kinetic energy $E_k = \frac{1}{2}mv^2$ gives $E_k \to -\frac{1}{2}mv_w^2 = -E_{kw}$. The Wick rotation for the full Hamiltonian is therefore $H = E_k + V(x) \to -H_w = -E_{kw} + V(x_w)$, so that the Wick rotated total energy is given by $H_w = E_{kw} - V(x_w)$. It follows from equation (3.98) that the Wick rotated Hamiltonian evaluated using the Euclidean trajectory obeys $\dot{H}_w = \mathrm{d}H_w/\mathrm{d}\tau = 0$.

For the simple case of tunneling through a rectangular barrier being considered here, the barrier of height V_0 and width L in real time becomes a well of depth $-V_0$ and width L in imaginary time, and in the region of the potential the energy becomes $E = \frac{1}{2}mv'^2 + V_o \to -\frac{1}{2}mv_w'^2 + V_o = -E_w$. The trajectories parameterized by τ that solve equation (3.98) also conserve the Wick rotated version of energy, allowing a simple characterization of the motion. As the particle crosses into the well its Wick rotated velocity changes as predicted by the conservation of energy, $E_w = \frac{1}{2}mv_w^2 = \frac{1}{2}mv_w'^2 - V_o$. For the case that x_i and x_f are on opposite sides of the well, the motion breaks into two segments. The first is motion at a constant Euclidean velocity v_w outside the well for a total time τ_o such that $v_w = (x_f - x_i - L)/\tau_o$. The second segment is when the particle is in the well, which occurs for the time $\tau_i = \beta - \tau_o$, during which its Euclidean velocity becomes $v_w' = \sqrt{v_w^2 + 2V_0/m}$ and gives $\tau_i = \beta - \tau_o = L/v_w'$. The exact values for τ_o and v_w' can be found since the trajectory must be such that $(x_f - x_i - L)/v_w + L/v_w' = \beta$. Using these simple results, the Wick rotated argument of the exponential in equation (3.84) becomes

$$
\begin{aligned}
\frac{i}{\hbar}(S_c + ET) &\to -\frac{1}{\hbar}(S_{cw} + E_w\beta) \\
&= -\frac{1}{\hbar}\int_0^{\beta} \mathrm{d}\tau\left(\frac{1}{2}m\left(\frac{\mathrm{d}x_c}{\mathrm{d}\tau}\right)^2 + V(x_c)\right) - \frac{E_w\beta}{\hbar} \\
&= -\frac{1}{\hbar}\left(\frac{1}{2}mv_w^2\tau_o + \left(\frac{1}{2}mv_w'^2 + V_o\right)(\beta - \tau_0) + E_w\beta\right) \\
&= -\frac{1}{\hbar}\left(\left(\frac{1}{2}mv_w^2 + E_w\right)\tau_o + \left(\frac{1}{2}mv_w'^2 + V_o + E_w\right)(\beta - \tau_0)\right) \\
&= -\frac{1}{\hbar}\left(mv_w^2\tau_o + mv_w'^2(\beta - \tau_0)\right) \\
&= -\frac{1}{\hbar}\left(mv_w(x_f - x_i - L) + mv_w'L\right).
\end{aligned}
\tag{3.99}
$$

It is important to remember that v_w and v_w' are functions of β, so that the integration over β to find $\tilde{G}_R$ will require these dependences. However, result (3.99) is often used to determine the relative probability of tunneling in the following way. Performing the inverse Wick rotation $\tau \to it$ gives $v_w \to -iv = -ip/m$ and $v_w' \to \sqrt{-v^2 + 2V_o/m} = -i\sqrt{v^2 - 2V_o/m} = -i\sqrt{p^2 - 2\,mV_o}/m$, where the minus sign is chosen for v_w' for consistency with v_w. The final result for the exponential is

$$
\exp\left\{\frac{i}{\hbar}(S_c + ET)\right\} = \exp\left\{\frac{i}{\hbar}\left(p(x_f - x_i - L) + L\sqrt{2\,m(E - V_o)}\right)\right\}, \tag{3.100}
$$

where $E = p^2/2m$. For the case that $E < V_o$ the square root becomes $i\sqrt{2\,m(V_o - E)}$, showing the exponential damping factor across the barrier found earlier using the method of steepest descent.

It must be stressed that this result is not the final result using this method. The next step in this approach is calculating the Wick rotated version of the prefactor (3.27). For a general potential $V(x)$ this is equivalent to finding the ratio of the functional determinants of the Wick rotated differential operators $\hat{D}_{w0}(\tau) = m\,\mathrm{d}^2/\mathrm{d}\tau^2$ and $\hat{D}_w(\tau) = m\,\mathrm{d}^2/\mathrm{d}\tau^2 - V''(x_w(\tau))$ in the prefactor (3.54). This can be done using Euclidean variants of the Gel'fand–Yaglom theorem. The final step is to perform the integral over β to obtain the Wick rotated version of $\tilde{G}_R$.

An area of considerable interest where this technique has been applied is the decay of a metastable quantum mechanical state. This occurs in a system which has a potential with two or more relative minima, so that $\partial V(x)/\partial x = 0$ has more than one solution. If a particle is localized in a minimum at a higher energy than another minimum, the particle will eventually tunnel the barrier between the two into the lower minimum. This phenomenon is known as *the decay of the false vacuum* and has been studied in the context of the path integral. Coleman examined the *persistence functional* for the false vacuum located at x_*, given by the path integral

$$P(x_*, T) = G(x_*, T, x_*, 0) = \int_{x_*}^{x_*} [\mathcal{D}x] \exp\left\{ i \int_0^T \mathrm{d}t\, \mathcal{L} \right\}. \tag{3.101}$$

The Wick rotated path integral has a set of classical solutions for use in equation (3.81), referred to as *instantons*, that describe the particle bouncing from the false ground state to the true ground state and back again. These solutions are available in the Wick rotated path integral since the potential is effectively inverted, as was shown in equations (3.97) and (3.98). Through analytic continuation Coleman demonstrated the existence of negative eigenvalues for the differential operators that determine the prefactor, and the square root of these negative eigenvalues create a damping factor of the form $\exp(-\Gamma t)$ in the persistence functional. The interested reader is referred to the original work of Coleman.

An extension of equation (2.27) can be expressed with the semiclassical approximation and used to analyze (3.101). The propagator can be rewritten as an integral, a variant of mathematical *convolution*, which takes the form in one dimension

$$\begin{aligned} \int \mathrm{d}x\, G(x_f, x, T-t)\, G(x, x_i, t) &= \int \mathrm{d}x\, \langle x_f | e^{-i\hat{H}(T-t)/\hbar} | x \rangle \langle x | e^{-i\hat{H}t/\hbar} | x_i \rangle \\ &= \langle x_f | e^{-i\hat{H}T/\hbar} | x_i \rangle = G(x_f, x_i, T), \end{aligned} \tag{3.102}$$

where t is *arbitrary* and the integration is over the physical domain of the system. Because t is arbitrary and does not appear on the right-hand side of equation (3.102), integrating both sides of equation (3.102) gives

$$G(x_f, x_i, T) = \frac{1}{T} \int_0^T \mathrm{d}t \int \mathrm{d}x\, G(x_f, x, T-t)\, G(x, x_i, t). \tag{3.103}$$

Each of the two propagators appearing in equation (3.102) can be given a semiclassical approximation using equation (3.75). By setting $x_i = x_f$, results such as equation (3.103) form a recent approach to analyzing equation (3.101) and the decay of metastable states. The reader is recommended to the review article by Andreassen *et al* that discusses open issues in the path integral approach to analyzing metastable states.

3.6 Coherent state path integrals and difference equations

The semiclassical method fails for the both the c-number and Grassmann coherent state path integrals given by equations (2.87) and (2.111). Concentrating on the c-number coherent state path integral (2.87), the time-sliced action, designated S_0, can be associated with a continuum Lagrangian by the usual recipe,

$$\begin{aligned} S_0 = \sum_{j=0}^{N-1} \epsilon\, \mathcal{L}_j &= \sum_{j=0}^{N-1} \epsilon \left(\frac{1}{2} i\hbar \lambda_{j+1}^* \dot{\lambda}_j - \frac{1}{2} i\hbar \dot{\lambda}_j^* \lambda_j - \hbar\omega \lambda_{j+1}^* \lambda_j - \frac{1}{2} \hbar\omega \right) \\ &\to \int_{t_i}^{t_f} \mathrm{d}t \left(\frac{1}{2} i\hbar \lambda^* \dot{\lambda} - \frac{1}{2} i\hbar \dot{\lambda}^* \lambda - \hbar\omega (\lambda^* \lambda + \frac{1}{2}) \right). \end{aligned} \tag{3.104}$$

The continuum Lagrangian of equation (3.104) gives the Euler–Lagrange equation of motion $\dot{\lambda} = -i\omega\lambda$ and its complex conjugate, $\dot{\lambda}^* = i\omega\lambda^*$. Since these equations are first order, they allow the choice of only one complex boundary condition, as opposed to the two complex initial and final conditions, λ_f and λ_i, associated with the transition element. Worse, substituting the equations of motion into the continuum action of equation (3.104) gives the action $-\frac{1}{2}\hbar\omega T$ *even if different boundary conditions are selected for* λ and λ^*. The root of the problem lies, once again, in treating the formal derivatives $\dot{\lambda}_j = (\lambda_{j+1} - \lambda_j)/\epsilon$ as true derivatives in the $\epsilon \to 0$ limit. In the case of the semiclassical approach for second order Lagrangians, the path integral measure could be altered so that the differentiable fluctuations alone gave the correct functional determinant and transition element. This solution fails for the coherent state path integral and so the *unruly paths* must be included. This means that terms $O(\epsilon^2)$ and higher *cannot be discarded* during the evaluation.

An approach similar to the semiclassical approximation for quadratic actions leaves the formal *difference derivatives* in the action for the two path integrals. It then sequentially decouples the variables of integration by solving the Euler–Lagrange equation in the absence of derivatives. This method is based on the property of the coherent state measures (2.80) and (2.107) that allows translation by *different complex functions* in both the c-number and Grassmann case. For example, the c-number Gaussian integral (1.99) can be combined with *two arbitrary complex numbers*, f and g, to give

$$\begin{aligned}\int \frac{d^2\tilde{\lambda}}{\pi}\, e^{-(\tilde{\lambda}^*+f^*)(\tilde{\lambda}+g)} &= \int_{-\infty}^{\infty} \frac{d\tilde{\lambda}_R\, d\tilde{\lambda}_I}{\pi}\, e^{-\tilde{\lambda}_R^2-\tilde{\lambda}_R(f^*+g)}e^{-\tilde{\lambda}_I^2-i\tilde{\lambda}_I(f^*-g)}e^{-f^*g} \\ &= e^{\frac{1}{4}(f^*+g)^2-\frac{1}{4}(f^*-g)^2-f^*g} = 1 \\ \Longrightarrow \int \frac{d^2\lambda}{\pi}\, e^{-\lambda^*\lambda} &= \int \frac{d^2\tilde{\lambda}}{\pi}\, e^{-(\tilde{\lambda}^*+f^*)(\tilde{\lambda}+g)}.\end{aligned} \tag{3.105}$$

A similar result holds for the Grassmann Gaussian integral (1.81).

The two coherent state path integrals to be evaluated, equations (2.87) and (2.111), are formally identical, with the only difference being c-number variables versus Grassmann variables. In what follows, the order of variables will be observed in both cases even though the c-number variable version will be analyzed. Evaluating the c-number path integral (2.87) begins by noting that the terms in the time-sliced Lagrangian have the form $\mathcal{L}_j = \mathcal{L}_j(\lambda_{j+1}, \lambda_j)$. This means that all references to the variables of integration λ_j and λ_j^* are contained in the sequential sum $\epsilon\mathcal{L}_j + \epsilon\mathcal{L}_{j-1} \equiv \mathcal{A}_j(\lambda_{j+1}, \lambda_j, \lambda_{j-1})$. The λ_j variables are decoupled by solving the Euler–Lagrange equations without derivatives, which give two *difference equations*,

$$\frac{\partial \mathcal{A}_j}{\partial \lambda_j} = 0, \quad \frac{\partial \mathcal{A}_j}{\partial \lambda_j^*} = 0, \tag{3.106}$$

and expanding $\mathcal{A}_j$ around the two solutions. The solutions to the respective equations of (3.106) are denoted $\lambda_j = \bar{\lambda}_j(\lambda_{j+1}, \lambda_{j-1})$ and $\lambda_j^* = \bar{\lambda}_j^*(\lambda_{j+1}, \lambda_{j-1})$ and the original variables of integration are *translated* according to $\lambda_j = \bar{\lambda}_j + \tilde{\lambda}_j$ and $\lambda_j^* = \bar{\lambda}_j^* + \tilde{\lambda}_j^*$. From equation (3.105) this is possible *even if* $\bar{\lambda}_j$ and $\bar{\lambda}_j^*$ are *not* complex conjugates of each other. The Taylor series for $\mathcal{A}_j$ is given by

$$\mathcal{A}_j = \mathcal{A}_j(\lambda_{j+1}, \bar{\lambda}_j, \bar{\lambda}_j^*, \lambda_{j-1}) + \tilde{\lambda}_j^* \tilde{\lambda}_j M(\bar{\lambda}_j, \bar{\lambda}_j^*) \tag{3.107}$$

where

$$M(\bar{\lambda}_j, \bar{\lambda}_j^*) = \left. \frac{\partial^2 \mathcal{A}_j}{\partial \lambda_j \, \partial \lambda_j^*} \right|_{\bar{\lambda}_j, \bar{\lambda}_j^*} . \tag{3.108}$$

The terms linear in $\tilde{\lambda}_j$ and $\tilde{\lambda}_j^*$ are absent in equation (3.107) by virtue of condition (3.106). Since the original action in equation (2.87) is quadratic, there are no higher order terms. The term proportional to $\tilde{\lambda}_j^* \tilde{\lambda}_j$ in the path integral can now be integrated using the measure from equation (2.80), $\mathrm{d}^2\tilde{\lambda} = \mathrm{d}\tilde{\lambda}_R \, \mathrm{d}\tilde{\lambda}_I$, with the result

$$\int \frac{\mathrm{d}^2 \tilde{\lambda}}{\pi} \exp\left\{ \frac{i}{\hbar} \tilde{\lambda}_j^* \tilde{\lambda}_j M(\bar{\lambda}_j, \bar{\lambda}_j^*) \right\} = i\hbar \, M^{-1}(\bar{\lambda}_j, \bar{\lambda}_j^*). \tag{3.109}$$

The remaining first term in equation (3.107) can now be combined with the next term in the Lagrangian sum, so that $\mathcal{A}_{j+1} = \mathcal{A}_j + \epsilon \mathcal{L}_{j+1}$, and the process is repeated. For a quadratic action this method is equivalent to sequentially completing the squares, as was done in equations (3.4) and (3.5). Translating the variables by the solutions to equation (3.106) provides a simple method to decouple the fluctuations $\tilde{\lambda}_j$ from each other and reduce the path integral to a sequence of Gaussians of the form (3.109). This method can be applied to all the quadratic path integrals considered so far.

Applying this method to the coherent state path integral action of equation (2.87) begins with the λ_1 variables, which are completely contained in $\mathcal{A}_1 = \epsilon \mathcal{L}_1 + \epsilon \mathcal{L}_0$, so that

$$\begin{aligned} \mathcal{A}_1 = & \frac{1}{2} i\hbar \lambda_2^* (\lambda_2 - \lambda_1) - \frac{1}{2} i\hbar (\lambda_2^* - \lambda_1^*) \lambda_1 + \frac{1}{2} i\hbar \lambda_1^* (\lambda_1 - \lambda_0) \\ & - \frac{1}{2} i\hbar (\lambda_1^* - \lambda_0^*) \lambda_0 - i\hbar D_1(\epsilon) \lambda_2^* \lambda_1 - i\hbar D_1(\epsilon) \lambda_1^* \lambda_0 - \epsilon \hbar \omega, \end{aligned} \tag{3.110}$$

where the initial coefficient $D_1(\epsilon) = -i\omega\epsilon$ is imaginary for notational convenience. Applying equation (3.106) to equation (3.110) yields the two difference equation solutions,

$$\begin{aligned} \frac{\partial \mathcal{A}_1}{\partial \lambda_1} &= i\hbar \lambda_1^* - i\hbar \lambda_2^* - i\hbar D_1(\epsilon) \lambda_2^* = 0 \Longrightarrow \bar{\lambda}_1^* = \lambda_2^* + D_1(\epsilon) \lambda_2^*, \\ \frac{\partial \mathcal{A}_1}{\partial \lambda_1^*} &= i\hbar \lambda_1 - i\hbar \lambda_0 - i\hbar D_1(\epsilon) \lambda_0 = 0 \Longrightarrow \bar{\lambda}_1 = \lambda_0 + D_1(\epsilon) \lambda_0. \end{aligned} \tag{3.111}$$

The second derivative gives $M(\bar{\lambda}_1, \bar{\lambda}_1^*) = \partial^2 \mathcal{A}_1/\partial\lambda_1\,\partial\lambda_1^* = i\hbar$, so that equation (3.109) becomes a factor of one. Inserting the solutions (3.111) into (3.110) gives

$$\mathcal{A}_1(\lambda_2, \bar{\lambda}_1, \bar{\lambda}_1^*, \lambda_0) = \frac{1}{2}i\hbar\lambda_2^*(\lambda_2 - \lambda_0) - \frac{1}{2}i\hbar(\lambda_2^* - \lambda_0^*)\lambda_0 - i\hbar D_2(\epsilon)\lambda_2^*\lambda_0 - \hbar\omega\epsilon, \quad (3.112)$$

where the new coefficient is given by $D_2(\epsilon) = (1 - i\omega\epsilon)D_1(\epsilon) - i\omega\epsilon$. Result (3.112) is identical to $\epsilon\mathcal{L}_1$ except that λ_2 has replaced λ_1 and $D_2(\omega)$ has replaced $D_1(\omega)$. This process can be repeated after combining equation (3.112) with $\epsilon\mathcal{L}_2$ and yields an identical result, with λ_3 replacing λ_2 and $D_3(\epsilon)$ replacing $D_2(\epsilon)$, where once again $D_3(\epsilon) = (1 - i\omega\epsilon)D_2(\epsilon) - i\omega\epsilon$. After the jth iteration the equation for $D_{j+1}(\epsilon)$ is again given by $D_{j+1} = (1 - i\omega\epsilon)D_j(\epsilon) - i\omega\epsilon$. This difference equation has the boundary condition $D_1(\epsilon) = -i\omega\epsilon$, but each successive iteration has higher powers of $\omega\epsilon$, with $D_j(\epsilon)$ containing a term proportional to $(\omega\epsilon)^j$. In the limit $\epsilon \to 0$ and j large this difference equation has the unique solution $D_j(\epsilon) = e^{-i\omega j\epsilon} - 1$. After all $N - 1$ possible iterations are complete, the original action sum $\sum_j \epsilon\mathcal{L}_j$ becomes a single term, $\mathcal{A}_{N-1}$, given by

$$\begin{aligned}\mathcal{A}_{N-1} &= \frac{1}{2}i\hbar\lambda_N^*(\lambda_N - \lambda_0) - \frac{1}{2}i\hbar(\lambda_N^* - \lambda_0^*)\lambda_0 - i\hbar D_N(\epsilon)\lambda_N^*\lambda_0 - \frac{1}{2}\hbar\omega N\epsilon \\ &= \frac{1}{2}i\hbar\lambda_f^*\lambda_f + \frac{1}{2}i\hbar\lambda_i^*\lambda_i - i\hbar\lambda_f^*\lambda_i e^{-i\omega T} - \frac{1}{2}\hbar\omega T.\end{aligned} \quad (3.113)$$

The path integral of equation (2.87) is then given by

$$\begin{aligned}G(\lambda_f, \lambda_i, t_f, t_i) &= \exp\left\{\frac{i}{\hbar}\mathcal{A}_{N-1}\right\} \\ &= \exp\left\{-\frac{1}{2}\lambda_f^*\lambda_f - \frac{1}{2}\lambda_i^*\lambda_i + \lambda_f^*\lambda_i e^{-i\omega T} - \frac{1}{2}i\omega T\right\}.\end{aligned} \quad (3.114)$$

Using equations (2.76), (2.77), and the harmonic oscillator energy eigenvalues gives

$$\begin{aligned}\langle\lambda_f|e^{-i\hat{H}T/\hbar}|\lambda_i\rangle &= e^{-\frac{1}{2}\lambda_f^*\lambda_f - \frac{1}{2}\lambda_i^*\lambda_i}e^{-\frac{1}{2}i\omega T}\sum_{n=0}^{\infty}\frac{(\lambda_f^*\lambda_i)^n e^{-in\omega T}}{n!} \\ &= \exp\left\{-\frac{1}{2}\lambda_f^*\lambda_f - \frac{1}{2}\lambda_i^*\lambda_i + \lambda_f^*\lambda_i e^{-i\omega T} - \frac{1}{2}i\omega T\right\},\end{aligned} \quad (3.115)$$

demonstrating that the operator approach yields a result identical to equation (3.114).

The steps used to evaluate the c-number coherent state path integral can be applied verbatim to the Grassmann coherent state path integral (2.111), yielding a formally similar result,

$$G(\xi_f, \xi_i, t_f, t_i) = \exp\left\{-\frac{1}{2}\xi_f^*\xi_f - \frac{1}{2}\xi_i^*\xi_i + \xi_f^*\xi_i e^{-i\omega T} + \frac{1}{2}i\omega T\right\}. \quad (3.116)$$

The sole differences are the Grassmann nature of ξ_f and ξ_i and the opposite sign of the ground state energy. An operator based derivation, like equation (3.115) but using Grassmann coherent states in terms of energy eigenstates (2.103), also yields equation (3.116).

It is important to extend the coherent state path integral to the case where a complex external source is present, so that the time-sliced Lagrangian in the c-number case is equation (2.90) and in the Grassmann case is given by equation (2.113). Since the source function is differentiable, it can be decoupled from the variables of integration by using *continuum methods*. Because the order of the variables will be important in this process, it will be performed for the Grassmann coherent state path integral. The procedure consists of translating the variables of integration by an *impulse function* according to $\xi_j \to \xi_j + I(t_j)$ and $\xi_j^* \to \xi_j^* + \mathcal{I}^*(t_j)$, where the two functions, $I(t_j)$ and $\mathcal{I}^*(t_j)$, are assumed to depend on the source functions $K(t)$ and $K^*(t)$ respectively, and will therefore behave as Grassmann variables. Because these functions are differentiable, result (3.20) can be extended to give $\dot{\xi}_j \to \dot{\xi}_j + \dot{I}(t_j)$ and $\dot{\xi}_j^* \to \dot{\xi}_j^* + \dot{\mathcal{I}}^*(t_j)$. There is an important subtlety in this statement, since only the variables of integration appearing in the action (2.90) can be translated, while ξ_i and ξ_f^* must remain unchanged. This means that $\dot{\xi}_0 \to (\xi_1 - \xi_i)/\epsilon + (I(t_1) - I(t_i))/\epsilon = \dot{\xi}_0 + \dot{I}(0)$, but *only if* $I(t_i) = 0$. Similarly, $\dot{\xi}_{N-1}^* \to (\xi_f^* - \xi_{N-1}^*)/\epsilon + (\mathcal{I}^*(t_f) - \mathcal{I}^*(t_{N-1}))/\epsilon = \dot{\xi}_{N-1}^* + \dot{\mathcal{I}}^*(t_{N-1})$, but *only if* $\mathcal{I}^*(t_f) = 0$. It is critical to this process that $\dot{\xi}_0^*$ and $\dot{\xi}_N$ do not occur in the action of equation (3.104), and therefore only one boundary condition is required for both I and $\mathcal{I}^*$. The formal derivatives of ξ and ξ^* become true derivatives of $I(t)$ and $\mathcal{I}^*(t)$ following the integration by parts derived in equation (3.11). Momentarily suppressing $\hbar$, the sum of the formal derivatives in equation (3.104) can be written in two equivalent forms,

$$\begin{aligned}\sum_{j=0}^{N-1} \epsilon\left(\frac{1}{2} i\dot{\xi}_{j+1}^* \xi_j - \frac{1}{2} i\xi_j^* \dot{\xi}_j\right) &= \frac{1}{2} i\xi_f^* \xi_f + \frac{1}{2} i\xi_i^* \xi_i - i\xi_1^* \xi_i - \sum_{j=1}^{N-1} \epsilon\ i\dot{\xi}_j^* \xi_j \\ &= \frac{1}{2} i\xi_f^* \xi_f + \frac{1}{2} i\xi_i^* \xi_i - i\xi_f^* \xi_{N-1} + \sum_{j=1}^{N-1} \epsilon\ i\xi_j^* \dot{\xi}_{j-1}.\end{aligned} \tag{3.117}$$

The two different forms can be used subsequent to translating the variables of integration. This replaces $\mathcal{I}^*(t_{j+1})\dot{\xi}_j$ and $-\dot{\xi}_j^* I(t_j)$ with $-\dot{\mathcal{I}}_{j+1}^* \xi_j$ and $\xi_j^* \dot{I}(t_j)$, *up to the surface terms* in equation (3.117). The translated action (2.90) is then given by

$$\begin{aligned}S \to\ & S_0 - i\hbar\, \xi_f^*\, I(t_{N-1}) - i\hbar\, \mathcal{I}^*(t_1)\, \xi_i \\ &+ \sum_{j=1}^{N-1} \epsilon\ \xi_j^*\big(i\hbar \dot{I}(t_{j-1}) - \hbar\omega I(t_{j-1}) + J(t_{j-1})\big) \\ &- \sum_{j=1}^{N-1} \epsilon\ \big(i\hbar \dot{\mathcal{I}}^*(t_j) + \hbar\omega \mathcal{I}^*(t_j) - J^*(t_j)\big)\xi_j \\ &+ \int_{t_i}^{t_f} \mathrm{d}t \left(\frac{1}{2} i\hbar \mathcal{I}^* \dot{I} - \frac{1}{2} i\hbar \dot{\mathcal{I}}^* I - \hbar\omega \mathcal{I}^* I + J^* I + J \mathcal{I}^*\right),\end{aligned} \tag{3.118}$$

where S_0 is the original action (3.104) without the source terms, while the Riemann sum of impulse function products becomes an integral. Because the impulse functions are differentiable it is possible to replace $\mathcal{I}^*(t_1)$ with $\mathcal{I}^*(t_i) + \epsilon\,\dot{\mathcal{I}}^*(t_i)$ and $I(t_{N-1})$ with $I(t_f) - \epsilon\,\dot{I}(t_f)$ and discard the term proportional to ϵ in both cases. The terms linear in ξ_j and ξ_j^* in equation (3.118) are removed by choosing the functions $\mathcal{I}^*(t)$ and $I(t)$ to satisfy the equations $i\hbar\dot{I}(t) - \hbar\omega I(t) + J(t) = 0$ and $i\hbar\dot{\mathcal{I}}^*(t) + \hbar\omega\mathcal{I}^*(t) - J^*(t) = 0$, along with the two previously determined boundary conditions $I(t_i) = \mathcal{I}^*(t_f) = 0$. Inserting the equation of motion for the impulse functions into equation (3.118) yields the resulting form of the translated action,

$$S \to S_0 - i\hbar\left(\xi_f^*\,I(t_f) + \mathcal{I}^*(t_i)\,\xi_i\right) + \int_{t_i}^{t_f} \mathrm{d}t \left(\frac{1}{2}\mathcal{I}^*(t)\,K(t) + \frac{1}{2}K^*(t)I(t)\right). \quad (3.119)$$

The two equations for $I(t)$ and $\mathcal{I}^*(t)$ are solved using Green's functions, where the step function (1.17) is used as it was in the impulse equation (3.14),

$$\begin{aligned} I(t) &= \frac{i}{\hbar}\int_{t_i}^{t_f} \mathrm{d}\tau\;\theta(t-\tau)\,e^{-i\omega(t-\tau)}K(\tau),\\ \mathcal{I}^*(t) &= \frac{i}{\hbar}\int_{t_i}^{t_f} \mathrm{d}\tau\;\theta(\tau-t)\,e^{-i\omega(\tau-t)}K^*(\tau). \end{aligned} \quad (3.120)$$

By construction, the impulse functions of equation (3.120) also satisfy the required boundary conditions $\mathcal{I}^*(t_f) = I(t_i) = 0$. Inserting the impulse functions (3.120) into equation (3.119) yields the contribution of the sources to the action,

$$\begin{aligned} S \to S_0 &+ \int_{t_i}^{t_f} \mathrm{d}t\left(e^{-i\omega(t_f-t)}\xi_f^*\,K(t) + e^{-i\omega(t-t_i)}K^*(t)\,\xi_i\right)\\ &+ \frac{i}{\hbar}\int_{t_i}^{t_f} \mathrm{d}t \int_{t_i}^{t_f} \mathrm{d}t'\;K^*(t)\,\theta(t-t')\,e^{-i\omega(t-t')}K(t')\\ &\equiv S_0 + F_\xi[K]. \end{aligned} \quad (3.121)$$

Since the action S_0 in equation (3.121) contains the variables of integration and is identical to equation (3.104), its evaluation will once again give the path integral of equation (3.114), denoted $G(\xi_f, \xi_i, t_f, t_i)$. Using the action (3.121) gives the final result for the path integral with source terms,

$$Z_{if}[K] = G(\xi_f, \xi_i, t_f, t_i)\exp\left\{\frac{i}{\hbar}F_\xi[K]\right\}, \quad (3.122)$$

which clearly reduces to equation (3.114) when $K = 0$. In the next section the order of the Grassmann source terms in equation (3.121) will be seen to be consistent with the time-ordering of fermionic operators discussed in equation (2.115).

Repeating these same steps for the c-number coherent state path integral with source terms (2.90), denoted $Z_{if}[J]$, yields an identical result to equation (3.122), with ξ_f, ξ_i, and $K(t)$ replaced with the c-number variables λ_f, λ_i, and $J(t)$, so that

$$Z_{if}[J] = G(\lambda_f, \lambda_i, t_f, t_i)\exp\left\{\frac{i}{\hbar}F_\lambda[J]\right\}, \tag{3.123}$$

where the functional $F_\lambda[J]$ is given by

$$\begin{aligned} F_\lambda[J] = &\frac{i}{\hbar}\int_{t_i}^{t_f} \mathrm{d}t \int_{t_i}^{t_f} \mathrm{d}t'\, J^*(t)\,\theta(t - t')\, e^{-i\omega(t-t')}J(t') \\ &+ \int_{t_i}^{t_f} \mathrm{d}t \left(e^{-i\omega(t_f-t)}\lambda_f^*\, J(t) + e^{-i\omega(t-t_i)}J^*(t)\,\lambda_i\right) \end{aligned} \tag{3.124}$$

The order of the c-number variables is irrelevant.

3.7 Generating functionals and perturbation theory

Path integrals with quadratic actions are exactly solvable. However, the path integrals for nonquadratic Hamiltonians are often as intractable as their wave mechanical and operator counterparts. The semiclassical approximation can be invoked if classical trajectories can be found. However, the path integral allows a relatively easy method to develop a perturbative approach to evaluating transition elements. The underlying ideas are simple to state. In a one-dimensional infinite volume system the propagator associated with the Hamiltonian $\hat{H} = \hat{P}^2/2\,m + V(\hat{X})$ is given by the configuration space path integral (2.13). This path integral cannot, in general, be directly evaluated. An associated path integral, with a *basis Lagrangian* given by $\mathcal{L}_0 + J(t)x$, is found that is exactly solvable and, loosely speaking, is *close* to the full Lagrangian $\mathcal{L}$ for the case that $J = 0$. The latter statement is vague and often the source of problems in perturbative approaches, since the choice of basis Lagrangian may give incomplete or misleading results.

The process begins by writing the Lagrangian in the time-sliced path integral (2.13) in arbitrary spatial dimensions as the sum of two terms, $\mathcal{L}(\dot{\boldsymbol{x}}, \boldsymbol{x}) = \mathcal{L}_0(\dot{\boldsymbol{x}}, \boldsymbol{x}) + \boldsymbol{J}\cdot\boldsymbol{x} + \mathcal{L}_I(\boldsymbol{x})$, where $\mathcal{L}_0(\dot{\boldsymbol{x}}, \boldsymbol{x}) + \boldsymbol{J}\cdot\boldsymbol{x}$ is the *basis Lagrangian* and is associated with an exactly solvable path integral. The term $\mathcal{L}_I(\boldsymbol{x})$ is *assumed* to create a *small perturbation* in the results obtained by evaluating the $\mathcal{L}_0$ path integral. This is often justified for the case that $\mathcal{L}_I$ has a small coefficient, or *weak coupling constant*, that is not expected to alter the energy level structure and values significantly. The basis Lagrangian is typically quadratic in $\dot{\boldsymbol{x}}$ and $\boldsymbol{x}$ and yields an exactly solvable path integral, denoted $Z_{if}^0[\boldsymbol{J}]$. The key to generating the perturbative representation of the propagator for the more complicated system including $\mathcal{L}_I$, referred to as the *perturbation*, is the presence of the source function $\boldsymbol{J}(t)$. Similarly to the linear term in equation (1.99), $\boldsymbol{J}(t)$ allows the more complicated system to be written in terms of functional derivatives,

$$G(\boldsymbol{x}_f, \boldsymbol{x}_i, t_f, t_i) = \exp\left\{\frac{i}{\hbar}\int_{t_i}^{t_f} \mathrm{d}t\, \mathcal{L}_I\left(\frac{\hbar}{i}\frac{\delta}{\delta \boldsymbol{J}(t)}\right)\right\} Z_{if}^0[\boldsymbol{J}]\Big|_{\boldsymbol{J}=0}. \tag{3.125}$$

Expanding the exponential in a Taylor series and applying the functional derivatives appearing in equation (3.125) to $Z^0_{if}[\boldsymbol{J}]$ and finally setting $\boldsymbol{J} = 0$ regenerates the original path integral for $\mathcal{L}$. The terms in the series will be proportional to powers of the coefficients that characterize $\mathcal{L}_I$, so that higher order terms are rapidly negligible as long as these coefficients are small, representing *weak forces*. This series yields a perturbative representation of the propagator for the full Lagrangian $\mathcal{L}_0 + \mathcal{L}_I$. This approach to perturbation theory was originally developed by Schwinger and is known as *source theory*.

The basis functional $Z_{if}[\boldsymbol{J}]$ is often referred to as a *generating functional* since its functional derivatives generate the expectation values for products of the Heisenberg picture observable to which $\boldsymbol{J}$ is coupled. This can be seen from the generating functionals evaluated so far, equation (3.17) for the free particle propagator, equation (3.66) for the harmonic oscillator propagator, and equation (3.122) for the harmonic oscillator coherent state transition element. For example, using the one-dimensional version of the free particle generating functional (3.17) and the definitions of the impulse functions (3.14) and (3.16) while retaining only terms that do not vanish when $J = 0$ gives

$$\begin{aligned}\langle \hat{X}_H(t)\rangle &= -i\hbar \frac{\delta Z^0_{if}[J]}{\delta J(t)}\bigg|_{J=0} = \left(x_i \frac{\delta I(t_i)}{\delta J(t)} + \frac{(x_f - x_i)}{T}\frac{\delta R_J}{\delta J(t)}\right) Z^0_{if}[J]\bigg|_{J=0} \\ &= \left(x_i + \frac{(x_f - x_i)}{T}(t - t_i)\right) Z^0_{if}[0],\end{aligned} \tag{3.126}$$

where $Z^0_{if}[0]$ is the free particle propagator (3.6). It should be noted that setting $J = 0$ in equation (3.126) makes contact with the source free theory, but *it is not required* to set J to zero. The Heisenberg picture position operator is found from its definition (1.48) by using the commutator (1.35), the Hamiltonian $\hat{P}^2/2m$, and the identity (1.54), which gives $\hat{X}_H(t) = \hat{X} + \hat{P}(t - t_i)/m$. Inserting the momentum projection operator $\hat{\mathbf{1}}_p$, as it was in equation (3.7), gives

$$\begin{aligned}\langle x_f| e^{-i\hat{H}T/\hbar}\hat{X}_H(t)|x_i\rangle &= \int_{-\infty}^{\infty} \frac{dp}{2\pi\hbar}\, e^{-ip^2T/2m\hbar} e^{ip(x_f - x_i)/\hbar}\left(x_i + \frac{p}{m}(t - t_i)\right) \\ &= \left(x_i + \frac{(x_f - x_i)}{T}(t - t_i)\right)\sqrt{\frac{m}{2\pi i\hbar T}}\exp\left\{\frac{i}{\hbar}\frac{1}{2}m\frac{(x_f - x_i)^2}{T}\right\},\end{aligned} \tag{3.127}$$

in complete agreement with equations (3.126) and (2.89).

Similarly, the harmonic oscillator coherent state path integral (3.122) can be used to generate the expectation value of creation and annihilation operators as well as their time-ordered products. For example, taking the functional derivative of equation (3.122) gives

$$\begin{aligned}\langle\lambda_f|U_J(t_f,t_i)\hat{a}_H^\dagger(t)|\lambda_i\rangle &= -i\hbar\frac{\delta Z_{if}[J]}{\delta J(t)} = \frac{\delta F_\lambda[J]}{\delta J(t)}Z_{if}[J] \equiv \langle\hat{a}_H^\dagger(t)\rangle Z_{if}[J] \\ &= \left(\lambda_f^* e^{-i\omega(t_f-t)} + \frac{i}{\hbar}\int_{t_i}^{t_f} d\tau\, \theta(\tau-t)e^{-i\omega(\tau-t)}J^*(\tau)\right)Z_{if}[J].\end{aligned} \tag{3.128}$$

The second derivative gives the time-ordered product,

$$\begin{aligned}\langle\lambda_f|U_J(t_f,t_i)\mathcal{T}\left\{\hat{a}_H(t')\,\hat{a}_H^\dagger(t)\right\}|\lambda_i\rangle &= (-i\hbar)^2\frac{\delta Z_{if}[J]}{\delta J^*(t')\,\delta J(t)} \\ &= \frac{\delta F_\lambda[J]}{\delta J^*(t')}\frac{\delta F_\lambda[J]}{\delta J(t)}Z_{if}[J] - i\hbar\frac{\delta^2 F_\lambda[J]}{\delta J^*(t')\,\delta J(t)}Z_{if}[J].\end{aligned} \tag{3.129}$$

The first term in the second line of equation (3.129) is $\langle\hat{a}_H(t')\rangle\langle\hat{a}_H^\dagger(t)\rangle/Z_{if}[J]$. The second term of equation (3.129) is given by

$$-i\hbar\frac{\delta^2 F_\lambda[J]}{\delta J^*(t)\,\delta J(t')}Z_{if}[J] = \theta(t-t')\,e^{-i\omega(t-t')}\,Z_{if}[J]. \tag{3.130}$$

It should be noted that equation (3.130) is nonzero *only if* $t > t'$, and this is consistent with the annihilation operator being at a later time than the creation operator. For such a case, the annihilation operator must be commuted with the creation operator and their commutation relation (2.72) causes the time-ordered product to be nonzero. The same result is obtained from the expression (3.122) for the Grassmann oscillator generating functional $Z_{if}[K]$. Applying result (2.115) shows that

$$i\hbar\,\frac{\delta^2 F_\xi[K]}{\delta K^*(t)\,\delta K(t')}Z_{if}[K] = \theta(t-t')\,e^{-i\omega(t-t')}\,Z_{if}[K] \tag{3.131}$$

The function appearing in equation (3.130) satisfies

$$\left(i\frac{d}{dt} - \omega\right)\left(\theta(t-t')e^{-i\omega(t-t')}\right) = i\,\delta(t-t'), \tag{3.132}$$

so that it is a Green's function for the differential operator appearing in the action of the path integral. In general, the derivatives of $Z_{if}[J]$ generate the Green's functions for the associated quantum theory as in the case (3.130). It is more common to consider the coherent state transition element for the case that the initial and final states are both the ground state, obtained by setting $\lambda_i = \lambda_f = 0$. The ground state is often referred to as the *vacuum* since there are no energy quanta present. This carries over into the field theory case, where the energy quanta are interpreted as particles and the absence of any particles is by definition the vacuum. As a result, the transition element from the initial to the final ground state with an external source J present is known as the *vacuum persistence functional.*

It is often useful to separate the expectation value of an operator from the transition element in which it is evaluated. This is done by defining the functional

$W_{if}[J] = -i\hbar \ln Z_{ij}[J]$. For the observable $\hat{O}$ coupled to J, the *connected part* of the expectation value of $\hat{O}_H(t)$, denoted $\langle \hat{O}(t) \rangle_c$, is obtained from $W_{if}[J]$,

$$\langle \hat{O}(t) \rangle_c = \frac{\delta W_{if}[J]}{\delta J(t)} = -\frac{i\hbar}{Z_{if}[J]} \frac{\delta Z_{if}[J]}{\delta J(t)}. \tag{3.133}$$

For example, using the free particle result (3.17) to define $W_{if}[J]$ gives

$$W_{if}[J] = x_i I(t_i) + \frac{m(x_f - x_i)^2}{2T} + \frac{(x_f - x_i)}{T} Q_J + \frac{1}{2\,mT} Q_J^2 - \frac{R_J}{2m}, \tag{3.134}$$

where $I(t)$, Q_J, and R_J are defined by equations (3.14) and (3.16). Evaluating equation (3.133) at $J = 0$ using equation (3.134) gives $\langle \hat{X}(t) \rangle_c = x_i + (x_f - x_i)(t - t_i)/T$, which is the expectation value without the transition element factor. This result extends to arbitrary time-ordered products. In general, the second derivative of $W_{if}[J]$ is given by

$$\frac{\delta^2 W_{if}[J]}{\delta J(t_1)\, \delta J(t_2)} = -\frac{i\hbar}{Z_{if}[J]} \frac{\delta^2 Z_{if}[J]}{\delta J(t_1)\, \delta J(t_2)} + \frac{i\hbar}{Z_{if}^2[J]} \frac{\delta Z_{if}[J]}{\delta J(t_1)} \frac{\delta Z_{if}[J]}{\delta J(t_2)}. \tag{3.135}$$

The second term in equation (3.135) subtracts the contribution of the expectation values from the Green's function, leaving only the contribution due to commutators between the time-ordered operators. This is apparent when the second functional derivative of equation (3.134) is taken, giving the connected two-point function $\langle \mathcal{T}\{\hat{X}_H(t)\hat{X}_H(t')\} \rangle_c$,

$$\frac{\delta^2 W_{if}[J]}{\delta J(t)\, \delta J(t')} = \frac{(t - t_i)(t' - t_i)}{mT} - \theta(t - t')\frac{(t' - t_i)}{m} - \theta(t' - t)\frac{(t - t_i)}{m}. \tag{3.136}$$

Result (1.17) shows that equation (3.136) is the free particle Green's function, satisfying

$$m\frac{\mathrm{d}^2}{\mathrm{d}t^2}\left(\frac{\delta^2 W_{if}[J]}{\delta J(t)\, \delta J(t')}\right) \equiv m\frac{\mathrm{d}^2}{\mathrm{d}t^2}\Delta(t, t') = \delta(t - t'), \tag{3.137}$$

as well as obeying the boundary conditions $\Delta(t_f, t') = \Delta(t_i, t') = 0$. The functional $W_{if}[J]$ is the generator of *connected* Green's functions, a term that will be clearer when its field theory counterpart is examined. A very useful property of the connected Green's functions is that the normalization factors for the path integral are not required since they are discarded by taking the logarithm. For example, in expression (3.133) the derivative creates a ratio to cancel divergent quantities.

A second important functional, the *effective action*, is defined by a functional Legendre transformation,

$$\Gamma_{if}[\langle \hat{O} \rangle] = W_{if}[J] - \int_{t_i}^{t_f} \mathrm{d}t\, \langle \hat{O}(t) \rangle\, J(t). \tag{3.138}$$

Using the functional chain rule (1.31) gives

$$\frac{\delta\Gamma_{if}[\langle\hat{O}\rangle]}{\delta\langle\hat{O}(\tau)\rangle} = \int_{t_i}^{t_f} \mathrm{d}t \, \frac{\delta W_{if}[J]}{\delta J(t)} \frac{\delta J(t)}{\delta\langle\hat{O}(\tau)\rangle} - \int_{t_i}^{t_f} \mathrm{d}t \, \langle\hat{O}(t)\rangle \frac{\delta J(t)}{\delta\langle\hat{O}(\tau)\rangle} - J(\tau), \quad (3.139)$$

and combining this with equation (3.133) gives

$$\frac{\delta\Gamma_{if}[\langle\hat{O}\rangle]}{\delta\langle\hat{O}(\tau)\rangle} = -J(\tau). \quad (3.140)$$

Result (3.140) gives a method to determine the expectation values that are consistent with the original source free theory. This idea plays a central role in the concept of spontaneously broken symmetry.

The effective action is more often used in the context of quantum field theory, but one of its important properties is also present in the quantum mechanical version. Using the functional chain rule gives

$$\begin{aligned} \delta(t - t') = \frac{\delta\langle\hat{O}(t)\rangle}{\delta\langle\hat{O}(t')\rangle} &= \int_{t_i}^{t_f} \mathrm{d}\tau \, \frac{\delta\langle\hat{O}(t)\rangle}{\delta J(\tau)} \frac{\delta J(\tau)}{\delta\langle\hat{O}(t')\rangle} \\ &= \int_{t_i}^{t_f} \mathrm{d}\tau \, \frac{\delta^2 W_{if}[J]}{\delta J(t)\, \delta J(\tau)} \frac{\delta^2\Gamma_{if}[\langle\hat{O}\rangle]}{\delta\langle\hat{O}(\tau)\rangle\, \delta\langle\hat{O}(t')\rangle}. \end{aligned} \quad (3.141)$$

This shows that the second functional derivative of the effective action is the *functional inverse* of the connected two-point Green's function. The Fourier transforms of the two functions appearing in equation (3.141) are the inverse of each other. The zeroes of the second derivative are therefore the poles of the connected two-point Green's function.

3.8 The partition function

The quantum mechanical partition function Z_β is central to modern statistical mechanics and its analysis is one of the important applications of the path integral. If $|E\rangle$ denotes an energy eigenstate, the partition function is given by

$$Z_\beta = \sum_E \langle E|e^{-\beta\hat{H}}|E\rangle = \sum_E e^{-\beta E} = \mathrm{Tr}\, e^{-\beta\hat{H}}, \quad (3.142)$$

where $\beta = 1/kT$ is the inverse temperature and the sum is over the complete set of energy states. The sum includes an integration over any of the continuous parts of the energy spectrum. The completeness of the c-number position eigenstates (1.33) allows the partition function to be written

$$\begin{aligned} Z_\beta &= \sum_E \langle E|e^{-\beta\hat{H}}|E\rangle = \sum_E \int \mathrm{d}\boldsymbol{x}\, \mathrm{d}\boldsymbol{x}' \, \langle E|\boldsymbol{x}'\rangle\langle\boldsymbol{x}'|e^{-\beta H}|\boldsymbol{x}\rangle\langle\boldsymbol{x}|E\rangle \\ &= \int \mathrm{d}\boldsymbol{x}\, \mathrm{d}\boldsymbol{x}' \, \sum_E \langle\boldsymbol{x}|E\rangle\langle E|\boldsymbol{x}'\rangle\langle\boldsymbol{x}'|e^{-\beta H}|\boldsymbol{x}\rangle = \int \mathrm{d}\boldsymbol{x}\, \mathrm{d}\boldsymbol{x}' \, \langle\boldsymbol{x}|\boldsymbol{x}'\rangle\langle\boldsymbol{x}'|e^{-\beta H}|\boldsymbol{x}\rangle \quad (3.143) \\ &= \int \mathrm{d}\boldsymbol{x} \, \langle\boldsymbol{x}|e^{-\beta H}|\boldsymbol{x}\rangle. \end{aligned}$$

The c-number partition function (3.143) is the path integral propagator continued to imaginary time, $t_f - t_i \to -i\beta\hbar$, and summed over *periodic* boundary conditions $\boldsymbol{x}_f = \boldsymbol{x}_i = \boldsymbol{x}$,

$$Z_\beta = \int \mathrm{d}\boldsymbol{x}\, \langle \boldsymbol{x}| e^{-i\hat{H}T/\hbar} |\boldsymbol{x}\rangle|_{T=-i\beta\hbar} = \int \mathrm{d}\boldsymbol{x}\; G(\boldsymbol{x}, \boldsymbol{x}, t_f - t_i = -i\beta\hbar). \tag{3.144}$$

This shows that the partition function of the quantum system is available from the Wick rotated path integral. Once Z_β is known it is then possible to find the Helmholtz free energy, $F_\beta = -\beta^{-1} \ln Z_\beta$. The Helmholtz free energy then gives all the thermodynamic quantities, such as the pressure, $P = -\partial F_\beta / \partial V$, where V is the volume of the system, and the entropy, $S = \beta^2\, \partial F_\beta / \partial \beta$. In addition, the ground state energy E_0 of the system dominates the partition function in the zero temperature limit, $\lim_{\beta\to\infty} Z_\beta = \exp(-\beta E_0)$. Therefore the ground state energy of the system emerges from the Wick rotated path integral in the limit $\beta \to \infty$.

Applying recipe (3.144) to the free particle propagator (3.6) gives

$$Z_\beta = \int_V \mathrm{d}x \sqrt{\frac{m}{2\pi\hbar^2\beta}} = V\sqrt{\frac{m}{2\pi\hbar^2\beta}}, \tag{3.145}$$

where V is the volume of the system. Using equation (3.145) gives $P = -\partial F_\beta / \partial V = 1/\beta V$, which is the perfect gas law, $PV = kT$. Applying recipe (3.144) to the coherent state path integral (3.114) and performing the two Gaussian integrations of equation (2.78) gives the familiar harmonic oscillator partition function,

$$\begin{aligned} Z_\beta &= \int \frac{\mathrm{d}^2\lambda_i}{\pi} \exp\left\{-(1 - e^{-\beta\hbar\omega})|\lambda_i|^2 - \frac{1}{2}\beta\hbar\omega\right\} \\ &= \frac{e^{-\frac{1}{2}\beta\hbar\omega}}{1 - e^{-\beta\hbar\omega}} = \frac{1}{2\sinh\frac{1}{2}\beta\hbar\omega}. \end{aligned} \tag{3.146}$$

Applying recipe (3.144) to the configuration space harmonic oscillator path integral (3.59) and performing the single Gaussian integration gives the same result,

$$Z_\beta = \int \mathrm{d}x \sqrt{\frac{m\omega}{2\pi\hbar \sinh(\beta\hbar\omega)}} \exp\left\{-\frac{m\omega x^2}{\hbar} \tanh\frac{1}{2}\beta\hbar\omega\right\} = \frac{1}{2\sinh\frac{1}{2}\beta\hbar\omega}. \tag{3.147}$$

Relation (3.144) is valid for c-number systems, but must be altered for Grassmann systems. This is due to result (2.108), which shows that the partition function of a Grassmann system is given by

$$
\begin{aligned}
Z_\beta &= \sum_E \langle E|e^{-\beta \hat{H}}|E\rangle = \int d^2\xi\, d^2\xi' \sum_E \langle E|\xi'\rangle\langle\xi'|e^{-\beta H}|\xi\rangle\langle\xi|E\rangle \\
&= \int d^2\xi\, d^2\xi' \sum_E \langle -\xi|E\rangle\langle E|\xi'\rangle\langle\xi'|e^{-\beta H}|\xi\rangle \\
&= \int d^2\xi\, d^2\xi'\, \langle -\xi|\xi'\rangle\langle\xi'|e^{-\beta H}|\xi\rangle = \int d^2\xi\, \langle -\xi|e^{-\beta H}|\xi\rangle.
\end{aligned}
\tag{3.148}
$$

The fermionic partition function is the propagator continued to imaginary time, $t_f - t_i \to -i\beta\hbar$, and summed over *antiperiodic* boundary conditions, $\xi_f = -\xi_i = -\xi$,

$$
Z_\beta = \int d^2\xi\, \langle -\xi|e^{-i\hat{H}T/\hbar}|\xi\rangle|_{T=-i\beta\hbar} = \int d^2\xi\, G(-\xi, \xi, t_f - t_i = -i\beta\hbar). \tag{3.149}
$$

Applying this recipe to the Grassmann coherent state integral (3.116) and using the rules of complex Grassmann integration, equations (1.78) and (1.79), gives

$$
\begin{aligned}
Z_\beta &= \int d\xi\, d\xi^*\, \exp\left(-(1 + e^{-\beta\hbar\omega})\xi^*\xi + \frac{1}{2}\beta\hbar\omega\right) \\
&= e^{\frac{1}{2}\beta\hbar\omega}(1 + e^{-\beta\hbar\omega}) = e^{\frac{1}{2}\beta\hbar\omega} + e^{-\frac{1}{2}\beta\hbar\omega} = 2\cosh\left(\frac{1}{2}\beta\hbar\omega\right).
\end{aligned}
\tag{3.150}
$$

Since the Grassmann oscillator has only the two states of equation (2.101), the partition function (3.150) reveals the two energy eigenvalues, $E_0 = -\frac{1}{2}\hbar\omega$ and $E_1 = \frac{1}{2}\hbar\omega$.

Rather than using the final result of evaluating the path integral to find the partition function, it is important to formulate the path integral version of the partition function. This allows both perturbative and semiclassical approximations for the partition function for more complicated potentials. The recipe for converting the path integral to the partition function consists of the Wick rotation to imaginary time, so that $\epsilon \to -i\epsilon$, where $\epsilon = \beta\hbar/N$. In turn, this means that the formal time derivatives become $\dot{x}_j = (x_{j+1} - x_j)/\epsilon \to -(x_{j+1} - x_j)/i\epsilon = i\dot{x}_j$, where the formal derivative is now with respect to imaginary time. If integrations over intermediate momenta are present these are unchanged by the Wick rotation. The final step is making the path integral periodic for c-numbers, $x_N = x_0$, or antiperiodic for Grassmann numbers, $\xi_N = -\xi_0$, and integrating over x_0 or ξ_0. This step adds x_0 or ξ_0 to the measure of the path integral.

For example, following this recipe for the c-number phase space path integral (2.11) yields the associated quantum mechanical partition function,

$$
Z_\beta = \int [\mathcal{D}x][\mathcal{D}p]\exp\left\{-\sum_{j=0}^{N-1}\frac{\epsilon}{\hbar}\left(\frac{1}{2m}p_j^2 - ip_j\dot{x}_j + V(x_j, t_j)\right)\right\}, \tag{3.151}
$$

where $x_N = x_0$ and the measure is symmetric over phase space,

$$
[\mathcal{D}x][\mathcal{D}p] = dp_0\, dp_1 \cdots dp_{N-1}\, dx_0\, dx_1 \cdots dx_{N-1}. \tag{3.152}
$$

Integrating the momenta in equation (3.151) gives the configuration space version of the partition function,

$$Z_\beta = \int [\tilde{\mathcal{D}}x] \exp\left\{-\sum_{j=0}^{N-1} \frac{\epsilon}{\hbar}\left(\frac{1}{2}m\dot{x}_j^2 + V(x_j, t_j)\right)\right\}, \tag{3.153}$$

where $x_N = x_0$ and the measure is given by

$$[\tilde{\mathcal{D}}x] = \left(\frac{m}{2\pi\epsilon\hbar}\right)^{N/2} \mathrm{d}x_0\, \mathrm{d}x_1 \cdots \mathrm{d}x_{N-1}. \tag{3.154}$$

The partition function of equation (3.153) can be analyzed by the Wick rotated version of the Fourier series method introduced in equation (3.30), writing the integration variables as

$$x_j = x_c(\tau_j) + \sqrt{\frac{2}{\beta\hbar}} \sum_{k=1}^{N-1} a_k \sin\left(\frac{k\pi\tau_j}{\beta\hbar}\right), \tag{3.155}$$

where $\tau_j = j\epsilon$ is the imaginary time. The function $x_c(\tau)$ is a solution to the Wick rotated equation of motion, $m\, \mathrm{d}^2x_c(\tau)/\mathrm{d}\tau^2 = \partial V(x_c(\tau))/\partial x_c(\tau)$, and is required to satisfy periodic boundary conditions $x_c(0) = x_c(\beta\hbar) = x_0$, where x_0 is *arbitrary*. This change of variables results in a factorization identical to the real time path integral (3.26) with the exception that $x_f = x_i = x_0$ and the associated classical action is evaluated at imaginary time. For example, in the case of the free particle the Wick rotated trajectory is $x_c(\tau) = x_0$, and using equation (3.155) shows that the partition function is the Wick rotated version of the fluctuation integral (3.38) with an additional integration over x_0. A similar result occurs for the harmonic oscillator, where the classical solution (3.55) can be Wick rotated and adapted to periodic boundary conditions. Because the real time applications of equation (3.155) have already been presented, attention will be turned elsewhere.

Before proceeding it is important to note that the thermal expectation values of a Wick rotated c-number Heisenberg picture operator, $\hat{O}(t \to -i\tau)$, is periodic in $\beta\hbar$. This follows from continuing relation (1.48) to imaginary time τ, giving

$$\begin{aligned} \langle\hat{O}(\tau)\rangle_\beta &= \sum_E \langle E|e^{-\beta\hat{H}}\hat{O}(\tau)|E\rangle = \sum_E \langle E|e^{-\beta\hat{H}}\hat{O}(\tau)e^{\beta\hat{H}}e^{-\beta\hat{H}}|E\rangle \\ &= \sum_E \langle E|\hat{O}(\tau + \beta\hbar)e^{-\beta\hat{H}}|E\rangle = \langle\hat{O}(\tau + \beta\hbar)\rangle_\beta. \end{aligned} \tag{3.156}$$

This requirement therefore applies to the integration variables in the path integral. It is satisfied explicitly by using a *Fourier cosine series* for x_j,

$$x_j = \sqrt{\frac{1}{\beta\hbar}}\, a_0 + \sqrt{\frac{2}{\beta\hbar}} \sum_{k=1}^{N-1} a_k \cos\left(\frac{2\pi k\tau_j}{\beta\hbar}\right). \tag{3.157}$$

Expression (3.157) is manifestly periodic over the interval $\beta\hbar$ and constitutes a change of all N of the x_j variables to the N Fourier variables a_k. The frequencies $f_k = 2\pi k/\beta\hbar$ appearing in equation (3.157) are twice those appearing in equation (3.155) for the fluctuations around $x_c(\tau)$, and are known as the periodic or *even Matsubara frequencies.* It is straightforward to show equation (3.157) yields the same result as equation (3.145) for the free particle by revisiting the analysis of equations (3.30)–(3.41). Using equations (3.157) and (3.35) in the Wick rotated action $\tilde{S}$ and setting $\tau_j = j\epsilon$ and $\beta\hbar = N\epsilon$ gives

$$\begin{aligned}\tilde{S} &= i\sum_{j=0}^{N-1}\frac{1}{2}m\frac{(x_{j+1}-x_j)^2}{\epsilon} = i\frac{2\,mN^2}{\beta^2\hbar^2}\sum_{k=1}^{N-1}a_k^2\sin^2\left(\frac{\pi k}{N}\right)\\ \Longrightarrow \exp\left(\frac{i}{\hbar}\tilde{S}\right) &= \exp\left(-\frac{2\,mN^2}{\beta^2\hbar^3}\sum_{k=1}^{N-1}a_k^2\sin^2\left(\frac{\pi k}{N}\right)\right).\end{aligned} \tag{3.158}$$

It is important to note that a_0 does not appear in the sum of equation (3.158) and is therefore playing the role of the integral over x_0 for the free particle. The series (3.157) gives the identification $\mathrm{d}a_0 = \mathrm{d}x_0\sqrt{\beta\hbar}$, so that the integration over a_0 simply creates the factor $V\sqrt{\beta\hbar}$ for the partition function. Repeating the steps that gave equation (3.33) shows that the Jacobian associated with equation (3.157) is $J = (N/\beta\hbar)^{N/2}$. Setting $\epsilon = \beta\hbar/N$ and using the Jacobian, the measure for the partition function is given by

$$[\mathcal{D}a] = V\sqrt{\beta\hbar}\left(\frac{mN^2}{2\pi\beta^2\hbar^3}\right)^{\frac{1}{2}N}\mathrm{d}a_1\,\mathrm{d}a_2\cdots\mathrm{d}a_{N-1}. \tag{3.159}$$

Performing the Gaussian integrals defined by the action in equation (3.158) yields the product of $N-1$ factors of the form $(\pi\beta^2\hbar^3/2\,mN^2\sin^2(\pi k/N))^{1/2}$. Using the identity equivalent to equation (3.41) for the doubled frequencies,

$$\prod_{k=1}^{N-1}\sin\left(\frac{\pi k}{N}\right) = \frac{N}{2^{N-1}}, \tag{3.160}$$

once again yields the one-dimensional free particle partition function (3.145),

$$Z_\beta = V\sqrt{\beta\hbar}\left(\frac{mN^2}{2\pi\beta^2\hbar^3}\right)^{\frac{1}{2}N}\left(\frac{\pi\beta^2\hbar^3}{2\,mN^2}\right)^{\frac{1}{2}(N-1)}\frac{2^{N-1}}{N} = V\sqrt{\frac{m}{2\pi\hbar^2\beta}}. \tag{3.161}$$

It was pointed out earlier that the semiclassical approximation fails for the coherent state path integral for the case of different initial and final coherent states. The reason is that there is no classical solution to the first order equations of motion that connects them. However, the method of smooth or differentiable fluctuations, which led to the prefactor expressed as the ratio of functional determinants in equation (3.54), can be adapted to evaluate the coherent state path integral for the partition function. The use of smooth fluctuations for the c-number coherent state partition function begins by defining the $\omega \to 0$ limit of the *continuum Lagrangian* given by equation (2.111), $\mathcal{L}_0 = \frac{1}{2} i\hbar(\lambda^* \dot{\lambda} - \dot{\lambda}^* \lambda)$, so that the Wick rotated version is given by $\tilde{\mathcal{L}}_0 = \frac{1}{2}\hbar(\lambda^* \dot{\lambda} - \dot{\lambda}^* \lambda)$. The λ variables are given a Fourier expansion using the even Matsubara frequencies $f_k = 2\pi k/\beta\hbar$ defined earlier. Fixing N to be even, the Fourier expansion is given by

$$\lambda(\tau_j) = \sqrt{\frac{1}{\beta\hbar}} \sum_{k=-\frac{1}{2}N}^{\frac{1}{2}N} \nu_k e^{if_k t_j} = \sqrt{\frac{1}{\beta\hbar}} \sum_{k=1}^{\frac{1}{2}N} (\nu_k e^{if_k t_j} + \nu_{-k} e^{-if_k t_j}). \tag{3.162}$$

The even Matsubara frequencies insure that $\lambda(\tau_0) = \lambda(\tau_N)$. The *zero mode* ν_0 is absent in equation (3.162) in order to match the N integration variables λ_j as well as to avoid an undefined integration in the partition function. The functions appearing in equation (3.162) are eigenfunctions of the differential operator $\hat{D}_0 = \mathrm{d}/\mathrm{d}\tau$ present in the continuum Lagrangian $\tilde{\mathcal{L}}_0$, and are associated with the eigenvalues $if_k = 2\pi ik/\beta\hbar$. Applying derivatives rather than differences to equation (3.162), the Wick rotated action is

$$\begin{aligned} -i\int_0^{\beta\hbar} \mathrm{d}\tau\, \tilde{\mathcal{L}}_0 &= -\frac{i}{\beta} \sum_{k,k'=-N/2}^{N/2} \nu_{k'}^* \nu_k \left(\frac{\pi i(k+k')}{\beta\hbar}\right) \int_0^{\beta\hbar} \mathrm{d}\tau \exp\left(\frac{2\pi i(k-k')\tau}{\beta\hbar}\right) \\ &= -i\hbar \sum_{k=-N/2}^{N/2} \nu_k^* \nu_k \frac{2\pi ik}{\beta\hbar} = -i\hbar \sum_{k=1}^{\frac{1}{2}N} (\nu_k^* \nu_k - \nu_{-k}^* \nu_{-k}) \frac{2\pi ik}{\beta\hbar}. \end{aligned} \tag{3.163}$$

The Jacobian associated with equation (3.162) is designated J, while applying derivatives to the fluctuations requires an additional normalization factor $\mathcal{N}_0$, as it did with equation (3.48). The coherent state partition function for $\tilde{\mathcal{L}}_0$ is a product of N Fresnel integrals,

$$\begin{aligned} Z_\beta(\omega = 0) &= \frac{\mathcal{N}_0 J}{\pi^N} \int \prod_{j=-\frac{1}{2}N}^{\frac{1}{2}N} \mathrm{d}\nu_j^* \mathrm{d}\nu_j \exp\left\{ \sum_{k=-\frac{1}{2}N}^{\frac{1}{2}N} \nu_k^* \nu_k \left(\frac{2\pi ik}{\beta\hbar}\right)\right\} \\ &= \mathcal{N}_0 J \prod_{k=1}^{\frac{1}{2}N} \left(-\frac{i\beta\hbar}{2\pi k}\right)\left(\frac{i\beta\hbar}{2\pi k}\right) = \mathcal{N}_0 J \prod_{k=1}^{N/2} \frac{\beta^2\hbar^2}{4\pi^2 k^2} = \frac{\mathcal{N}_0 J}{\det \hat{D}_0}, \end{aligned} \tag{3.164}$$

where $\hat{D}_0 = \mathrm{d}/\mathrm{d}\tau$. The normalization factor $\mathcal{N}_0$ for the smooth fluctuation method can now be chosen to give the correct coherent state partition function for the case that $\omega \approx 0$. Result (3.146) gives $Z_\beta(\omega \approx 0) = 1/\beta\hbar\omega$. The normalization factor for the coherent state partition function must therefore be chosen to be

$$\mathcal{N}_0 = \frac{1}{J\beta\hbar\omega} \det \hat{D}_0. \tag{3.165}$$

Result (3.165) allows a relatively easy analysis of the partition function for the full Wick rotated Lagrangian $\tilde{\mathcal{L}}_\omega = \frac{1}{2}\hbar\lambda^*\dot{\lambda} - \frac{1}{2}\hbar\dot{\lambda}^*\lambda - \hbar\omega(\lambda^*\lambda + \frac{1}{2})$. The differential operator $\hat{D}_\omega = \mathrm{d}/\mathrm{d}\tau - \omega$ has the eigenvalues $(2\pi ik/\beta\hbar) - \omega$ when applied to the periodic functions in the expansion (3.162). The product of these eigenvalues is

$$\det \hat{D}_\omega = \prod_{k=-N/2}^{N/2}\left(\frac{2\pi ik}{\beta\hbar} - \omega\right) = \prod_{k=1}^{N/2}\left(\frac{4\pi^2k^2}{\beta^2\hbar^2} + \omega^2\right), \tag{3.166}$$

where the $k = 0$ mode is absent since it is absent in equation (3.162). The Jacobian J is the same as that for $\tilde{\mathcal{L}}_0$ since the change of variables is identical. However, the normalization factor $\mathcal{N}_\omega$ for the $\tilde{\mathcal{L}}_\omega$ partition function is given by $\mathcal{N}_\omega = \mathcal{N}_0 e^{\frac{1}{2}\beta\hbar\omega}$, which obeys $\lim_{\omega\to 0}\mathcal{N}_\omega = \mathcal{N}_0$ and cancels the ground state energy in the action. In the $N \to \infty$ limit the partition function path integral becomes

$$\begin{aligned} Z_\beta(\omega) &= \frac{\mathcal{N}_\omega J e^{-\frac{1}{2}\beta\hbar\omega}}{\det D_\omega} = \frac{1}{\beta\hbar\omega}\frac{\det \hat{D}_0}{\det \hat{D}_\omega} = \frac{1}{\beta\hbar\omega}\prod_{k=1}^{N/2}\frac{4\pi^2k^2/\beta^2\hbar^2}{\omega^2 + (4\pi^2k^2/\beta^2\hbar^2)} \\ &= \frac{1}{\beta\hbar\omega}\prod_{k=1}^{N/2}\frac{1}{1 + (\beta^2\omega^2\hbar^2/4\pi^2k^2)} = \frac{1}{2\sinh\left(\frac{1}{2}\beta\hbar\omega\right)}, \end{aligned} \tag{3.167}$$

which is the same result as equation (3.146). Continuum methods can therefore be applied to coherent state path integrals with periodic boundary conditions. This result will be employed with coherent state path integrals in quantum field theory.

The Grassmann oscillator partition function can also be evaluated with similar continuum methods. As before, the use of smooth fluctuations for the Grassmann coherent state partition function begins by defining the $\omega \approx 0$ form of the *continuum Lagrangian* from (2.111), $\mathcal{L}_0 = \frac{1}{2}i\hbar(\xi^*\dot{\xi} - \dot{\xi}^*\xi)$, so that the Wick rotated version is given by $\tilde{\mathcal{L}}_0 = \frac{1}{2}\hbar(\xi^*\dot{\xi} - \dot{\xi}^*\xi)$. The Fourier expansion of the Grassmann variables is written by choosing an even value for N and using the antiperiodic or *odd Matsubara frequencies* $\tilde{f}_k = \pi(2k - 1)/\beta\hbar$,

$$\xi(\tau_j) = \sqrt{\frac{1}{\beta\hbar}}\sum_{k=1}^{N/2}\left(\zeta_k e^{i\tilde{f}_k\tau_j} + \zeta_{-k}e^{-i\tilde{f}_k\tau_j}\right) \implies \xi(\tau_N) = -\xi(\tau_0), \tag{3.168}$$

which is antiperiodic by virtue of $\exp(\pm\pi i) = -1$ and, like the c-number version (3.162), does not include a zero mode. The functions appearing in the expansion (3.168) are antiperiodic orthonormal eigenfunctions of the differential operator $\hat{D}_0 = \mathrm{d}/\mathrm{d}\tau$ associated with the eigenvalues $i\tilde{f}_k = \pi i(2k - 1)/\beta\hbar$. Because the functions in equation (3.168) are orthonormal, the action given by $\tilde{\mathcal{L}}_0$ is formally identical to equation (3.163) expressed in the Grassmann modes ζ_k and the odd Matsubara frequencies,

$$-i\int_0^{\beta\hbar} \mathrm{d}\tau\, \tilde{\mathcal{L}}_0 = -i\hbar \sum_{k=1}^{\frac{1}{2}N} (\zeta_k^* \zeta_k - \zeta_{-k}^* \zeta_{-k}) \left(\frac{\pi i(2k-1)}{\beta\hbar} \right). \tag{3.169}$$

The Jacobian associated with equation (3.168) is designated $\tilde{J}$, while the normalization factor that compensates for using derivatives is denoted $\tilde{\mathcal{N}}_0$. After changing variables and using derivatives rather than differences, the Grassmann oscillator partition function for $\tilde{\mathcal{L}}_0$ is obtained from the rules of Grassmann integration,

$$\begin{aligned} Z_\beta(\omega=0) &= \tilde{\mathcal{N}}_0 \tilde{J} \int \prod_{j=-N/2}^{N/2} \mathrm{d}\zeta_j \mathrm{d}\zeta_j^* \exp\left\{ \sum_{k=1}^{N/2} (\zeta_k^*\zeta_k - \zeta_{-k}^*\zeta_{-k}) \left(\frac{\pi i(2k-1)}{\beta\hbar} \right) \right\} \\ &= \tilde{\mathcal{N}}_0 \tilde{J} \prod_{k=1}^{N/2} \frac{\pi^2(2k-1)^2}{\beta^2\hbar^2} = \tilde{\mathcal{N}}_0 \tilde{J} \,\tilde{\det}\, \hat{D}_0, \end{aligned} \tag{3.170}$$

where $\hat{D}_0 = \mathrm{d}/\mathrm{d}\tau$ and $\tilde{\det}$ refers to the product of the odd Matsubara frequencies. The normalization factor $\tilde{\mathcal{N}}_0$ for the smooth fluctuation method must be chosen to give the coherent state partition function for $\omega \approx 0$. From equation (3.150) it follows that $Z_\beta(\omega \approx 0) = 2$. As a result, the normalization factor for the Grassmann coherent state partition function is given by

$$\tilde{\mathcal{N}}_0 = \frac{2}{\tilde{J}\,\tilde{\det}\, \hat{D}_0}. \tag{3.171}$$

The full Wick rotated Lagrangian $\mathcal{L}_\omega = \frac{1}{2}\hbar\xi^*\dot{\xi} - \frac{1}{2}\hbar\dot{\xi}^*\xi - \hbar\omega(\xi^*\xi - \frac{1}{2})$ can now be analyzed using the same expansion since the differential operator $\hat{D}_\omega = \mathrm{d}/\mathrm{d}\tau + \omega$ has the eigenvalues $\omega + \pi i(2k+1)/\beta\hbar$ for the orthonormal functions used in the expansion. Similarly to $\tilde{\det}\, \hat{D}_0$, the product of the eigenvalues gives

$$\tilde{\det}\, \hat{D}_\omega = \prod_{k=-N/2}^{N/2} \left(\frac{\pi i(2k-1)}{\beta\hbar} + \omega \right) = \prod_{k=1}^{N/2} \left(\omega^2 + \frac{\pi^2(2k-1)^2}{\beta^2\hbar^2} \right). \tag{3.172}$$

As before, the normalization $\tilde{\mathcal{N}}_\omega$ is modified to cancel the ground state energy, so that $\tilde{\mathcal{N}}_\omega = \tilde{\mathcal{N}}_0 e^{-\frac{1}{2}\beta\hbar\omega}$. Using equation (3.165) the partition function for $\tilde{\mathcal{L}}_\omega$ is given by

$$\begin{aligned} Z_\beta(\omega) = \tilde{\mathcal{N}}_\omega \tilde{J}\,\tilde{\det}\, D_\omega e^{\frac{1}{2}\beta\hbar\omega} &= 2\,\frac{\tilde{\det}\, \hat{D}_\omega}{\tilde{\det}\, \hat{D}_0} \\ &= 2\prod_{k=1}^{N/2} \frac{\omega^2 + (\pi^2(2k-1)^2/\beta^2\hbar^2)}{\pi^2(2k-1)^2/\beta^2\hbar^2} \\ = 2\prod_{k=1}^{N/2} \left(1 + \frac{\beta^2\hbar^2\omega^2}{\pi^2(2k-1)^2} \right) &= 2\cosh\left(\frac{1}{2}\beta\hbar\omega \right), \end{aligned} \tag{3.173}$$

where the Euler identity reproduces the Grassmann partition function (3.150).

3.9 Symmetry and canonical transformations

The role of symmetry has grown greatly in importance in physics, particularly in modeling subatomic processes. Symmetries of the action are associated with conservation laws, and in some cases dictate the nature of the dynamics present in the quantum system. The generating functional approach to calculating quantum processes provides a convenient tool to analyze this combination. In classical mechanics an infinitesimal symmetry is defined as a variation of the coordinates $\boldsymbol{x} \to \boldsymbol{x} + \delta\boldsymbol{x}$ that leaves the Lagrangian $\mathcal{L}(\dot{\boldsymbol{x}}, \boldsymbol{x})$ invariant for the case that the variation is evaluated using a solution of the Euler–Lagrange equation. Unlike the variational principle, the symmetry variation $\delta\boldsymbol{x}$ does not necessarily vanish at the endpoints. Adapting equations (1.4) and (1.6), the action of the symmetry gives

$$\delta S = \int_{t_i}^{t_f} \mathrm{d}t \left(\frac{\mathrm{d}}{\mathrm{d}t}\left(\frac{\partial \mathcal{L}}{\partial \dot{\boldsymbol{x}}} \cdot \delta\boldsymbol{x} \right) + \left(\frac{\partial \mathcal{L}}{\partial \boldsymbol{x}} - \frac{\mathrm{d}}{\mathrm{d}t}\frac{\partial \mathcal{L}}{\partial \dot{\boldsymbol{x}}} \right) \cdot \delta\boldsymbol{x} \right) = 0. \tag{3.174}$$

The second term vanishes when evaluated using a solution of the Euler–Lagrange equation (1.7), so that the first must also vanish. This is true if and only if

$$\frac{\mathrm{d}}{\mathrm{d}t}\left(\frac{\partial \mathcal{L}}{\partial \dot{\boldsymbol{x}}} \cdot \delta\boldsymbol{x} \right) = \frac{\mathrm{d}}{\mathrm{d}t}(\boldsymbol{p} \cdot \delta\boldsymbol{x}) \equiv \frac{\mathrm{d}G}{\mathrm{d}t} = 0, \tag{3.175}$$

where $\boldsymbol{p} = \partial\mathcal{L}/\partial\dot{\boldsymbol{x}}$ is the momentum along the classical trajectory. For example, if the symmetry of the system is *translational invariance*, then the Lagrangian is invariant under the change of coordinates $\delta\boldsymbol{x} = \boldsymbol{a}$, where $\boldsymbol{a}$ is a constant vector in the direction the origin can be translated. The Hamiltonian is said to be *cyclic* in the coordinate parallel to $\boldsymbol{a}$. For such a case, the conserved quantity is $G = \boldsymbol{p} \cdot \boldsymbol{a} = p_{\|}\, a$, where $p_{\|}$ is the momentum parallel to $\boldsymbol{a}$. Since a is a constant, equation (3.175) states that $\dot{p}_{\|} = 0$, so the momentum in the direction of $\boldsymbol{a}$ is conserved.

The classical symmetry transformation may also leave the quantum mechanical path integral action invariant. However, in order to leave the path integral invariant, it must simultaneously leave the action invariant *and* the measure invariant. For example, the free particle path integral (3.1) is invariant under the simultaneous translations $x_i \to x_i + a$ and $x_f \to x_f + a$ since the intermediate variables of integration (3.2) can also be translated to compensate. For a infinitesimal this gives

$$\begin{aligned} G(x_f + a, x_i + a, t_f, t_i) - G(x_f, x_i, t_f, t_i) &= a\left(\frac{\partial}{\partial x_i} + \frac{\partial}{\partial x_f} \right) G(x_f, x_i, t_f, t_i) \\ &= a\frac{\partial}{\partial(x_i + x_f)} G(x_f, x_i, t_f, t_i) = 0. \end{aligned} \tag{3.176}$$

The translational symmetry of the action for the free particle shows that the path integral is a function of $x_f - x_i$ with no occurrence of the variable combination $x_f + x_i$. This will occur anytime the Hamiltonian is cyclic in a particular coordinate. Another example is rotational invariance, which leads to conservation of angular momentum.

There is a second manifestation of symmetry in the path integral that is quantum mechanical in nature. The coherent state generating functional is invariant under the simultaneous transformations $\lambda_j \to e^{i\alpha}\lambda_j$ and $J(t_j) \to e^{i\alpha}J(t_j)$ along with their complex conjugates. Designating $\lambda_f^{*'} = e^{-i\alpha}\lambda_f^*$ and $\lambda_i' = e^{i\alpha}\lambda_i$, this gives

$$\begin{aligned} Z_{i'f'}[Je^{i\alpha}, J^*e^{-i\alpha}] - Z_{if}[J, J^*] &= i\alpha \int_{t_i}^{t_f} \mathrm{d}t \left(\frac{\delta}{\delta J(t)} - \frac{\delta}{\delta J^*(t)}\right) Z_{if}[J, J^*] \\ &= -\frac{\alpha}{\hbar}\int_{t_i}^{t_f} \mathrm{d}t\, (\langle \hat{a}_H^\dagger(t)\rangle - \langle \hat{a}_H(t)\rangle) = \frac{\alpha}{\sqrt{2\, m\omega\hbar}}\int_{t_i}^{t_f} \mathrm{d}t\, \langle \hat{P}_H(t)\rangle = 0, \end{aligned} \tag{3.177}$$

where the definition (2.73) of $\hat{a}$ was used. Since t_i and t_f are arbitrary, this symmetry is equivalent to $\langle \hat{P}_H(t)\rangle = 0$.

A related topic is the use of canonical transformations to change phase space coordinates from p and x to P and X in such a way that Hamilton's equations (1.11) are preserved. At the classical level, a canonical transformation of the third kind is achieved by choosing a generating function $F(p, X, t)$. After an integration by parts on the term $p\dot{x}$, the phase space action is then transformed so that

$$\int_{t_i}^{t_f} \mathrm{d}t\, (-x\dot{p} - H(p, x)) = \int_{t_i}^{t_f} \mathrm{d}t \left(P\dot{X} - H(P, X) + \frac{\mathrm{d}}{\mathrm{d}t}F(p, X, t)\right). \tag{3.178}$$

Using $\dot{F} = (\partial F(p, X, t)/\partial X)\, \dot{X} + (\partial F(p, X, t)/\partial p)\, \dot{p} + \partial F(p, X, t)/\partial t$, the two sides can be identical up to a surface term only if

$$x = -\frac{\partial F(p, X, t)}{\partial p}, \quad P = -\frac{\partial F(p, X, t)}{\partial X}. \tag{3.179}$$

It is assumed that equation (3.179) can be solved to specify unique forms for $X(x, p)$ and $P(x, p)$. For example, choosing $F(p, X) = -p^2 \tan X/2\, m\omega$ gives $X = \arctan(m\omega x/p)$ and $\omega P = p^2/2\, m + \frac{1}{2}m\omega^2 x^2$. This transformation therefore turns the harmonic oscillator Hamiltonian into the cyclic Hamiltonian $H = \omega P$.

One of the important properties of a canonical transformation is that it preserves the volume of phase space by giving a Jacobian that is unity. For a one-dimensional system the Jacobian J of the canonical transformation is given by

$$J^{-1} = \frac{\partial X}{\partial x}\frac{\partial P}{\partial p} - \frac{\partial X}{\partial p}\frac{\partial P}{\partial x}. \tag{3.180}$$

It is straightforward to prove $J = 1$ by finding the partial derivatives of P from $P = -\partial F(p, X, t)/\partial X$ using the chain rule and combining them with the identity

$$1 = \frac{\partial x}{\partial x} = -\frac{\partial^2 F(p, X, t)}{\partial p\, \partial X}\frac{\partial X}{\partial x}. \tag{3.181}$$

At first glance, it would seem that applying canonical transformations to phase space path integrals with symmetric measures, such as the mixed propagator (2.15), would offer an alternative method for evaluating more complicated quantum systems. The problem is that $\dot{x}_j = (x_{j+1} - x_j)/\epsilon \equiv \Delta x_j/\epsilon$ is not a true time derivative, as has been pointed out numerous times. Changing the integration variables using the *classical* canonical transformation, $x_j = x(X_j, P_j)$ and $p_j = p(X_j, P_j)$, preserves the form of the measure, $[\mathcal{D}x][\mathcal{D}p] \rightarrow [\mathcal{D}X][\mathcal{D}P]$ since the Jacobian is one. However, it does *not necessarily give* $p_j\, \Delta x_j \rightarrow P_j\, \Delta X_j$, instead giving $p_j\, \Delta x_j = -p_j \partial F(p_{j+1}, X_{j+1}, t)/\partial p_{j+1} + p_j \partial F(p_j, X_j, t)/\partial p_j$. This is similar to the ordering problems that occur when canonical transformations are applied in the operator formulation of quantum mechanics.

Fukutaka and Kashiwa defined a canonical transformation that is consistent with the formal time derivatives that appear in equation (3.185). The canonical transformation of the third kind for path integrals is defined as

$$\begin{aligned} -P_j\,(X_{j+1} - X_j) &= F(p_j, X_{j+1}) - F(p_j, X_j), \\ -x_{j+1}\,(p_{j+1} - p_j) &= F(p_{j+1}, X_{j+1}) - F(p_j, X_{j+1}). \end{aligned} \tag{3.182}$$

The definitions of equation (3.182) give the desired result for the path integral,

$$-x_{j+1}(p_{j+1} - p_j) = P_j(X_{j+1} - X_j) + F(p_{j+1}, X_{j+1}) - F(p_j, X_j). \tag{3.183}$$

The last two terms in equation (3.183) act like the total time derivative of equation (3.178) when summed in the action of the path integral, becoming surface terms

$$\sum_{j=-1}^{N-1} \Big(F(p_{j+1}, X_{j+1}) - F(p_j, X_j)\Big) = F(p_f, X_f) - F(p_i, X_i), \tag{3.184}$$

where the identifications $X_{-1} = X_i$ and $X_N = X_f$ have been made.

The first step in implementing this canonical transformation requires the integration by parts (2.18) in the mixed propagator (2.15), which gives

$$p_i x_0 + \sum_{j=0}^{N-1} p_j(x_{j+1} - x_j) = p_f x_f - \sum_{j=-1}^{N-1} x_{j+1}(p_{j+1} - p_j), \tag{3.185}$$

where the notation $p_{-1} = p_i$, $p_N = p_f$, $x_{-1} = x_i$, and $x_N = x_f$ is in effect. As discussed earlier, only two of these variables, p_i and x_f, are defined in the symmetric transition element (2.15), leaving p_f and x_i undefined pseudovariables in the path integral action. At the moment, all the terms involving p_f and x_i that occur in equation (3.185) cancel among each other and so their value is irrelevant. However, the action of a canonical transformation results in new variables $X(p, x)$ and $P(p, x)$ that depend on *pairs* of the original variables. At the classical level this is not a difficulty, since the new variables $X(p, x)$ and $P(p, x)$ will satisfy the two Hamilton equations (1.11). The solutions of these two equations will give two relations for $X_f(p_f, x_f)$ and $P_f(p_f, x_f)$ in terms of $X_i(p_i, x_i)$ and $P_i(p_i, x_i)$, thereby determining x_i and p_f in terms of the two original boundary conditions x_f and p_i. However, the presence of p_f and x_i in the quantum mechanical context is problematic since the classical solutions appear only in the semiclassical approximation, which in turn is exact only for quadratic Hamiltonians. As a result, the expression $X_i = X(p_i, x_i)$ is ambiguous, with one argument, p_i, well defined and the other, x_i, undefined. The same problem besets the other boundary values X_f, P_f, and P_i. The resolution to this problem presented here consists of imposing additional boundary conditions on the new variables, and these additional conditions will be determined in the course of developing the canonical transformation.

The final form for equation (3.185) after the canonical transformation of equation (3.182) is

$$-\sum_{j=-1}^{N-1} x_{j+1}(p_{j+1} - p_j) = \sum_{j=-1}^{N-1} P_j(X_{j+1} - X_j) + F(p_f, X_f) - F(p_i, X_i), \tag{3.186}$$

which mimics the action of the classical canonical transformation. As promised, there are two additional undefined quantities, $P_{-1} \equiv P_i$ and $P_N \equiv P_f$, that have emerged from these formal manipulations.

The new variables X_j and P_j defined by equation (3.183) will typically not be solely dependent on x_j and p_j. This possibility becomes apparent when the definitions (3.182) are expanded in a Taylor series. The first three terms give

$$\begin{aligned} P_j &\approx -\frac{\partial F(p_j, X_j)}{\partial X_j} - \frac{1}{2}\frac{\partial^2 F(p_j, X_j)}{\partial X_j^2}\Delta X_j - \frac{1}{6}\frac{\partial^3 F(p_j, X_j)}{\partial X_j^3}(\Delta X_j)^2, \\ x_j &\approx -\frac{\partial F(p_j, X_j)}{\partial p_j} - \frac{1}{2}\frac{\partial^2 F(p_j, X_j)}{\partial p_j^2}\Delta p_j - \frac{1}{6}\frac{\partial^3 F(p_j, X_j)}{\partial p_j^3}(\Delta p_j)^2, \end{aligned} \tag{3.187}$$

where $\Delta X_j = X_{j+1} - X_j$ and $\Delta p_j = p_j - p_{j-1}$. The first term in both expansions reproduces the classical transformations of equation (3.179). However, the other terms in the expansions of equation (3.187) cannot be ignored in the transformed Hamiltonian unless ΔX_j and Δp_j are effectively $O(\epsilon)$. If they are $O(\epsilon)$, then they are suppressed because of the overall factor of ϵ in the expression $\epsilon H(P_j, X_j, \Delta P_j, \Delta X_j)$. The validity of discarding these terms depends on the properties of the transformed Hamiltonian. In particular, if the transformed Hamiltonian is cyclic up to terms of $O(\Delta X_j)$, then these $O(\Delta X_j)$ terms are suppressed. This is reasonably easy to demonstrate by using an action in the path integral of the form

$$\tilde{S} = F(p_f, X_f) - F(p_i, X_i) + \sum_{j=-1}^{N-1}\left(P_j\Delta X_j - \epsilon(\tilde{H}(P_j) + K(P_j)\Delta X_j)\right), \tag{3.188}$$

with the assumption that the P_j are variables of integration with an infinite range. Simply translating the integration variables according to $P_j \to P_j + \epsilon K(P_j)$ removes the term $\epsilon K(P_j)\Delta X_j$ and results in $\epsilon\tilde{H}(P_j) \to \epsilon\tilde{H}(P_j + \epsilon K(P_j))$. This shows that the effect of the original term $\epsilon K(P_j)\Delta X_j$ is $O(\epsilon^2)$ and can be ignored.

Unfortunately, the technical difficulties with canonical transformations also extend into the path integral measure, where the ΔX_j and ΔP_j terms prevent the

Jacobian from being unity *even if* they are $O(\epsilon)$. Using the Taylor series for P_j from equation (3.187), combining it with $\partial\Delta X_j/\partial p_j = -\partial X_j/\partial p_j$, and retaining only terms $O(\Delta X_j)$ gives

$$
\begin{aligned}
\frac{\partial P_j}{\partial p_j} &= -\frac{\partial^2 F(p_j, X_j)}{\partial p_j\, \partial X_j} - \frac{1}{2}\frac{\partial^2 F(p_j, X_j)}{\partial X_j^2}\frac{\partial X_j}{\partial p_j} \\
&\quad -\left(\frac{1}{2}\frac{\partial^3 F(p_j, X_j)}{\partial p_j \partial X_j^2} + \frac{1}{6}\frac{\partial^3 F(p_j, X_j)}{\partial X_j^3}\frac{\partial X_j}{\partial p_j}\right)\Delta X_j, \\
\frac{\partial P_j}{\partial x_j} &= -\frac{1}{2}\frac{\partial^2 F(p_j, X_j)}{\partial X_j^2}\frac{\partial X_j}{\partial x_j} - \frac{1}{6}\frac{\partial^3 F(p_j, X_j)}{\partial X_j^3}\frac{\partial X_j}{\partial x_j}\Delta X_j.
\end{aligned}
\tag{3.189}
$$

This gives the Jacobian for the canonical transformation,

$$
\begin{aligned}
J^{-1} &= \prod_{j=0}^{N-1}\left[\frac{\partial X_j}{\partial x_j}\frac{\partial P_j}{\partial p_j} - \frac{\partial X_j}{\partial p_j}\frac{\partial P_j}{\partial x_j}\right] \\
&= \prod_{j=0}^{N-1}\left[-\frac{\partial^2 F(p_j, X_j)}{\partial p_j\, \partial X_j}\frac{\partial X_j}{\partial x_j} - \frac{1}{2}\frac{\partial^3 F(p_j, X_j)}{\partial p_j\, \partial X_j^2}\frac{\partial X_j}{\partial x_j}\Delta X_j\right].
\end{aligned}
\tag{3.190}
$$

The final step is to combine equation (3.190) with the $O(\Delta p_j)$ identity similar to equation (3.181),

$$
1 = \frac{\partial x_j}{\partial x_j} = -\frac{\partial^2 F(p_j, X_j)}{\partial p_j\, \partial X_j}\frac{\partial X_j}{\partial x_j} - \frac{1}{2}\frac{\partial^3 F(p_j, X_j)}{\partial p_j^2\, \partial X_j}\frac{\partial X_j}{\partial x_j}\Delta p_j. \tag{3.191}
$$

To $O(\Delta X_j)$ and $O(\Delta p_j)$ the final result for the Jacobian of equation (3.182) is

$$
\begin{aligned}
J &= \prod_{j=0}^{N-1}\left(1 + A_j\Delta X_j + B_j\Delta p_j\right)^{-1} = \exp\left\{-\sum_{j=0}^{N-1}\ln(1 + A_j\Delta X_j + B_j\Delta p_j)\right\} \\
&\approx \exp\left\{\frac{i}{\hbar}\sum_{j=-1}^{N-1}\left(i\hbar A_j\,\Delta X_j + i\hbar B_j\,\Delta p_j\right)\right\},
\end{aligned}
\tag{3.192}
$$

where the terms of $O(\epsilon)$ following from $\ln(1 + x) \approx x$ are given by

$$
A_j = -\frac{1}{2}\frac{\partial^3 F(p_j, X_j)}{\partial p_j\, \partial X_j^2}\frac{\partial X_j}{\partial x_j}, \quad B_j = \frac{1}{2}\frac{\partial^3 F(p_j, X_j)}{\partial p_j^2\, \partial X_j}\frac{\partial X_j}{\partial x_j}. \tag{3.193}
$$

The sum has been extended to include the $j = -1$ for later simplicity of evaluation. Doing so requires that $A_{-1} = A_i$ vanishes since $\Delta X_{-1} = X_0 - X_i$ does not. However, $\Delta p_{-1} = \Delta p_i = 0$ by definition, so B_{-1} is unconstrained. The Jacobian of the transformation (3.182) effectively contributes two additional terms to the path integral action that are $O(\hbar)$. It is possible to relate these terms to the commutators $[\hat{x}, \hat{p}] = i\hbar$ that occur in the quantum mechanical version of the canonical transformations. Whereas the classical canonical transformation is a symmetry of phase space, leaving it invariant, the quantum canonical transformation is *not* a symmetry of phase space. The failure of the measure of a path integral to respect a *classical symmetry* is often referred to as an *anomaly*. In this case, the anomaly is induced by the formal time derivatives of the action.

These abstract considerations are clarified by considering the symmetric path integral (2.15) for the harmonic oscillator potential so that $H = p^2/2\,m + \frac{1}{2}m\omega^2x^2$. Applying the integration by parts of equation (2.18), the symmetric path integral becomes

$$\begin{aligned} G(x_f, p_i, t_f, t_i) = & \int [\bar{\mathcal{D}}p][\bar{\mathcal{D}}x]\times \\ & \times \exp\left\{\frac{i}{\hbar}\left(p_f x_f - \sum_{j=-1}^{N-1} x_{j+1}(p_{j+1} - p_j) + \sum_{j=0}^{N-1} \epsilon\, H(p_j, x_j)\right)\right\}, \end{aligned} \tag{3.194}$$

where $p_{-1} = p_i$, $x_N = x_f$, and the measure is given by equation (2.16). The pseudovariable p_f has appeared in the action of equation (3.194), but there is no longer any reference to the pseudovariable $x_{-1} = x_i$ after the integration by parts. The path integral (3.194) is given the canonical transformation generated by $F(p_j, X_j) = -(p_j^2 \tan X_j)/2\,m\omega$. Suppressing the terms of $O(\Delta X_j)$ and $O(\Delta P_j)$ under the assumption they are $O(\epsilon)$ yields the new canonical variables $P_j = p_j^2/2\,m\omega + \frac{1}{2}m\omega x_j^2$ and $X_j = \arctan(m\omega x_j/p_j)$. Similarly, ignoring $O(\Delta X)$ terms results in the transformed harmonic oscillator Hamiltonian $H(p_j, x_j) = \tilde{H}(P_j) = \omega P_j$. Denoting the Jacobian (3.192) as J, the path integral for a symmetric phase space measure becomes

$$G(x_f, p_i, T) = \exp\left(\frac{i}{\hbar}S_c\right)\int [\bar{\mathcal{D}}X][\bar{\mathcal{D}}P]\, J \exp\left\{\frac{i}{\hbar}\tilde{S}(P, X)\right\}, \tag{3.195}$$

where

$$\begin{aligned} \tilde{S}(P, X) &= \sum_{j=-1}^{N-1} P_j(X_{j+1} - X_j) - \sum_{j=0}^{N-1} \epsilon\, \omega P_j \\ S_c &= p_f x_f + F(p_f, X_f) - F(p_i, X_i). \end{aligned} \tag{3.196}$$

It is important to note that the sum over the new variables in the action (3.196) includes $P_{-1} = P_i$, $X_N = X_f$, and $X_{-1} = X_i$.

The next step is to evaluate the measure anomaly by retaining only terms of $O(\Delta X_j)$ and $O(\Delta P_j)$ and using the expansion

$$\Delta p_j = \frac{\partial p_j}{\partial X_j}\Delta X_j + \frac{\partial p_j}{\partial P_j}\Delta P_j. \tag{3.197}$$

Using equation (3.193) and the specific forms for $F(p_j, X_j)$, X_j, and P_j gives, after considerable algebra,

$$A_j\,\Delta X_j + B_j\,\Delta p_j = \frac{1}{2}\tan X_j\,\Delta X_j - \frac{1}{4}\frac{\Delta P_j}{P_j}. \tag{3.198}$$

To $O(\Delta X_j)$ and $O(\Delta P_j)$ the anomaly terms of equation (3.198) can be written

$$A_j\,\Delta X_j + B_j\,\Delta p_j = \frac{1}{2}\left(\ln\cos X_{j+1} - \ln\cos X_j\right) - \frac{1}{4}(\ln P_{j+1} - \ln P_j), \tag{3.199}$$

so that the sum that yields the total anomaly J becomes

$$-\sum_{j=-1}^{N-1}\left(A_j\,\Delta X_j + B_j\,\Delta p_j\right) = -\frac{1}{2}\ln\frac{\cos X_f}{\cos X_i} + \frac{1}{4}\ln\frac{P_f}{P_i}. \tag{3.200}$$

The utility of having extended the sum in equation (3.192) to include the $j = -1$ term is apparent, since otherwise $\cos X_0$, involving a variable of integration, would have occurred in the anomaly instead of $\cos X_i$, and this would have complicated evaluation of the path integral considerably. However, the price to be paid for this simplification is the requirement that $A_i = \tan X_i = 0$. This requirement is met if $X_i = \arctan(m\omega x_i/p_i) = 0$, so that the pseudovariable x_i must be set to zero at the end of the calculation.

The next step is to evaluate the remaining path integral. This begins by rewriting the action $\tilde{S}$ in equation (3.196) with an integration by parts,

$$\tilde{S} = P_{N-1}X_f - P_iX_i - \sum_{j=0}^{N-1}\left(X_j(P_j - P_{j-1}) + \epsilon\,\omega P_j\right). \tag{3.201}$$

The integrals over the X_j variables reduce the path integral (3.195) to

$$G(x_f, p_i, t_f, t_i) = \int[\mathcal{D}P]J\prod_{k=0}^{N-1}\delta(P_k - P_{k-1})\exp\left\{\frac{i}{\hbar}(S_c + \tilde{S}')\right\}, \tag{3.202}$$

where S_c is given by equation (3.196). After the X_j integrations, the remaining integration variable terms in the action, denoted $\tilde{S}'$, are given by

$$\tilde{S}' = P_{N-1}X_f - P_iX_i - \sum_{j=0}^{N-1}\epsilon\,\omega P_j, \tag{3.203}$$

and the measure is now given by

$$[\mathcal{D}P] = \frac{1}{\sqrt{2\pi\hbar}}\,\mathrm{d}P_0\cdots\mathrm{d}P_{N-1}. \tag{3.204}$$

It is important to note that the integrals over the P_j have the limits $[0, \infty]$ and would not yield Dirac deltas as the integrals over the X_j did. Using $X_f = \arctan(m\omega x_f/p_f)$

and $X_i = \arctan(m\omega x_i/p_i)$ gives $F(p_f, X_f) = -\frac{1}{2}p_f x_f$ and $F(p_i, X_i) = -\frac{1}{2}p_i x_i$. Performing the integrals over the P_j and using $\sum_{j=0}^{N-1}\epsilon = N\epsilon = T$ results in

$$G(x_f, p_i, T) = \frac{J}{\sqrt{2\pi\hbar}} \exp\left\{\frac{i}{\hbar}\left(\frac{1}{2}(p_f x_f + p_i x_i) + P_i(X_f - X_i - \omega T)\right)\right\}. \quad (3.205)$$

The final step is to identify the pseudovariables by using the classical equation of motion. The solutions for $p_f = p_c(t_f)$ and $x_i = x_c(t_i)$ are found from equation (3.55), which gives $x_i = (x_f - p_i \sin\omega T/m\omega)/\cos\omega T$ and $p_f = (p_i - m\omega x_f \sin\omega T)/\cos\omega T$. It follows that $X_f(x_f, p_f) - X_i(x_i, p_i) = \omega T$ and $P_f(x_f, p_f) = P_i(p_i, x_i)$, so that

$$\begin{aligned} S_c(x_f, p_i, T) &= \frac{1}{2}(p_f x_f + p_i x_i) + P_i(X_f - X_i - \omega T) \\ &= -\frac{1}{2}m\omega x_f^2 \tan\omega T - \frac{p_i^2}{2\,m\omega}\tan\omega T + \frac{p_i x_f}{\cos\omega T}. \end{aligned} \quad (3.206)$$

Using $P_i = P_f$ in the anomaly factor (3.200) gives

$$J = \exp\left\{-\frac{1}{2}\ln\left(\frac{\cos X_f}{\cos X_i}\right) + \frac{1}{4}\ln\left(\frac{P_f}{P_i}\right)\right\} = \sqrt{\frac{\cos X_i}{\cos X_f}}. \quad (3.207)$$

The requirement that $X_i = 0$ gives $\cos X_i = 1$ and $\cos X_f = \cos\omega T$. Unfortunately, this requires that $x_i = 0$, so that the two variables p_i and x_f must satisfy $m\omega x_f = p_i \sin\omega T$, reducing the generality of the result. The final result of the canonical transformation evaluation of the harmonic oscillator transition element gives

$$G(x_f, p_i, T) = \sqrt{\frac{1}{2\pi\hbar\cos\omega T}} \exp\left\{\frac{i}{\hbar}S_c(x_f, p_i, T)\right\}. \quad (3.208)$$

This is the same result that is obtained from performing the integral over x_0 in equation (2.15) using the path integral result (3.59) for the harmonic oscillator,

$$\begin{aligned} &\int \frac{dx_0}{\sqrt{2\pi\hbar}} \sqrt{\frac{m\omega}{2\pi i\hbar\sin\omega T}} \exp\left\{\frac{i}{\hbar}\left(p_i x_0 + \frac{1}{2}m\omega\left[(x_f^2 + x_0^2)\cot\omega T - \frac{2x_0 x_f}{\sin\omega T}\right]\right)\right\} \\ &= \sqrt{\frac{1}{2\pi\hbar\cos\omega T}} \exp\left\{\frac{i}{\hbar}S_c(x_f, p_i, T)\right\} = G(x_f, p_i, T), \end{aligned} \quad (3.209)$$

where S_c is given by equation (3.206). Result (3.209) does not have the limiting restriction $m\omega x_f = p_i \sin\omega T$ imposed by setting $X_i = 0$. Nevertheless, canonical transformations of path integrals are far more difficult to implement than might be expected.

3.10 Implementing constraints

A *quantum mechanical constraint* is understood as a relation that is placed on the position and momentum eigenvalues $\boldsymbol{x}$ and $\boldsymbol{p}$ available to the system in n-dimensions. Although constraints are arguably of greater importance in gauge field theory, it is useful to understand how constraints are implemented in quantum mechanical path integrals. Cartesian coordinates will be used to avoid the coordinate singularities associated with spherical coordinates, and the Cartesian components are denoted x_a and p_b. For the case of ℓ classical constraints, they are stated as $f^j(\boldsymbol{p}, \boldsymbol{x}, t) = 0$, where $j = 1, \ldots, \ell$. The system to be constrained is governed by the Lagrangian $\mathcal{L}_0(\boldsymbol{p}, \boldsymbol{x})$. In classical mechanics the ℓ constraints are themselves added to the action by using ℓ Lagrange multipliers λ_j and modifying the basic action,

$$\mathcal{L} = \mathcal{L}_0(\boldsymbol{p}, \boldsymbol{x}) + \sum_{j=1}^{\ell} \lambda_j f^j(\boldsymbol{p}, \boldsymbol{x}, t). \tag{3.210}$$

The constraints then become part of the set of equations of motion, where the ℓ equations $\partial\mathcal{L}/\partial\lambda^\ell = 0$ are solved simultaneously with the $2n$ Hamilton equations that also involve the ℓ Lagrange multipliers. Combining these $2n + \ell$ equations with the boundary conditions is assumed to allow a unique solution for the time behavior of all $2n + \ell$ unknowns, which are the $2n$ canonical variables $\boldsymbol{p}$ and $\boldsymbol{x}$ and the ℓ Lagrange multipliers λ_j. The physical interpretation of the Lagrange multipliers requires detailed knowledge of the specific system, but they are often associated with forces required or created by the constraint. The ℓ *primary constraints* $f^j = 0$ may give rise to *secondary constraints* if they are not time-independent. If $g^j = \dot{f}^j$ is not trivially zero, then g^j must be added to the list of constraints. In turn the new constraints may generate yet more constraints through their time-dependence. It is assumed this process eventually terminates.

For simplicity, constraints in the path integral formalism will be developed initially for the simple case of a single constraint in an n-dimensional system of the form $f(\boldsymbol{x}) = 0$ with no explicit time-dependence. It is assumed there is a unique solution to the classical constraint equation that allows the nth Cartesian component x_n to be written $x_n = c(\boldsymbol{x}_\perp)$, so that x_n becomes a function of the orthogonal subspace $\boldsymbol{x}_\perp$ spanned by x_1 through x_{n-1}. Implementing this primary constraint at the quantum mechanical level requires that the eigenvalue of the position operator $\hat{X}_n$ coincides with $c(\boldsymbol{x}_\perp)$, although subtleties may occur due to operator ordering in more complicated constraints. The constrained position states are denoted $|\tilde{\boldsymbol{x}}\rangle \equiv |\boldsymbol{x}_\perp, c(\boldsymbol{x}_\perp)\rangle$, so that the primary constraint $\hat{X}_n|\tilde{\boldsymbol{x}}\rangle = c(\boldsymbol{x}_\perp)|\tilde{\boldsymbol{x}}\rangle$ is satisfied. However, this constraint gives rise to the secondary constraint $\mathrm{d}\hat{X}_n/\mathrm{d}t = i[\hat{H}, \hat{X}_n]/\hbar = \hat{P}_n/m = 0$ for the case that the Hamiltonian contains the term $\hat{\boldsymbol{P}}^2/2m$. The constrained momentum states, $|\tilde{\boldsymbol{p}}\rangle = |\boldsymbol{p}_\perp, 0\rangle$, satisfy $\hat{P}_n|\tilde{\boldsymbol{p}}\rangle = 0$. These two constraints reduce the independent degrees of freedom or *dimensions* of the system's phase space since $\hat{X}_n$ and $\hat{P}_n$ no longer have arbitrary eigenvalues. At the quantum mechanical level this creates the additional issue that the standard commutator $[\hat{X}_n, \hat{P}_n] = i\hbar$ is contradicted by its matrix element $\langle\tilde{p}\,|[\hat{X}_n, \hat{P}_n]|\tilde{p}\,'\rangle = 0$ in the constrained set of states. This is resolved by replacing the standard commutators of quantum mechanics by *Dirac commutators* in a constrained system. The interested reader is recommended to the literature for further details.

These considerations complicate deriving the path integral by the time-slicing approach, since the constrained position and momentum states are no longer orthogonal in the Dirac delta sense. The constrained form of the position and momentum states, $|\tilde{x}\rangle = |x_\perp, c(x_\perp)\rangle$ and $|\tilde{p}\rangle = |p_\perp, 0\rangle$ inherit the original inner product. When combined with the property $\delta(x-a)\,\delta(x-a) = \delta(x-a)\,\delta(0)$, this gives

$$\begin{aligned}\langle\tilde{x}|\tilde{x}'\rangle &= \delta^{n-1}(x_\perp - x_\perp')\,\delta(c(x_\perp) - c(x_\perp')) = \delta^{n-1}(x_\perp - x_\perp')\,\delta(0),\\ \langle\tilde{p}|\tilde{p}\,'\rangle &= \delta^{n-1}(p_\perp - p_\perp')\,\tilde{\delta}(0),\end{aligned} \tag{3.211}$$

$$\langle\tilde{x}|\tilde{p}\rangle = \frac{1}{(2\pi\hbar)^{n/2}}\exp\left\{\frac{i}{\hbar}p_\perp\cdot x_\perp\right\}, \tag{3.212}$$

where $2\pi\hbar\,\delta(0) = \int \mathrm{d}p_n$ and $2\pi\hbar\,\tilde{\delta}(0) = \int \mathrm{d}x_n$ are the infinite one-dimensional position and momentum volumes being removed from the phase space of the system by the constraint. This shows that the constrained position and momentum states possess an infinite norm, and this is problematic for interpreting the inner product as a probability amplitude. However, the inner product (3.212) remains well defined, and this provides a solution to the dilemma.

The first step is to introduce a normalizing projection operator consistent with the constraint. This projection operator is defined as

$$\hat{\mathbf{P}}_p = 2\pi\hbar\int \mathrm{d}^n p\;\delta(p_n)\,|p\rangle\langle p| = 2\pi\hbar\int \mathrm{d}^{n-1}p_\perp\;|p_\perp, 0\rangle\langle p_\perp, 0|, \tag{3.213}$$

and it has the critical property that it normalizes the constrained inner product,

$$\langle\tilde{x}|\hat{\mathbf{P}}_p|\tilde{x}'\rangle = \int\frac{\mathrm{d}^{n-1}p_\perp}{(2\pi\hbar)^{n-1}}\exp\left\{\frac{i}{\hbar}p_\perp\cdot(x_\perp - x_\perp')\right\} = \delta^{n-1}(x_\perp - x_\perp'). \tag{3.214}$$

However, $\hat{\mathbf{P}}_p$ is not idempotent, instead satisfying $\hat{\mathbf{P}}_p\cdot\hat{\mathbf{P}}_p = 2\pi\hbar\,\tilde{\delta}(0)\,\hat{\mathbf{P}}_p$. It has the units of length, as it must to remove the factor of $\delta(0)$ from the inner product. The value of equation (3.214) is that it allows the definition of the *normalized transition element* for two constrained position states,

$$G_c(\tilde{x}_f, \tilde{x}_i, t_f, t_i) \equiv \langle\tilde{x}_f|\hat{\mathbf{P}}_p\hat{U}(t_f, t_i)|\tilde{x}_i\rangle. \tag{3.215}$$

In the limit $t_f \to t_i$ the transition element defined by equation (3.215) becomes the correctly normalized inner product (3.214), and this allows a probabilistic interpretation to the transition element for the constrained *quantum system*.

The time-sliced path integral representation of equation (3.215) can now be derived using the constrained projection operator $\hat{\mathbf{P}}^j_{xp}$, given by

$$\hat{\mathbf{P}}^{j}_{xp} = \int \mathrm{d}^n \boldsymbol{x}_j \frac{\mathrm{d}^n \boldsymbol{p}_{j-1}}{(2\pi\hbar)^{\frac{1}{2}n-1}} \delta(x_{j,n} - c_j)\, \delta(p_{j-1,n})\, e^{\frac{i}{\hbar}\boldsymbol{x}_j\cdot\boldsymbol{p}_{j-1}} |\boldsymbol{x}_j\rangle\langle \boldsymbol{p}_{j-1}|. \tag{3.216}$$

Using equation (3.211) shows that $\hat{\mathbf{P}}^{j}_{xp} \cdot \hat{\mathbf{P}}^{j}_{xp} = \hat{\mathbf{P}}^{j}_{xp}$ and $\hat{\mathbf{P}}_{p} \cdot \hat{\mathbf{P}}_{xp} = \hat{\mathbf{P}}_{p}$. The projection operator $\hat{\mathbf{P}}_{xp}$ is therefore idempotent in the subspace determined by the constraint. The key step in demonstrating this is first to perform the integrations over the Dirac deltas to recreate the constrained states of equation (3.211), allowing the product of projection operators to be evaluated using equation (3.211). The indices on the dummy variables of integration have been chosen in order to reproduce the previous time-sliced path integral (2.11) for the case that the constraint is present. For consistency of notation, the dummy momentum variables of integration in $\hat{\mathbf{P}}_p$ are relabeled $\boldsymbol{p}_{N-1}$. The position constraint in equation (3.216) can be rewritten by expanding the original constraint around the solution $c(\boldsymbol{x}_\perp)$, giving $f(\boldsymbol{x}) \approx (x_n - c(\boldsymbol{x}_\perp))\, \partial f(\boldsymbol{x}_\perp, c)/\partial c$, where the partial derivative is evaluated at $c = c(\boldsymbol{x}_\perp)$. Assuming the constraint solution is unique, using equation (1.19) allows the delta function in equation (3.216) to be written

$$\delta(x_{jn} - c_j) = \left| \frac{\partial f(\boldsymbol{x}_{j\perp}, c_j)}{\partial c_j} \right| \delta(f(\boldsymbol{x}_j)), \tag{3.217}$$

where $c_j = c(\boldsymbol{x}_{j\perp})$. The final step uses (1.25) to introduce the functional equivalent of Lagrange multipliers, writing the projection operator of equation (3.216) as

$$\begin{aligned} \hat{\mathbf{P}}^{j}_{xp} = \int \mathrm{d}^n \boldsymbol{x}_j \frac{\mathrm{d}^n \boldsymbol{p}_{j-1}}{(2\pi\hbar)^{\frac{1}{2}n-1}} \frac{\mathrm{d}\lambda_j}{2\pi\hbar} \frac{\mathrm{d}\nu_{j-1}}{2\pi\hbar} \epsilon^2 \left| \frac{\partial f(\boldsymbol{x}_{j\perp}, c_j)}{\partial c_j} \right| \\ \times \exp\left\{ \frac{i}{\hbar}\Big(\epsilon\lambda_j f(\boldsymbol{x}_j) + \epsilon\nu_{j-1} p_{j-1,n} + \boldsymbol{x}_j \cdot \boldsymbol{p}_{j-1} \Big) \right\} |\boldsymbol{x}_j\rangle\langle \boldsymbol{p}_{j-1}|, \end{aligned} \tag{3.218}$$

where ϵ is the infinitesimal time element. For the case that $f(\boldsymbol{x})$ has the units of length, the units of λ_j correspond to force, while those of ν_k always correspond to velocity. Combining equation (3.218) with equation (2.5) gives the phase space path integral for the transition between constrained states from the time-sliced evolution operator $\hat{U}$,

$$\begin{aligned} G_c(\tilde{\boldsymbol{x}}_f, \tilde{\boldsymbol{x}}_i, t_f, t_i) &= \langle \tilde{\boldsymbol{x}}_f | \mathbf{P}_p \hat{U}_{N-1} \mathbf{P}^{N-1}_{xp} \hat{U}_{N-2} \mathbf{P}^{N-2}_{xp} \dots \hat{U}_1 \hat{\mathbf{P}}^{1}_{xp} \hat{U}_0 | \tilde{\boldsymbol{x}}_i \rangle \\ &= \int [\mathcal{D}^c \boldsymbol{x}][\mathcal{D}^c \boldsymbol{p}] \exp\left\{ \frac{i}{\hbar} \sum_{j=0}^{N-1} \epsilon\Big(\boldsymbol{p}_j \cdot \dot{\boldsymbol{x}}_j - H(\boldsymbol{p}_j, \boldsymbol{x}_j) + \lambda_j f(\boldsymbol{x}_j) + \nu_j p_{jn} \Big) \right\}, \end{aligned} \tag{3.219}$$

where $\boldsymbol{x}_N = \tilde{\boldsymbol{x}}_f$, $\boldsymbol{x}_0 = \tilde{\boldsymbol{x}}_i$, and $\lambda_0 = 0$. The variables of integration, λ_j and ν_j, appear in the action in the same manner as the Lagrange multipliers in equation (3.210). The constraint measure is given by

$$[\mathcal{D}^c\boldsymbol{x}][\mathcal{D}^c\boldsymbol{p}] = \prod_{j=0}^{N-1}\left(\frac{\mathrm{d}^n\boldsymbol{p}_j}{(2\pi\hbar)^{n-1}}\,\epsilon\frac{\mathrm{d}\nu_j}{2\pi\hbar}\right)\prod_{k=1}^{N-1}\left(\mathrm{d}^n\boldsymbol{x}_k\,\epsilon\frac{\mathrm{d}\lambda_k}{2\pi\hbar}\left|\frac{\partial f_k}{\partial c_k}\right|\right), \tag{3.220}$$

where $f_k = f(\boldsymbol{x}_{k\perp}, c_k)$. Since $\lambda_k f_k$ and $\nu_j p_j^n$ have the units of energy, it follows that the measure of equation (3.220) has the units of inverse length to the power $n - 1$, which matches the units of the *constrained transition element* after the infinite spatial volume $2\pi\hbar\,\tilde{\delta}(0)$ has been cancelled by $\hat{\mathbf{P}}_p$.

It is useful to see how the constrained path integral works in the simple case of the two-dimensional free particle in the configuration space (x, y) with the constraint $f(x, y) = y = 0$. The constrained time-sliced Lagrangian can be rewritten as

$$\begin{aligned}\mathcal{L}_j &= p_{jx}\dot{x}_j + p_{jy}\dot{y}_j - \frac{p_{jx}^2}{2m} - \frac{p_{jy}^2}{2m} + \lambda_j y_j + \nu_j p_{jy}\\ &= p_{jx}\dot{x}_j - \frac{p_{jx}^2}{2m} - \frac{\left(p_{jy} - m(\dot{y}_j + \nu_j)\right)^2}{2m} + \frac{m}{2}(\nu_j + \dot{y}_j)^2 + \lambda_j y_j.\end{aligned} \tag{3.221}$$

The integrals over p_{jy} and ν_j are now Gaussian, and performing them reduces the constrained path integral to

$$G_c(\tilde{\boldsymbol{x}}_f, \tilde{\boldsymbol{x}}_i, t_f, t_i) = \int [\tilde{\mathcal{D}}^c\boldsymbol{x}][\tilde{\mathcal{D}}^c p_x]\exp\left\{\frac{i}{\hbar}\sum_{j=0}^{N-1}\epsilon\left(p_{jx}\dot{x}_j - \frac{p_{jx}^2}{2m} + \lambda_j y_j\right)\right\}, \tag{3.222}$$

where the remaining measure is given by

$$[\tilde{\mathcal{D}}^c\boldsymbol{x}][\tilde{\mathcal{D}}^c p_x] = \prod_{j=0}^{N-1}\left(\frac{\mathrm{d}p_{jx}}{2\pi\hbar}\right)\prod_{k=1}^{N-1}\left(\mathrm{d}x_k\,\mathrm{d}y_k\,\epsilon\frac{\mathrm{d}\lambda_k}{2\pi\hbar}\right). \tag{3.223}$$

Recalling that $\lambda_0 = 0$, the integrations over the λ_k yield $N - 1$ delta functions of the form $2\pi\,\delta(\epsilon y_k/\hbar) = 2\pi\hbar\,\delta(y_k)/\epsilon$. The remaining integrations over the $N - 1$ y_k variables become trivial, so that the final path integral is given by

$$G_c(\tilde{\boldsymbol{x}}_f, \tilde{\boldsymbol{x}}_i, t_f, t_i) = \int [\mathcal{D}x][\mathcal{D}p_x]\exp\left\{\frac{i}{\hbar}\sum_{j=0}^{N-1}\epsilon\left(p_{jx}\dot{x}_j - \frac{p_{jx}^2}{2m}\right)\right\}, \tag{3.224}$$

with the measure

$$[\mathcal{D}x][\mathcal{D}p_x] = \frac{\mathrm{d}p_{0x}}{2\pi\hbar}\cdots\frac{\mathrm{d}p_{(N-1)x}}{2\pi\hbar}\,\mathrm{d}x_1\cdots\mathrm{d}x_{N-1}. \tag{3.225}$$

Results (3.224) and (3.225) are identical to the one-dimensional free particle path integral and measure derived in equation (2.11).

There is an important type of constraint that occurs in gauge field theory and that can be examined in a quantum mechanical context. For the case of a system with n degrees of freedom, where the Cartesian position variables are denoted $\boldsymbol{x}$, the classical Lagrangian may be independent of $\dot{x}_c$, where x_c is one of the Cartesian coordinates $\boldsymbol{x}$. For such a case the momentum p_c canonically conjugate to x_c does not exist since $p_c = \partial\mathcal{L}/\partial\dot{x}_c = 0$. However, Hamilton's equations may give $\partial H/\partial x_c = \dot{p}_c \neq 0$, depending on the form for H. If so, the simple requirement that $p_c = 0$ becomes inconsistent with the equations of motion.

This problem can be treated as the enforcement of *two constraints*, $p_c = 0$ and $\dot{p}_c = -\partial H/\partial x_c = 0$. However, it is first necessary *to define* p_c in order to implement the constraint enforcement process. In order to do this the original Lagrangian is modified by adding the term $G(\boldsymbol{x}, \dot{x}_c)$, so that $\mathcal{L} \to \mathcal{L}_G = \mathcal{L} + G$. It is assumed that G has no explicit time-dependence and has been chosen so that $p_c = \partial G(\boldsymbol{x}, \dot{x}_c)/\partial\dot{x}_c$ is now a well-defined mathematical quantity. It is *also assumed* that the canonical momentum $p_c = \partial G(\boldsymbol{x}, \dot{x}_c)/\partial\dot{x}_c$ can be solved to express the previously absent $\dot{x}_c$ in terms of the canonically conjugate coordinates $\boldsymbol{x}$ and p_c, so that $\dot{x}_c = \dot{x}_c(\boldsymbol{x}, p_c)$. If so, the standard recipe for the modified Hamiltonian, $H_G(\boldsymbol{x}, \boldsymbol{p}) = \boldsymbol{p}\cdot\dot{\boldsymbol{x}} - \mathcal{L}_G$, yields a well-defined expression for the term $p_c\dot{x}_c(\boldsymbol{x}, p_c)$. As part of this process the additional term $G(\boldsymbol{x}, \dot{x}_c)$ is also re-expressed in terms of the canonical momentum p_c, so that the Hamiltonian now has the term $G_p(\boldsymbol{x}, p_c) \equiv G(\boldsymbol{x}, \dot{x}_c(\boldsymbol{x}, p_c))$. While the specific form for the function G is somewhat arbitrary and is often chosen for convenience of calculation, it must result in the relation $|G_p(\boldsymbol{x}, p_c)_{p_c=0} = 0$. This condition is required in order for H_G to coincide with the original Hamiltonian H when the constraint $p_c = 0$ is enforced. This means that G must be chosen so that $G_p(\boldsymbol{x}, p_c)$ has the Taylor series expansion around $p_c = 0$ given by

$$G_p(\boldsymbol{x}, p_c) = \left.\frac{\partial G_p(\boldsymbol{x}, p_c)}{\partial p_c}\right|_{p_c=0} p_c + \frac{1}{2}\left.\frac{\partial^2 G_p(\boldsymbol{x}, p_c)}{\partial p_c^2}\right|_{p_c=0} p_c^2 + \dots \tag{3.226}$$

It is assumed that G_p has no roots other than $p_c = 0$, so that equation (3.226) vanishes if and only if $p_c = 0$. In this sense, the constraint $G_p(\boldsymbol{x}, p_c) = 0$ is equivalent to the original constraint $p_c = 0$.

The choice of the term G is referred to as *gauge fixing* in gauge field theory. The condition $G = 0$ is often enforced *a priori* as a primary constraint and is typically referred to as a *gauge condition*. The field theory aspect of this is discussed in greater detail in the next chapter. At the classical level enforcing the gauge condition is achieved by solving $G_p(\boldsymbol{x}, p_c) = 0$ and using the result in the other equations of motion. At the quantum mechanical level the gauge condition is elevated to an operator statement and the gauge condition is enforced by defining the *physical subspace* as the set of states $|\Psi_p\rangle$ for which $\hat{G}_p(\hat{\boldsymbol{X}}, \hat{P}_c)|\Psi_p\rangle = 0$. This statement can be subject to operator ordering ambiguities since $\hat{X}_c$ and $\hat{P}_c$ do not commute, and so it is assumed that the operator version of $\hat{G}_p$ has the ordering given by equation (3.226). The physical subspace therefore obeys $\hat{P}_c|\Psi_p\rangle = 0$ in exact agreement with the original formulation of the problem. However, the Hilbert space now includes an *unphysical subspace*, denoted $|\Phi_u\rangle$, that corresponds to $\hat{G}_p|\Phi_u\rangle \neq 0$. Assuming $\hat{G}_p^{-1}$ exists, the physical and unphysical subspaces are orthogonal since

$$\langle\Phi_u|\Psi_p\rangle = \langle\Phi_u|\left(\hat{G}_p^{-1}\hat{G}_p\right)|\Psi_p\rangle = \langle\Phi_u|\hat{G}_p^{-1}\left(\hat{G}_p|\Psi_p\rangle\right) = 0. \tag{3.227}$$

If the system is prepared initially in the physical subspace, it is critical that the state does not evolve into the unphysical subspace where the required constraint no longer holds. *If the gauge condition commutes with the evolution operator*, then this failure is prevented since

$$\langle\Phi_u|\hat{U}(t_f, t_i)|\Psi_p\rangle = \langle\Phi_u|\hat{G}_p^{-1}\hat{U}(t_f, t_i)\left(\hat{G}_p|\Psi_p\rangle\right) = 0. \tag{3.228}$$

However, if $\hat{G}_p$ does not commute with the evolution operator, then there is a secondary constraint on the physical subspace,

$$\frac{d\hat{G}_p}{dt}|\Psi_p\rangle = \frac{i}{\hbar}[\hat{H}_G, \hat{G}_p]|\Psi_p\rangle = 0. \tag{3.229}$$

For the choice of ordering given by equation (3.226), the lowest order term in equation (3.226) gives

$$\frac{d\hat{G}_p}{dt} = \frac{i}{\hbar}[\hat{H}_G, \hat{G}_p] = \frac{i}{\hbar}([\hat{H}_G, \hat{D}(\hat{X})]\hat{P}_c + \hat{D}(\hat{X})[\hat{H}_G, \hat{P}_c]), \tag{3.230}$$

where $\hat{D}(\hat{X})$ is the momentum derivative factor $\partial G_p(\boldsymbol{x}, p_c)/\partial p_c|_{p_c=0}$ in equation (3.226) with $\boldsymbol{x}$ elevated to the operators $\hat{X}$. The first term in equation (3.230) vanishes when applied to the physical subspace since $\hat{P}_c|\Psi_p\rangle = 0$. However, if $[\hat{H}_G, \hat{P}_c] \neq 0$, the second term in equation (3.230) will result in a secondary constraint. For simplicity it is *assumed* that $i[\hat{H}_G, \hat{P}_c]/\hbar \equiv \hat{F}(\hat{X})$ is a function solely of the position operators $\hat{X}$. It is straightforward to see that the higher order terms in equation (3.226) will vanish for the case that both $\hat{P}_c$ and $\hat{F}(\hat{X})$ yield zero when acting on the physical states. Because $\hat{X}_c$ and $\hat{P}_c$ define a single physical degree of freedom, the secondary constraint $\hat{F}(\hat{X})|\Phi_p\rangle = 0$ must be implemented by a constraint on $\hat{X}_c$. Such a constraint is implemented so that $\hat{P}_c|\Psi_p\rangle = 0$, making it consistent with the gauge fixing procedure.

In deriving the path integral for this system, the Hamiltonian H_G is used to define the evolution operator, while the physical momentum and position projection operators are assumed to obey the operator statements $\hat{P}_c = 0$ and $\hat{F}(\hat{X}) = 0$ as primary and secondary constraints. As in the previous derivation of the constrained path integral, it is assumed that the spatial constraint $\hat{F}(\hat{X})|\boldsymbol{x}\rangle = 0$ can be satisfied by solving it for the eigenvalues of $\hat{X}_c$ that are the roots of the classical constraint $F(\boldsymbol{x}) = 0$, once again denoted $x_c = c(\boldsymbol{x}_\perp)$. The constrained projection operator that is used to derive the time-sliced phase space path integral is then identical in form to equation (3.216), using the delta functions $\delta(p_{cj})$ and $\delta(x_{cj} - c_j)$, where $c_j = c(\boldsymbol{x}_{\perp j})$ is the solution to the time-sliced constraint $F(\boldsymbol{x}_j) = 0$. It is possible there are a set of solutions to this equation, but that will not be considered here. It is useful to rewrite the product of the two delta functions appearing in the constrained measure using the property of the Dirac delta given by equation (1.19), the Taylor series expansion of G_p around $p_c = 0$ given by equation (3.226), and a similar Taylor expansion of F around its root $x_c = c$, which allows the time-sliced delta functions to be written

$$\delta(p_{cj})\,\delta(x_{cj} - c_j) = \left.\frac{\partial F(\boldsymbol{x}_j)}{\partial x_{cj}}\right|_{x_{cj}=c} \left.\frac{\partial G_p(\boldsymbol{x}_j, p_{cj})}{\partial p_{cj}}\right|_{p_{cj}=0} \delta(\,G_p(\boldsymbol{x}_j, p_{cj}))\,\delta(F(\boldsymbol{x}_j)). \tag{3.231}$$

The utility of expressing the constraints on the measure in this fashion is seen when there is a set of ℓ generalized gauge constraints that can be expressed as $G_j = G_j(\boldsymbol{x}, \boldsymbol{p}) = \sum_{k=1}^{\ell} M_{jk}(\boldsymbol{x})\,p_k = 0$, where $\boldsymbol{p}$ are the ℓ undefined momenta while $\boldsymbol{x}$ may include all the position variables. In this expression the $M_{jk}(\boldsymbol{x})$ are the elements of an $\ell \times \ell$ matrix $\mathbf{M}(\boldsymbol{x})$ that is *assumed to be invertible*, so that the set of constraints can be written $\mathbf{G} = \mathbf{M}(\boldsymbol{x}) \cdot \boldsymbol{p} = 0$, where the G_j are treated as the elements of the vector $\mathbf{G}$. Since $\mathbf{M}(\boldsymbol{x})$ is invertible, its determinant is nonzero and it therefore has no zero eigenvalue. This means that there is no nontrivial eigenvector $\boldsymbol{p}$ that would satisfy the constraint $\mathbf{M}(\boldsymbol{x}) \cdot \boldsymbol{p} = 0$. As a result, all ℓ constraints are solved uniquely by $\boldsymbol{p} = 0$, which means that the constraint is equivalent to $p_k = 0$ for all $k = 1, \ldots, \ell$. Defining the variable $\lambda_j = \sum_{k=1}^{\ell} M_{jk}(\boldsymbol{x})\,p_k$ shows that

$$1 = \int \mathrm{d}^\ell \lambda \prod_{j=1}^{\ell} \delta(\lambda_j) = \int \mathrm{d}^\ell p \det \mathbf{M}(\boldsymbol{x}) \prod_{j=1}^{\ell} \delta\left(\sum_{k=1}^{\ell} M_{jk}(\boldsymbol{x})\,p_k\right). \tag{3.232}$$

Simply relabeling $\lambda_j \to p_j$ in the first integral shows that equation (3.232) gives the equality

$$\prod_{j=1}^{\ell} \delta(p_j) = \det \mathbf{M}(\boldsymbol{x}) \prod_{j=1}^{\ell} \delta\left(\sum_{k=1}^{\ell} M_{jk}(\boldsymbol{x})\,p_k\right) = \det\left(\frac{\partial \mathbf{G}}{\partial \boldsymbol{p}}\right) \prod_{j=1}^{\ell} \delta\big(G_j\big), \tag{3.233}$$

where $\partial G_j / \partial p_k$ is the jk element of the matrix $\partial \mathbf{G} / \partial \boldsymbol{p}$ whose determinant appears in equation (3.233). A similar results follows for the n secondary constraints $\mathbf{F}(\boldsymbol{x})$ obtained from the primary constraints $\mathbf{G}(\boldsymbol{x}, \boldsymbol{p})$ using the same argument as equation (3.230). Using $\det \mathbf{A} \det \mathbf{B} = \det(\mathbf{A} \cdot \mathbf{B})$, the primary and secondary constraints can be written

$$\prod_{j=1}^{n} \delta(p_j)\,\delta(x_j - c_j) = \det\left(\frac{\partial \mathbf{G}}{\partial \boldsymbol{p}} \cdot \frac{\partial \mathbf{F}}{\partial \boldsymbol{x}}\right) \prod_{j=1}^{\ell} \delta\big(G_j\big)\,\delta(F_j), \tag{3.234}$$

where it has been *assumed* that $F_j(\boldsymbol{x}) = 0$ has the solution $x_j = c_j$. The product of the two matrices in equation (3.234) has the jk element

$$\sum_{i=1}^{n} \frac{\partial G_j}{\partial p_i} \cdot \frac{\partial F_k}{\partial x_i} = \big\{F_k,\, G_j\big\}. \tag{3.235}$$

where it was *assumed* that $\partial F_j/\partial p_i = 0$. The final form for the generalized constraint can be written using the Poisson bracket as

$$\prod_{k=1}^{\ell} \delta(p_k)\,\delta(x_k - c_k) = \det\{\mathbf{F}, \mathbf{G}\} \prod_{k=1}^{\ell} \delta(G_k)\,\delta(F_k). \tag{3.236}$$

The form (3.236) is used in deriving the time-sliced path integral for the case of ℓ gauge conditions in a manner identical to the case of the single constraint (3.219). The constraints on the measure can be incorporated into the path integral action by using the integral representation (1.26) of the delta functions. Introducing the notation $\mathrm{d}^\ell\boldsymbol{\lambda}_j = \prod_{k=1}^{\ell} \mathrm{d}\lambda_{kj}$ and $\mathrm{d}^\ell\boldsymbol{\nu}_j = \prod_{k=1}^{\ell} \mathrm{d}\nu_{kj}$, the product of the two Dirac deltas can be written

$$\begin{aligned}
&\prod_{k=1}^{\ell} \delta\Big(G_k(\boldsymbol{x}_j, \boldsymbol{p}_j)\Big)\,\delta(F_k(\boldsymbol{x}_j)) \\
&= \int \epsilon^{2\ell} \frac{\mathrm{d}^\ell\boldsymbol{\lambda}_j}{(2\pi\hbar)^\ell} \frac{\mathrm{d}^\ell\boldsymbol{\nu}_j}{(2\pi\hbar)^\ell} \exp\left\{\frac{i}{\hbar}\left(\sum_{k=1}^{\ell} \epsilon(\lambda_{kj} G_k(\boldsymbol{x}_j, \boldsymbol{p}_j) + \nu_{kj} F_k(\boldsymbol{x}_j))\right)\right\} \qquad (3.237) \\
&\equiv \int \epsilon^{2\ell} \frac{\mathrm{d}^\ell\boldsymbol{\lambda}_j}{(2\pi\hbar)^\ell} \frac{\mathrm{d}^\ell\boldsymbol{\nu}_j}{(2\pi\hbar)^\ell} \exp\left\{\frac{i}{\hbar} S_{cj}\right\}.
\end{aligned}$$

The term $S_{cj} = \sum_{k=1}^{\ell} \epsilon(\lambda_{kj} G_k(\boldsymbol{x}_j, \boldsymbol{p}_j) + \nu_{kj} F_k(\boldsymbol{x}_j))$ represents the contribution of the constraints to the action of the theory. This result generalizes the path integral (3.219) for a single constraint to the case of several gauge constraints,

$$\begin{aligned}
&G_c(\tilde{\boldsymbol{x}}_f, \tilde{\boldsymbol{x}}_i, t_f, t_i) \\
&= \int [\mathcal{D}^c\boldsymbol{x}][\mathcal{D}^c\boldsymbol{p}] \exp\left\{\frac{i}{\hbar} \sum_{j=0}^{N-1} \epsilon\Big(\boldsymbol{p}_j \cdot \dot{\boldsymbol{x}}_j - H(\boldsymbol{p}_j, \boldsymbol{x}_j) + S_{cj}(\boldsymbol{x}_j, \boldsymbol{p}_j)\Big)\right\}.
\end{aligned} \tag{3.238}$$

the constraint measure in equation (3.238) is given by

$$\begin{aligned}
[\mathcal{D}^c\boldsymbol{x}][\mathcal{D}^c\boldsymbol{p}] = &\prod_{j=0}^{N-1} \left(\frac{\mathrm{d}^n\boldsymbol{p}_j}{(2\pi\hbar)^{n-1}}\,\epsilon^\ell \frac{\mathrm{d}^\ell\boldsymbol{\nu}_j}{(2\pi\hbar)^\ell}\right) \\
&\times \prod_{k=1}^{N-1} \left(\mathrm{d}^n\boldsymbol{x}_k\,\epsilon^\ell \frac{\mathrm{d}^\ell\boldsymbol{\lambda}_k}{(2\pi\hbar)^\ell} \det\left\{F(\boldsymbol{x}_k), G(\boldsymbol{x}_k, \boldsymbol{p}_k)\right\}\right).
\end{aligned} \tag{3.239}$$

In the constrained measure of equation (3.239) the 2ℓ Lagrange multipliers are integrated over two ℓ-dimensional volumes in each of the time slices.

As a final step, the Grassmann result (1.129) can be used to write the determinants appearing in equation (3.239) as a Grassmann Gaussian integral,

$$\det \mathbf{M}(\boldsymbol{x}_j, \boldsymbol{p}_j) = \int \mathrm{d}c_{1j}\, \mathrm{d}\bar{c}_{1j} \cdots \mathrm{d}c_{\ell j}\, \mathrm{d}\bar{c}_{\ell j} \exp \left\{ \bar{c}_j^T \cdot \mathbf{M}(\boldsymbol{x}_j, \boldsymbol{p}_j) \cdot c_j \right\}, \tag{3.240}$$

where the $\bar{c}_j$ and c_j are two sets of ℓ anticommuting Grassmann variables. The Grassmann measure of equation (3.240) is added to the path integral measure for each time slice while the Grassmann term $-i\hbar\, \bar{c}_j^T \cdot \mathbf{M}(\boldsymbol{x}_j, \boldsymbol{p}_j) \cdot c_j$ becomes part of the path integral action sum. In the general case both the primary and secondary gauge constraints may contribute to the Grassmann component of the constrained action via form (3.240) and the determinant appearing in equation (3.236). However, in field theory it is the determinant associated with the secondary constraint that contributes to the action via Grassmann variables, where it is known as the *Faddeev–Popov determinant*.

Further reading

There are many texts and monographs where quantum mechanical path integrals are evaluated. In order of original publication date these include

- R P Feynman, A R Hibbs, and D Styer 2010 *Quantum Mechanics and Path Integrals* (New York: Dover)
- L S Schulman 2005 *Techniques and Applications of Path Integration* (New York: Dover)
- H Kleinert 2004 *Path Integrals in Quantum Mechanics, Statistics, Polymer Physics, and Financial Markets* (Singapore: World Scientific)
- M S Swanson 2014 *Path Integrals and Quantum Processes* (New York: Dover)
- C Grosche and F Steiner 1998 *Handbook of Feynman Path Integrals* (Berlin: Springer)

In addition, there are online reviews and articles that develop many of the introductory concepts of the path integral. These include

- C Grosche 1993 *An Introduction Into the Feynman Path Integral* (arXiv:hep-th/9302097v1)
- R Mackenzie 2000 *Path Integral Methods and Applications* (arXiv:quant-ph/0004090v1)

The use of Lanczos convergence factors in Fourier series is discussed in the monograph by Arfken. The Maslov index is discussed in

- V P Maslov and M V Fedoriuk 1981 *Semi-Classical Approximations in Quantum Mechanics* (Dordrecht: Reidel)

The Gel'fand–Yaglom theorem appeared in

- I M Gel'fand and A M Yaglom 1960 *J. Math. Phys.* **1** 48

and is discussed extensively in the previously referenced monograph by Kleinert. It is applied in both quantum mechanics and quantum field theory in the review by

- G Dunne 2008 *Functional Determinants in Quantum Field Theory*, Lectures at the 14th WE Heraeus Saalburg Summer School in Wolfersdorf, Thuringia (2008)

The use of instantons to calculate the decay rate of metastable states originated with

- S Coleman 1977 *Phys. Rev. D* **15** 2929

and is presented in the review by

- A Andreassen, D Farhi, W Frost and M Schwartz 2017 (arXiv:1604.06090v2 [hep-th])

Generating functionals in quantum mechanics and quantum field theory originated with

- J Schwinger 1951 *Phys. Rev.* **82** 664
- K Symanzik 1954 *Z. Natürforsch. A* **9** 10
- G Jona-Lasinio 1964 *Nuovo Cimento* **34** 1790
- J Schwinger 1969 *Particles and Sources* (London: Gordon and Breach)

and the previously referenced monograph by Fried. Statistical mechanics is presented in

- W Greiner, L Neise and H Stöcker 1995 *Thermodynamics and Statistical Mechanics* (Berlin: Springer)
- D McQuarrie 1973 *Statistical Thermodynamics* (Mill Valley, CA: University Science Books)

Canonical transformations of the type analyzed in this monograph originated with

- H Fukutaka and T Kashiwa 1988 *Ann. Phys.* **185** 301

the associated anomaly was analyzed in

- M Swanson 1993 *Phys. Rev. A* **47** 2431

and generalized to field theory in

- M Blasone, P Jizba and L Smaldone 2017 *J. Phys.: Conf. Ser.* **804** 012006

Dirac commutators as well as many aspects of constrained systems are discussed in

- K Sundermeyer 1983 *Constrained Dynamics* (Berlin: Springer).

IOP Publishing

Path Integral Quantization

Mark S Swanson

Chapter 4

Quantum field theory and path integrals

Quantum mechanics is a profoundly successful theory for systems where the number of point particles is fixed and their velocities remain small compared to the speed of light c. It was primarily to remedy these two shortcomings that relativistic quantum field theory was first developed. At first glance, fields appear to be fundamentally different from pointlike Newtonian particles since the field exists everywhere in space and changes in time. However, relativistic quantum field theory reveals the particle content of a field in a manner consistent with both quantum mechanics and special relativity. In this chapter the path integral formulation of relativistic quantum field theory is developed. As in the case of quantum mechanical path integrals, the meaning of the field theory path integral is greatly clarified by deriving it from the operator formulation of quantum field theory. In what follows it is assumed that the reader is familiar with special relativity, the rudiments of group theory, and the action formulation for scalar, spinor, and vector fields. For the sake of clarifying notation, these topics are briefly reviewed, but a great many important details will be omitted. The reader needing further insights into these details should consult the classical field theory references listed in the bibliography.

Although field theories can be defined in any spatial dimension, the presentation will target a flat four-dimensional spacetime. Following a brief review of classical fields the canonical quantization of free scalar, spinor, and gauge fields is presented in order to clarify their particle content. The LSZ formalism is then developed to describe scattering and particle production processes associated with interacting fields. The evolution operator for quantum field theory is then given an interaction picture representation. Once these essential elements of quantized fields are developed, the path integral for field theory is derived by extending the coherent state formalism introduced in chapter 2 where it was used to define the quantum mechanical coherent state path integral. This is done first for the simpler cases of scalar and spinor fields. The gauge field theory path integral is defined using the

doi:10.1088/978-0-7503-3547-8ch4

constraint method for dealing with gauge conditions and is shown to be equivalent to the more commonly used Faddeev–Popov method for factorizing gauge volumes.

4.1 Special relativity and relativistic notation

Einstein's special theory of relativity postulates that the speed of light c is constant for all observers. Accommodating this postulate requires time t to be elevated from the parameter of Newtonian physics and placed on an equal footing with the spatial coordinates. A point x in *Minkowski space* or *spacetime* is given by $x = (x^0, \boldsymbol{x})$, where $x^0 = ct$ while $\boldsymbol{x}$ is the spatial position. Points in spacetime are referred to as *events* since they have a specific spatial location and a time of occurrence. The components of x are written x^μ, where $\mu = 0, \ldots 3$. Relativity is concerned with how the coordinates of a point in Minkowski space change or *transform* when a different frame of reference is chosen. The change of frame is represented by an *invertible change* of coordinates to a new set of coordinates $x'^\mu(x)$. The chain rule shows that the differentials in the two coordinate systems are related by

$$\mathrm{d}x'^\mu = \sum_{\nu=0}^{3} \frac{\partial x'^\mu}{\partial x^\nu}\, \mathrm{d}x^\nu \equiv \frac{\partial x'^\mu}{\partial x^\nu}\, \mathrm{d}x^\nu. \tag{4.1}$$

Statement (4.1) has adopted the *summation convention*, in which repeated indices are understood to be summed. If the repeated indices are Greek, such as μ, ν, and ρ, they are summed from 0 to 3, so that all spacetime components are included in the sum. If the repeated indices are Latin, such as i, j, or k, they are summed from 1 to 3, so that only spatial indices are included in the sum. In other cases the repeated indices are summed over a range that is determined contextually. For example, the action of a 2×2 matrix σ on a 2×1 vector ξ results in another 2×1 vector $\chi = \sigma \cdot \xi$. The elements of χ are given by $\chi_a = \sigma_{ab}\, \xi_b$, where the repeated index b is summed from 1 to 2.

In general, the *contravariant components of any spacetime four-vector* are indicated with a superscript as $V^\mu(x)$. These components are assumed to transform identically to $\mathrm{d}x^\mu$, so that $V'^\mu(x') = (\partial x'^\mu/\partial x^\nu) V^\nu(x)$. On the other hand, inverting the transformation so that $x^\mu = x^\mu(x')$, the chain rule also shows that the components of the gradient ∂, defined as $\partial_\mu = \partial/\partial x^\mu$, transform according to

$$\partial'_\mu \equiv \frac{\partial}{\partial x'^\mu} = \frac{\partial x^\nu}{\partial x'^\mu}\, \partial_\nu. \tag{4.2}$$

These are referred to as the *covariant components of the gradient vector*, and are distinguished from the contravariant components by using a subscript. These two transformation properties are shared by all *spacetime tensors*. In component notation, a general spacetime tensor is written $T^{\mu\nu\ldots}_{\rho\sigma\ldots}(x)$, so that it may have a mixture of contravariant and covariant components. Under a change of coordinates the contravariant and covariant components transform according to equations (4.1) and (4.2).

It is an important observation that the differential, given by the summed product $\mathrm{d} = \mathrm{d}x^\mu\partial_\mu$, is *invariant* under a transformation since the chain rule gives

$$\begin{aligned}\mathrm{d}' = \mathrm{d}x'^\mu\partial'_\mu &= \frac{\partial x^\nu}{\partial x'^\mu}\frac{\partial x'^\mu}{\partial x^\rho}\,\mathrm{d}x^\nu\partial_\rho = \frac{\partial x^\nu}{\partial x^\rho}\mathrm{d}x^\nu\partial_\rho \\ &= \delta^\nu{}_\rho\,\mathrm{d}x^\nu\partial_\rho = \mathrm{d}x^\nu\partial_\nu = \mathrm{d},\end{aligned} \tag{4.3}$$

where $\delta^\nu{}_\rho$ is the Kronecker delta (1.67) in four dimensions. A summed product of contravariant and covariant indices such as equation (4.3) is referred to as an *index contraction*, and the contraction is unchanged under coordinate transformations. A quantity such as equation (4.3) that is invariant under coordinate transformations is referred to as a *scalar* and has a value that all observers can agree upon. A scalar spacetime tensor, denoted $\varphi(x)$, has no contravariant or covariant components. It transforms according to $\varphi'(x') = \varphi(x)$, where x' and x are different coordinates for the *same event*. Because the choice of coordinate systems is arbitrary, Lorentz or relativistic scalars are the building blocks of all physically meaningful relativistic field theories, since all observers will agree on their value at a spacetime point. Combining spacetime tensors into scalars, e.g. combining a vector field into terms such as $A_\mu(x)A^\mu(x)$, allows theories of more complicated tensor fields to be built from Lorentz scalars.

If the speed of light c is invariant for all observers, the *d'Alembertian wave operator* $\Box \equiv \partial_0^2 - \nabla^2$ that governs light waves must be invariant under a coordinate change. This requires that $\Box = \partial_\mu\partial^\mu$, and in turn this requires that the contravariant components of ∂ are given by $\partial^\mu = \eta^{\mu\nu}\partial_\nu$, where η can be viewed a 4×4 matrix η with the elements given by $\eta^{00} = 1$ and $\eta^{ij} = -\delta^{ij}$, all others vanishing. This gives $\partial^0 = \partial_0$ and $\partial^i = -\partial_i$, so that $\partial_\mu\partial^\mu = \partial_0^2 - \partial_i\partial_i = \partial_0^2 - \nabla^2$. The quantity η is known as the *metric tensor*. The elements of η^{-1} are designated $\eta_{\mu\nu}$ and obey the matrix multiplication equation $\eta^{\rho\nu}\eta_{\nu\mu} = \delta^\rho{}_\mu$. The inverse is designated with subscripts since it yields $\eta_{\mu\rho}\partial^\rho = \eta_{\mu\rho}\eta^{\rho\nu}\partial_\nu = \delta_\mu{}^\nu\partial_\nu = \partial_\mu$, providing the covariant components of ∂ from its contravariant components. The metric tensor and its inverse can therefore be used to raise and lower the indices of spacetime tensors. In addition, the generalization of spatial momentum $\boldsymbol{p}$ and kinetic energy E leads to their combination into a four-vector $p = (E/c, \boldsymbol{p})$. For a particle with inertial mass m, the four-momentum satisfies the famous relation $p^2 = p_\mu p^\mu = E^2/c^2 - \boldsymbol{p}\cdot\boldsymbol{p} = m^2c^2$.

It is the central postulate of special relativity that η takes the same form for *all inertial observers*. This property of the metric tensor serves to determine the transformations that relate *inertial frames*. The required invariance of the d'Alembertian and η gives

$$\eta^{\rho\sigma}\partial_\rho\partial_\sigma = \eta^{\mu\nu}\partial'_\mu\partial'_\nu = \eta^{\mu\nu}\frac{\partial x'^\rho}{\partial x^\mu}\frac{\partial x'^\sigma}{\partial x^\nu}\partial_\rho\partial_\sigma \Longrightarrow \eta^{\mu\nu}\frac{\partial x^\rho}{\partial x'^\mu}\frac{\partial x^\sigma}{\partial x'^\nu} = \eta^{\rho\sigma}. \tag{4.4}$$

Because η is independent of x, differentiating equation (4.4) shows that $\partial^2 x'^\rho/\partial x^\mu\,\partial x^\alpha = 0$ or $\partial x'^\rho/\partial x^\mu = \Lambda^\rho{}_\mu$, where $\Lambda^\rho{}_\mu$ is independent of x and is referred to as a *Lorentz transformation*. The infinitesimal Lorentz transformation can be written $\Lambda^\rho{}_\mu = \delta^\rho{}_\mu + \lambda^\rho{}_\mu$, where $O(\lambda^2) \approx 0$. Substituting this into the relation (4.4) and using $\eta^{\mu\nu}\lambda^\sigma{}_\nu\delta^\rho{}_\mu = \lambda^{\sigma\rho}$ and $\eta^{\mu\nu}\lambda^\rho{}_\mu\delta^\sigma{}_\nu = \lambda^{\rho\sigma}$ gives $\lambda^{\rho\sigma} + \lambda^{\sigma\rho} = 0$ to $O(\lambda)$. The 4×4 matrix λ is antisymmetric and therefore has only six independent components. As a result, a general Lorentz transformation possesses six independent degrees of freedom or parameters, corresponding to the three angles θ^k of a spatial rotation and the three components of the rapidity $\phi^i = v^i/c$, where v^i is the relative velocity of two different inertial observers who agree that $x = x' = 0$ describes the same event. If the two observers have different origins, the full solution to equation (4.4) is given by a *Poincaré transformation*, $x'^\mu = \Lambda^\mu{}_\nu x^\nu + b^\mu$, where b^μ is a constant four-vector representing translation of the coordinate system's origin in the new inertial frame. The Lorentz transformations form a *group* since their products yield another Lorentz transformation. They are *orthogonal matrices* since $\Lambda^\mu{}_\rho\Lambda_\mu{}^\sigma x_\sigma x^\rho = x_\sigma x^\sigma$ gives $\Lambda^\mu{}_\rho\Lambda_\mu{}^\sigma = \delta_\rho{}^\sigma$. In matrix notation this gives

$$\mathbf{\Lambda}^T \cdot \mathbf{\Lambda} = \mathbf{I} \implies \det{}^2 \mathbf{\Lambda} = 1 \implies \det \mathbf{\Lambda} = \pm 1. \tag{4.5}$$

Those transformations that satisfy $\det \mathbf{\Lambda} = 1$ are referred to as the *proper Lorentz group*. The transformations satisfying $\det \mathbf{\Lambda} = -1$ allow a change in parity or handedness, $\boldsymbol{x} \to -\boldsymbol{x}$, or time reversal, $x_0 \to -x_0$. They are often referred to as *improper Lorentz transformations.* However, their inclusion is critical to understanding the symmetries of quantum field theory. The six parameter *proper Lorentz group* is known as SO(3, 1), the *special orthogonal group in* 3 + 1 *dimensions.*

Spacetime tensors such as scalar fields $\varphi(x)$ and vector fields $A^\mu(x)$ do not exhaust the types of fields that transform using the Lorentz group. The Lorentz group can be adapted to act on a second type of spacetime object known as a *spinor*. The *Weyl spinor* $\xi(x)$ is a two component complex field with the components designated as $\xi^a(x)$, where $a = 1, 2$. The Weyl spinor transforms as $\xi'(x') = \mathbf{U} \cdot \xi(x)$, where $\mathbf{U}$ is a 2×2 complex matrix representing the action of the Lorentz transformation on the spinor, which is given by matrix multiplication of the spinor, $\mathbf{U} \cdot \xi(x)$. The individual components of the spinor transformation can be written using the summation convention as $\xi'^a(x') = U^{ab}\xi^b(x)$, where the sum runs over 1 and 2 for a Weyl spinor. Since $\mathbf{U}$ is a 2×2 complex valued matrix, it has a total of eight free parameters, while the Lorentz transformation has six parameters. Therefore, two additional conditions are required to reduce the number of free parameters in $\mathbf{U}$ down to the six parameters of the Lorentz group, and these two conditions are given by demanding that $\det \mathbf{U} = 1$. This requires the imaginary part of $\det \mathbf{U}$ to vanish, while the real part must be one. It also guarantees that $\mathbf{U}$ has an inverse, which is necessary to represent the Lorentz group. If both $\mathbf{U}_1$ and $\mathbf{U}_2$ have determinant unity, then

$$\mathbf{U} = \mathbf{U}_1\mathbf{U}_2 \implies \det \mathbf{U} = \det(\mathbf{U}_1\mathbf{U}_2) = \det \mathbf{U}_1 \det \mathbf{U}_2 = 1. \tag{4.6}$$

As a result, these matrices form a group known as SL(2, C), the *special linear group* of 2×2 complex matrices with determinant one.

A representation of the Lorentz group for use with the Weyl spinor begins with the 2×2 *Pauli spin matrices* σ^i, given by

$$\sigma^1 = \begin{pmatrix} 0 & 1 \\ 1 & 0 \end{pmatrix}, \quad \sigma^2 = \begin{pmatrix} 0 & -i \\ i & 0 \end{pmatrix}, \quad \sigma^3 = \begin{pmatrix} 1 & 0 \\ 0 & -1 \end{pmatrix}. \tag{4.7}$$

These matrices have the properties $\frac{1}{2}\{\sigma^i, \sigma^j\} = \delta^{ij}\mathbf{1}$ and $\frac{1}{2}[\sigma^i, \sigma^j] = i\varepsilon^{ijk}\sigma^k$. They are combined with the identity matrix $\sigma^0 = \mathbf{1}$ into two sets of 2×2 matrices $\sigma^\mu = (\sigma^0, \boldsymbol{\sigma})$ and $\bar{\sigma}^\mu = (\sigma^0, -\boldsymbol{\sigma})$. Both of these combinations are Hermitian, $\sigma^{\mu\dagger} = \sigma^\mu$ and $\bar{\sigma}^{\mu\dagger} = \bar{\sigma}^\mu$. These two sets of matrices satisfy the critical anticommutator identity $\frac{1}{2}\{\bar{\sigma}^\mu, \sigma^\nu\} = \eta^{\mu\nu}\,\mathbf{1}$. These two sets of matrices can be contracted with a covariant four-vector x_μ to define two distinct 2×2 matrices $x_\mu\sigma^\mu = x_0\sigma^0 + x_i\sigma^i$ and $x_\mu\bar{\sigma}^\mu = x_0\sigma^0 - x_i\sigma^i$. The product $x_\mu\sigma^\mu x_\nu\bar{\sigma}^\nu$ is given by

$$\begin{aligned} \sigma^\mu x_\mu \bar{\sigma}^\nu x_\nu &= \sigma^0\sigma^0 x_0 x_0 + [\sigma^0, \sigma^i]x_0 x_i - \frac{1}{2}\{\sigma^i, \sigma^j\}x_i x_j - \frac{1}{2}[\sigma^i, \sigma^j]x_i x_j \\ &= (x_0 x_0 - x_i x_i)\mathbf{1} = x_\mu x^\mu \mathbf{1}, \end{aligned} \tag{4.8}$$

where the properties of the Pauli spin matrices were used. The explicit form of the Pauli spin matrices gives the important result

$$\det(x_\mu\sigma^\mu) = \det\begin{pmatrix} x_0 + x_3 & x_1 - ix_2 \\ x_1 + ix_2 & x_0 - x_3 \end{pmatrix} = x_0^2 x_0 - \mathbf{x}^2 = x^\mu x_\mu, \tag{4.9}$$

along with the identical result $\det(x_\mu\bar{\sigma}^\mu) = x^\mu x_\mu$. Both determinants therefore yield expressions invariant under Lorentz transformations.

The *spinor Lorentz transformation* $\mathbf{U}$ corresponding to the inverse spacetime Lorentz transformation $(\Lambda^{-1})^\nu_\mu$ is defined as the member of SL(2, C) that satisfies the relation

$$\mathbf{U}^\dagger\sigma^\mu\mathbf{U} = \sigma^\nu(\Lambda^{-1})^\mu_\nu \implies \mathbf{U}^\dagger\sigma^\mu x'_\mu\mathbf{U} = \sigma^\nu(\Lambda^{-1})^\mu_\nu x'_\mu = \sigma^\nu x_\nu. \tag{4.10}$$

The expression (4.10) is understood as matrix multiplication, so that $\mathbf{U}\sigma^\mu = \mathbf{U}\cdot\sigma^\mu$. The elements of the matrix product can be expressed using the summation convention as $(\mathbf{U}\sigma^\mu)_{ab} = \mathbf{U}_{ac}\sigma^\mu_{cb}$. Combining equations (4.9) and (4.10) with $\det\mathbf{U} = \det\mathbf{U}^\dagger = 1$ shows that

$$\begin{aligned} x_\mu x^\mu &= \det(x_\mu\sigma^\mu) = \det(\mathbf{U}^\dagger x'_\mu\sigma^\mu\mathbf{U}) \\ &= \det\mathbf{U}^\dagger \det(x'_\mu\sigma^\mu)\det\mathbf{U} = \det(x'_\mu\sigma^\mu) = x'_\mu x^{\mu'}. \end{aligned} \tag{4.11}$$

This shows that the action of an SL(2, C) transformation leaves $x^\mu x_\mu$ invariant, and so it corresponds to a Lorentz transformation. For a given $\mathbf{U}$, the associated Lorentz transformation is found by combining equation (4.10) with $\frac{1}{2}\mathrm{Tr}(\bar{\sigma}^\mu\sigma^\nu) = \eta^{\mu\nu}$, $\bar{\sigma}_\alpha = \eta_{\alpha\beta}\bar{\sigma}^\beta$, and $\frac{1}{2}\mathrm{Tr}(\sigma^\mu\bar{\sigma}_\nu) = \delta^\mu{}_\nu$ to find

$$(\Lambda^{-1})^\nu_\mu x_\nu = \frac{1}{2}\,\mathrm{Tr}\left(\mathbf{U}^\dagger\sigma^\nu\mathbf{U}\bar{\sigma}_\mu\right)x_\nu \implies (\Lambda^{-1})^\nu_\mu = \frac{1}{2}\,\mathrm{Tr}(\bar{\sigma}_\mu\mathbf{U}^\dagger\sigma^\nu\mathbf{U}). \tag{4.12}$$

This constitutes a *mapping* from SL(2, C) to the Lorentz group. Using the rapidity $\phi^i = v^i/c$ and the three angles θ^k of a rotation, a general member of SL(2, C) can be written as $\mathbf{U}(\phi, \theta) = \exp\left\{\frac{1}{2}i\sigma_j(\theta^j + i\phi^j)\right\}$.

These results show that the bilinear form $\xi^\dagger\sigma^\mu\partial_\mu\xi$ is Lorentz invariant since equation (4.10) gives

$$\begin{aligned}\xi^{\dagger'}(x')\sigma^\mu\partial'_\mu\xi'(x') &= \xi^\dagger(x)\mathbf{U}^\dagger\sigma^\mu\partial'_\mu\mathbf{U}\xi(x) \\ &= \xi^\dagger(x)\sigma^\nu(\Lambda^{-1})^\mu_\nu\partial'_\nu\xi(x) \\ &= \xi^\dagger(x)\sigma^\nu\partial_\nu\xi(x)\end{aligned} \tag{4.13}$$

However, $\mathbf{U}$ is not unitary since $\mathbf{U}^\dagger = \exp\left\{-\frac{1}{2}i\sigma_j(\theta^j - i\phi^j)\right\} \neq \mathbf{U}^{-1}$. As a result, the bilinear form $\xi^\dagger\xi$ is not Lorentz invariant. It will be seen in the next section that this renders it impossible to associate a mass with a Weyl spinor. The second is that the choice $\sigma^\mu = (\sigma^0, \boldsymbol{\sigma})$ can be replaced by $\bar{\sigma}^\mu = (\sigma^0, -\boldsymbol{\sigma})$ to obtain *another* representation of the Lorentz group. These two sets of matrices are related by a parity transformation, $\boldsymbol{x} \to -\boldsymbol{x}$. It can be shown that there is no 2×2 matrix P that gives $P^\dagger\sigma^i P = -\sigma^i$ for all i, so these two matrices are *not* unitarily equivalent. This means that there is no transformation from a right to a left-handed coordinate system when Weyl spinors are used. As a result, Weyl spinors are limited to modeling massless left- or right-handed particles, but not particles that are both.

Incorporating parity transformations and a mass term requires a *bispinor*, ψ, often referred to as a *Dirac spinor* whose components are designated ψ^a, where $a = 1, \ldots, 4$. The σ^μ are combined with the 2×2 zero matrix $\mathbf{0}$ to give four 4×4 *Dirac matrices* γ^μ,

$$\psi = \begin{pmatrix}\psi^1 \\ \vdots \\ \psi^4\end{pmatrix}, \quad \gamma^0 = \begin{pmatrix}\sigma^0 & \mathbf{0} \\ \mathbf{0} & -\sigma^0\end{pmatrix}, \quad \gamma^i = \begin{pmatrix}\mathbf{0} & \sigma^i \\ -\sigma^i & \mathbf{0}\end{pmatrix}. \tag{4.14}$$

This choice for the γ^μ is referred to as the *Dirac representation* and is not unique since the γ^μ can be redefined with a unitary transformation. However, it is assumed that all observers agree upon the choice of representation and so the γ^μ are identical in all reference frames. The properties of the Pauli spin matrices show that the Dirac matrices satisfy $\gamma^{i\dagger} = -\gamma^i$, $\gamma^{0\dagger} = \gamma^0$, $\mathrm{Tr}\gamma^\mu = 0$, and $\frac{1}{4}\gamma_\mu\gamma^\mu = \mathbf{1}$, as well as the critical identity $\frac{1}{2}\{\gamma^\mu, \gamma^\nu\} = \eta^{\mu\nu}\mathbf{1}$, all of which are invariant under a unitary transformation. Contracting the spacetime vector ∂_μ with the γ^μ gives a spinor valued differential operator, denoted $\gamma^\mu\partial_\mu \equiv \not{\partial}$. It follows that

$$\not{\partial}\not{\partial} = \frac{1}{2}\{\gamma^\mu, \gamma^\nu\}\partial_\mu\partial_\nu + \frac{1}{2}[\gamma^\mu, \gamma^\nu]\partial_\mu\partial_\nu = \mathbf{1}\eta^{\mu\nu}\partial_\mu\partial_\nu = 1\Box, \tag{4.15}$$

where $[\gamma^\mu, \gamma^\nu]\partial_\mu\partial_\nu = [\gamma^\nu, \gamma^\mu]\partial_\nu\partial_\mu = -[\gamma^\mu, \gamma^\nu]\partial_\mu\partial_\nu$ shows that the anticommutator term vanishes. The spinor differential operator transforms under the Lorentz group similarly to the Weyl spinor (4.10),

$$\not{\partial} = \mathbf{U}^{-1}\not{\partial}'\mathbf{U} = (\Lambda^{-1})^{\nu}_{\mu}\gamma^{\mu}\partial'_{\nu} \tag{4.16}$$

where $\mathbf{U}$ is a 4×4 matrix satisfying the required group property $\det \mathbf{U} = 1$. The relation (4.16) has the important difference that $\mathbf{U}^{-1}$ is present, rather than $\mathbf{U}^{\dagger}$, allowing the Lorentz transformation to be unitary. It can be shown that equation (4.16) is consistent with $\mathbf{U} = \exp(\frac{1}{4}i\lambda_{\mu\nu}\sigma^{\mu\nu})$. In this expression the $\lambda_{\mu\nu}$ are the elements of an antisymmetric matrix encapsulating the six Lorentz group parameters found earlier for the infinitesimal case, while $\sigma^{\mu\nu}$ is the antisymmetric 4×4 matrix $\sigma^{\mu\nu} = [\gamma^{\mu}, \gamma^{\nu}]$. Because the γ^{μ} are traceless, it follows that $\det \mathbf{U} = 1$. This is proved using the matrix identity $\det \mathbf{U} = \exp\{\mathrm{Tr} \ln \mathbf{U}\}$. It can be shown that combining $\gamma^{j\dagger}\gamma^{0} = \gamma^{0}\gamma^{j}$ with the definition of $\mathbf{U}$ gives the important property $\mathbf{U}^{\dagger}\gamma^{0} = \gamma^{0}\mathbf{U}^{-1}$.

The Dirac spinor transforms as $\psi'(x') = \mathbf{U}\psi(x) \equiv \mathbf{U} \cdot \psi(x)$, where the dot for matrix multiplication will be suppressed going forward. The *conjugate spinor* is defined as $\bar{\psi}(x) = \psi^{\dagger}(x)\gamma^{0}$, and it transforms as $\bar{\psi}'(x') = \psi^{\dagger}(x)\mathbf{U}^{\dagger}\gamma^{0} = \bar{\psi}(x)\mathbf{U}^{-1}$. As a result, the bilinear $\bar{\psi} \cdot \psi \equiv \bar{\psi}\psi = \bar{\psi}^{a}\psi^{a}$ (summation convention) is a Lorentz invariant quantity. Like the Weyl spinor, the Lorentz transformation is given by $(\Lambda^{-1})^{\mu}{}_{\nu} = \frac{1}{4}\mathrm{Tr}(\mathbf{U}^{-1}\gamma^{\mu}\mathbf{U}\gamma_{\nu})$, and this results in $i\bar{\psi}\not{\partial}\psi$ being a Lorentz scalar,

$$i\bar{\psi}'\not{\partial}'\psi' = i\bar{\psi}\mathbf{U}^{-1}\gamma^{\mu}\mathbf{U}\partial'_{\mu}\psi = i\bar{\psi}\gamma^{\nu}(\Lambda^{-1})^{\mu}{}_{\nu}\partial'_{\mu}\psi = i\bar{\psi}\not{\partial}\psi. \tag{4.17}$$

Using both $\bar{\psi}\psi$ and $i\,\bar{\psi}\not{\partial}\psi$ allows a theory of massive spinor particles to be discussed in the next section.

4.2 Action functionals and relativistic fields

The action functional (1.1) for particles can be generalized to fields by treating the field at each point as an independent degree of freedom. A generic spacetime field is designated as $\varphi^{\alpha}(x)$, where α represents a set of tensor or spinor indices with well defined transformation properties under the Lorentz group. The action functional $S[\varphi, \partial\varphi]$ is written as a spacetime integral, and for one generic field in four-dimensional spacetime it takes the form

$$S[\varphi, \partial\varphi] = \int \mathrm{d}^4x\ \mathcal{L}(\varphi^{\alpha}(x), \partial_{\mu}\varphi^{\alpha}(x)). \tag{4.18}$$

For the moment, the two constants $\hbar$ and c will be kept in the following definitions in order to clarify units. The Lagrangian density is constructed so that the units of S are the same as $\hbar$, and so its units are those of $\hbar$ over four-dimensional volume. It is possible to develop Lorentz transformations in arbitrary dimensions, but that will not be considered until chapter 5. It is important to note that the spacetime volume element of the action integral is invariant under a Lorentz transformation since $\mathrm{d}^4x' = \mathrm{d}^4x \det \mathbf{\Lambda} = \mathrm{d}^4x$. Specific forms for the Lagrangian density $\mathcal{L}$ will be analyzed, but all of them will have the general property that only the fields and their gradient are present in $\mathcal{L}$.

Like the classical mechanics case (1.4), the Euler–Lagrange equation for fields is found by demanding the variation of the action around the classical solution φ_c^{α} vanishes, where the variation is defined by $\varphi^{\alpha} \to \varphi_c^{\alpha} + \delta\varphi^{\alpha}$ with $\delta\varphi^{\alpha}$ infinitesimal and vanishing at the boundaries of the spacetime volume. Using $\delta(\partial_\mu\varphi^\alpha) = \partial_\mu\delta\varphi^\alpha$ and integrating by parts, this requirement gives

$$\delta S = S[\varphi^\alpha + \delta\varphi^\alpha] - S[\varphi^\alpha] = \int \mathrm{d}^4x\, \delta\varphi^\alpha \left(\frac{\partial\mathcal{L}}{\partial\varphi^\alpha} - \partial_\mu \frac{\partial\mathcal{L}}{\partial(\partial_\mu\varphi^\alpha)} \right)\Bigg|_{\varphi_c^\alpha} = 0. \tag{4.19}$$

Since $\delta\varphi^\alpha$ is arbitrary, this yields the Euler–Lagrange equation, given by

$$\frac{\partial\mathcal{L}}{\partial\varphi^\alpha} - \partial_\mu \frac{\partial\mathcal{L}}{\partial(\partial_\mu\varphi^\alpha)} = 0, \tag{4.20}$$

which must hold at every spacetime point.

Treating $\varphi^\alpha(\boldsymbol{x}, t)$ as an independent degree of freedom for each spatial point $\boldsymbol{x}$ allows the *momentum density* canonically conjugate to $\varphi^\alpha(\boldsymbol{x}, t)$ to be defined similarly to the classical mechanics case,

$$\pi_\alpha(\boldsymbol{x}, t) = \frac{\delta S}{\delta\dot{\varphi}^\alpha(\boldsymbol{x}, t)}. \tag{4.21}$$

where the relativistic field notation $\dot{\varphi}^\alpha = \partial\varphi^\alpha/\partial x^0 = \partial\varphi^\alpha/\partial(ct)$ is now in use. It should be noted that equation (4.21) can differ from $\partial\mathcal{L}/\partial\dot{\varphi}^\alpha$ since the definition of the functional derivative (1.28) incorporates terms that are related by an integration by parts. It is assumed that equation (4.21) can be solved to give $\dot{\varphi}^\alpha = \dot{\varphi}^\alpha(\varphi^\alpha, \pi_\alpha)$ at each spacetime point. Similarly to the classical Hamiltonian, the associated Hamiltonian density $\mathcal{H}$ or energy per unit spatial volume is defined as

$$\mathcal{H}(\varphi^\alpha, \nabla\varphi^\alpha, \pi^\alpha) = c\, \pi_\alpha\, \dot{\varphi}^\alpha(\varphi^\alpha, \pi^\alpha) - c\, \mathcal{L}(\varphi^\alpha, \nabla\varphi^\alpha, \dot{\varphi}^\alpha(\varphi^\alpha, \pi^\alpha)), \tag{4.22}$$

where there is an implicit sum over the repeated α index in the product $\pi_\alpha\dot{\varphi}^\alpha$. The factor of c is needed in order that $\mathcal{H}$ has the units of a spatial energy density. This follows from noting that $\mathcal{L}$ has the units of $\hbar$ over the fourth power of length in four dimensions and so $c\,\mathcal{L}$ has the units of energy per three-dimensional volume. The Hamiltonian H is then found by integrating the Hamiltonian density over the spatial volume V, so that

$$H = \int_V \mathrm{d}^3x\, \mathcal{H}(\varphi^\alpha, \nabla\varphi^\alpha, \pi^\alpha). \tag{4.23}$$

If equation (4.23) is evaluated with a solution to the Euler–Lagrange equation (4.20) that vanishes at the boundary of the spatial volume V, then it is straightforward to show that $\dot{H} = 0$ and the energy of the field configuration is conserved. Varying $\mathcal{L} = c\, \pi_\alpha\dot{\varphi}^\alpha - \mathcal{H}$ yields the functional version of Hamilton's equations (1.11),

$$\frac{\delta H}{\delta\pi_\alpha(x)} = c\, \dot{\varphi}^\alpha(x), \qquad \frac{\delta H}{\delta\varphi^\alpha(x)} = -c\, \dot{\pi}_\alpha(x). \tag{4.24}$$

The functional derivative reduces the units of the Hamiltonian by a factor of spatial volume as well as by the units of the field quantity, either φ^α or π_α.

A *free field* obeys the law of superposition and therefore satisfies a linear version of the Euler–Lagrange equation. The solutions to the free field equations are often used to construct a perturbative solution to related *nonlinear* equations of motion, and these solutions are interpreted as modeling *interacting fields*. There are three basic Lagrangians that define free scalar, bispinor, and vector fields. These Lagrangians are constructed using quadratic Lorentz scalars built from the respective fields since this will give an Euler–Lagrange equation that is both linear *and covariant*, taking the same form in any inertial frame. The latter property is the basic requirement for a fundamental law of nature consistent with special relativity.

4.2.1 The free scalar field

The Lagrangian for a free real scalar field $\varphi(x)$ associated with the inertial mass m is given by combining two relativistic scalars,

$$\mathcal{L}(\varphi,\, \partial\varphi) = \hbar\left(\frac{1}{2}\eta^{\mu\nu}\partial_\mu\varphi\, \partial_\nu\varphi - \frac{m^2c^2}{2\hbar^2}\varphi^2\right). \tag{4.25}$$

In four dimensions the units associated with φ must be inverse length in order that both terms in the parentheses have the units of inverse length to the fourth power. The Euler–Lagrange equation associated with equation (4.25) is given by the linear partial differential equation

$$\partial_\mu\frac{\partial\mathcal{L}}{\partial(\partial_\mu\varphi)} - \frac{\partial\mathcal{L}}{\partial\varphi} = \hbar\left(\eta^{\mu\nu}\partial_\mu\partial_\nu + \frac{m^2c^2}{\hbar^2}\right)\varphi = \hbar\left(\Box + \frac{m^2c^2}{\hbar^2}\right)\varphi = 0, \tag{4.26}$$

which is known as the *Klein–Gordon equation*. Using the definition (4.21), the momentum canonically conjugate to φ is given by $\pi = \hbar\dot{\varphi}$. The Hamiltonian density (4.22) is therefore

$$\mathcal{H}(\varphi,\, \nabla\varphi,\, \pi) = \frac{c}{2\hbar}\pi^2 + \frac{\hbar c}{2}\nabla\varphi\cdot\nabla\varphi + \frac{m^2c^3}{2\hbar}\varphi^2. \tag{4.27}$$

The signs in the Lagrangian (4.25) were chosen so that equation (4.27) is positive-definite. Combining equation (4.24) with $\pi(x) = \hbar\dot{\varphi}$ gives

$$c\,\dot{\pi}(x) = \hbar c\,\ddot{\varphi} = -\frac{\delta H}{\delta\varphi(x)} = \hbar c\,\nabla^2\varphi - \frac{m^2c^3}{\hbar}\varphi, \tag{4.28}$$

which reproduces equation (4.26).

4.2.2 The free bispinor field

The Lagrangian density for the free bispinor field $\psi(x)$ of equation (4.14) associated with the inertial mass m is given by combining the two previously discussed relativistic scalars that are quadratic in the bispinor fields,

$$\mathcal{L}(\bar{\psi}, \psi, \partial\bar{\psi}, \partial\psi) = \hbar\left(\frac{1}{2}i\bar{\psi}\gamma^{\mu}(\partial_{\mu}\psi) - \frac{1}{2}i(\partial_{\mu}\bar{\psi})\gamma^{\mu}\psi - \frac{mc}{\hbar}\bar{\psi}\psi\right). \tag{4.29}$$

In four dimensions the units of $\bar{\psi}\psi$ are inverse spatial volume. The first two terms involving the spinor gradient are related by an integration by parts. Both are present in order that equation (4.29) is real valued. The demonstration that equation (4.29) is real begins by noting $\bar{\psi}^{\dagger} = (\psi^{\dagger}\gamma^{0})^{\dagger} = \gamma^{0\dagger}\psi = \gamma^{0}\psi$. Combining this result with $\gamma^{i\dagger}\gamma^{0\dagger} = -\gamma^{i}\gamma^{0} = \gamma^{0}\gamma^{i}$ and $\gamma^{0\dagger}\gamma^{0\dagger} = \gamma^{0}\gamma^{0}$ shows that $(i\bar{\psi}\gamma^{\mu}\partial_{\mu}\psi)^{*} = -i\partial_{\mu}\psi^{\dagger}\gamma^{\mu\dagger}\gamma^{0\dagger}\psi = -i(\partial_{\mu}\bar{\psi})\gamma^{\mu}\psi$. The Euler–Lagrange equation gives

$$\frac{\partial\mathcal{L}}{\partial\bar{\psi}} - \partial_{\mu}\frac{\partial\mathcal{L}}{\partial(\partial_{\mu}\bar{\psi})} = \hbar\left(i\gamma^{\mu}\partial_{\mu}\psi - \frac{mc}{\hbar}\psi\right) = 0, \tag{4.30}$$

and the conjugate equation

$$\frac{\partial\mathcal{L}}{\partial\psi} - \partial_{\mu}\frac{\partial\mathcal{L}}{\partial(\partial_{\mu}\psi)} = \hbar\left(-i\partial_{\mu}\bar{\psi}\gamma^{\mu} - \frac{mc}{\hbar}\bar{\psi}\right) = 0, \tag{4.31}$$

both of which are known as the *Dirac equation*. The conjugate equation can be obtained from equation (4.30) by taking its Hermitian adjoint. The momentum density canonical to ψ is given by $\pi = \frac{1}{2}i\hbar\psi^{\dagger}$, while the momentum density canonically conjugate to $\bar{\psi}$ is $\bar{\pi} = -\frac{1}{2}i\hbar\gamma^{0}\psi$. It follows that $\bar{\pi}^{\dagger}\gamma^{0} = \frac{1}{2}i\hbar\psi^{\dagger} = \pi$. It is not possible to solve these relations to express $\dot{\psi}$ in terms of π and ψ. However, since the Lagrangian (4.29) is linear in the spinor gradient, the Hamiltonian density is independent of time derivatives and is given by

$$\begin{aligned}\mathcal{H} = c\pi(\partial_{0}\psi) + c(\partial_{0}\bar{\psi})\bar{\pi} - c\mathcal{L} = -\frac{1}{2}i\hbar c\; \bar{\psi}\gamma^{j}(\partial_{j}\psi) \\ + \frac{1}{2}i\hbar c\; (\partial_{j}\bar{\psi})\gamma^{j}\psi + mc^{2}\bar{\psi}\psi.\end{aligned} \tag{4.32}$$

The Hamiltonian density of equation (4.32) is real valued. Like the scalar field, Hamilton's equations (4.24) reproduce the equations of motion (4.30) and (4.31) by treating ψ and $\bar{\psi}$ as independent degrees of freedom.

4.2.3 The free vector or gauge field

The third basic field is the vector or *gauge field* $a^{\mu}(x)$, which has the units of inverse length. The free field Lagrangian is constructed using the associated *antisymmetric second rank tensor* $F_{\mu\nu} = \partial_{\mu}a_{\nu} - \partial_{\nu}a_{\mu}$,

$$\mathcal{L} = -\frac{1}{4}\hbar F_{\mu\nu}F^{\mu\nu} = -\frac{1}{2}\hbar\; \partial_{\mu}a_{\nu}\; \partial^{\mu}a^{\nu} + \frac{1}{2}\hbar\; \partial_{\mu}a_{\nu}\partial^{\nu}a^{\mu}. \tag{4.33}$$

The contravariant components a^{μ} are given by $(a_{0}, \boldsymbol{a})$, where the electric field components are identified as $\mathrm{E}_{i} = -E^{i} = -\partial_{i}a^{0} - \partial_{0}a^{i} = \partial^{i}a^{0} - \partial^{0}a^{i} = F^{i0}$ and the magnetic field components are $B_{i} = \varepsilon_{ijk}\partial_{j}a^{k} = -\varepsilon_{ijk}\partial_{j}a_{k}$, so that $F_{ij} = -\varepsilon_{ijk}B_{k}$. Both

$F_{\mu\nu}$ and the Lagrangian (5.51) are *gauge invariant*, which means that they are unchanged by an *abelian gauge transformation*, $a_\mu \to a_\mu - \partial_\mu \Lambda$, where Λ is an arbitrary dimensionless spacetime scalar function satisfying $\partial_\mu \partial_\nu \Lambda = \partial_\nu \partial_\mu \Lambda$ and vanishing at the boundaries of spacetime. It is possible to add a Lorentz invariant mass term proportional to $a_\mu a^\mu$, but such a term is *not gauge invariant* and creates difficulties in the quantum version of the theory. The equation of motion is

$$\frac{\partial \mathcal{L}}{\partial a^\nu} - \partial_\mu \frac{\partial \mathcal{L}}{\partial(\partial_\mu a^\nu)} = \hbar\left(\Box a_\nu - \partial_\nu \partial_\mu a^\mu\right) = 0, \tag{4.34}$$

where the identification $a_\mu = \eta_{\mu\nu} a^\nu$ was used in the Euler–Lagrange equation. The canonical momentum is given by $\pi_\mu = -\hbar(\partial_0 a_\mu - \partial_\mu a_0) = -\hbar F_{0\mu}$, which vanishes for $\mu = 0$ and leaves π_0 undefined. The Hamiltonian density is found by combining $\dot{a}_i = \partial_i a_0 - \pi_i/\hbar$ with $\mathcal{H} = c\,\pi_\mu \dot{a}^\mu - c\,\mathcal{L} = c\,\pi_i \dot{a}^i - c\,\mathcal{L} = -c\,\pi_i \dot{a}_i - c\,\mathcal{L}$ to find

$$\mathcal{H} = -c\,\pi_i \dot{a}_i - c\,\mathcal{L} = \frac{c}{2\hbar}\pi_i \pi_i + \frac{\hbar c}{2}\left(\partial_j a^i \partial_j a^i - \partial_j a^i \partial_i a^j\right) - c\,\pi_i \partial_i a_0. \tag{4.35}$$

The first of Hamilton's equations of motion (4.24) gives

$$\frac{\delta H}{\delta \pi_i} = \frac{c}{\hbar}\pi_i - c\,\partial_i a_0 = -c\,(\dot{a}_i - \partial_i a_0) - c\,\partial_i a_0 = -c\,\dot{a}_i = c\,\dot{a}^i, \tag{4.36}$$

which is the predicted result. Enforcing the second of Hamilton's equations gives

$$\dot{\pi}_i = -\frac{\delta H}{\delta a^i} \implies -\hbar c(\ddot{a}_i - \partial_i \dot{a}_0) = -\hbar c\left(\nabla^2 a_i + \partial_i \partial_j a^j\right), \tag{4.37}$$

which can be written $\Box a_i - \partial_i \partial_\mu a^\mu = 0$, the equation of motion (4.34) for a_i. The first of Hamilton's equations gives $\delta H/\delta \pi_0 = 0$, which requires $\dot{a}_0 = 0$. However, the second of Hamilton's equations, $\dot{\pi}_0 = -\delta H/\delta a_0$, breaks down since π_0 is undefined, and it is therefore not possible to derive the equation of motion as was done in equation (4.37). Setting $\dot{\pi}_0 = 0$ gives

$$c\,\dot{\pi}_0 = -\frac{\delta H}{\delta a^0} = -c\,\partial_i \pi_i = \hbar c(\partial_i \dot{a}_i - \nabla^2 a_0) = \hbar c\left(\Box a_0 - \partial_0 \partial_\mu a^\mu\right) = 0, \tag{4.38}$$

which is the correct equation of motion (4.34) for a_0.

The combination of gauge invariance and π_0 being undefined means that the solutions of the equations of motion for a_μ *are not unique*. The simultaneous solution to both of these problems is to add a gauge fixing term to the Lagrangian (5.51). The addition of a gauge fixing term defines π_0 in terms of the a_μ, allowing $\dot{\pi}_0 = -\delta H/\delta a_0$ to be a well defined equation of motion. Equally important, it creates a subsidiary or gauge condition, $\pi_0 = 0$, and this will render the solutions to the equations of motion unique by restricting the types of gauge transformations that may be performed. A commonly used gauge fixing term is referred to as the *Landau gauge* and invokes the *Lorenz condition*. It is given by adding the manifestly invariant term $G = -\frac{1}{2}\hbar(\partial_\mu a^\mu)^2$ to the action, which becomes

$$\mathcal{L}_G = -\frac{1}{4}F_{\mu\nu}F^{\mu\nu} - \frac{1}{2}(\partial_\mu a^\mu)^2. \tag{4.39}$$

The modified Lagrangian $\mathcal{L}_G$ now gives $\pi_0 = \delta S/\delta \dot{a}^0 = -\hbar\, \partial_\mu a^\mu$, while the equation of motion (4.34) becomes $\Box a_\mu = 0$ for all four components. However, using $\dot{a}^0 = -\partial_i a^i - \pi_0/\hbar$ in the term $c\, \pi_0 \dot{a}^0$ shows that the Hamiltonian density associated with the gauge fixed Lagrangian is

$$\mathcal{H}_G = \frac{c}{2\hbar}(\pi_i\pi_i - \pi_0\pi_0) + \frac{\hbar c}{2}\left(\partial_j a^i \partial_j a^i - \partial_j a^i \partial_i a^j\right) - c\, \pi_i \partial_i a^0 - c\, \pi_0 \partial_i a^i. \tag{4.40}$$

Taking care to recall that $a^i = -a_i$, it is straightforward to show that the Hamilton equations obtained from the Hamiltonian density (4.40) now give $\delta H_G/\delta\pi_\mu = c\, \dot{a}^\mu$ and $\Box a_\mu = 0$ for all four components. However, the Hamiltonian density (4.40) is *no longer bounded from below* unless the primary *gauge constraint* $\pi_0 = -\hbar\, \partial_\mu a^\mu = 0$ is enforced. Hamilton's second equation gives $\dot{\pi}_0 = -\delta H_G/\delta a_0 = -\partial_j \pi_j$. Enforcing $\dot{\pi}_0 = -\partial_j\pi_j = 0$ as a secondary constraint is necessary for consistency with the primary gauge constraint $\pi_0 = 0$. The secondary constraint can be written

$$G = \dot{\pi}_0 = -\partial_j\pi_j = -\nabla^2 a_0 - \nabla \cdot \dot{a} = 0, \tag{4.41}$$

which is simply Gauss's law in the absence of charges. The method for enforcing these two constraints was developed in the context of the quantum mechanical path integral with a gauge condition and resulted in the path integral (3.238). This approach will be further developed and applied to abelian gauge theory in the final section of this chapter.

4.3 Canonical quantization of free fields

Although this is a monograph on path integral methods, it is important to understand the connection between the functional formulation of field processes and their canonical quantization counterparts. This is of particular importance in gauge theory, where the gauge constraint discovered in the previous section is required to exclude the existence of zero and negative norm states in the quantum theory. Expressing this in the functional formalism achieves a greater clarity when the connection to the underlying Hilbert space is understood. Unfortunately, the need for brevity requires the omission of many details, and the reader is recommended to texts where these details are presented more thoroughly.

Quantizing the three types of free fields is achieved by extending quantum mechanical techniques to the case of fields in a relativistic manner. As mentioned earlier, each point in space $\boldsymbol{x}$ is treated as an individual quantum mechanical degree of freedom playing the role of position. The quantum mechanical quantization condition $[\hat{X}, \hat{P}] = i\hbar$ is adapted by turning the fields into operators and enforcing an *equal time (anti)commutation relation*,

$$[\hat{\varphi}^\alpha(\boldsymbol{x}, t), \hat{\pi}_\beta(\boldsymbol{y}, t)]_\pm = i\hbar f^\alpha{}_\beta\, \delta^3(\boldsymbol{x} - \boldsymbol{y}), \tag{4.42}$$

where it was assumed the fields exist in a three-dimensional space. Since φ^α has the units of inverse length. while π_β has the units of $\hbar$ per unit length squared, expression (4.42) has the correct units as long as the $f^\alpha{}_\beta$ are dimensionless constants. The coefficients $f^\alpha{}_\beta$ and commutation or anticommutation relation are chosen to be consistent with the type of field and the need for relativistic covariance. For the case of a single scalar field commutation relations are chosen and $f^\alpha{}_\beta = 1$. For the case of a single bispinor field anticommutation relations are chosen and $f^\alpha{}_\beta = \delta^\alpha{}_\beta$, where α and β are the spinor indices. For the case of a single gauge field commutation relations are chosen and $f^\alpha{}_\beta = \eta^\alpha{}_\beta$. In all three cases the $f^\alpha{}_\beta$ are relativistically invariant.

In the remainder of this monograph *natural or Planck units* will be employed. Along with c and $\hbar$, the gravitational constant G, the Coulomb constant K, and Boltzmann's constant k_B are typically set to unity. This greatly simplifies formulas such as the relativistic mass and energy formula, which becomes $E^2 = \boldsymbol{p}^2c^2 + m^2c^4 \to \boldsymbol{p}^2 + m^2$. In such units mass and momentum are typically measured in units of energy or inverse length. For example, mass m and inverse length $1/L$ are related through the Compton wavelength formula, $m = \hbar/Lc$, so that setting $\hbar = c = 1$ gives $m = 1/L$. This allows masses to be measured in inverse lengths or energies and lengths and times to be measured in masses or energies. If e is the fundamental unit of electric charge, then the quantity $Ke^2/\hbar c \to e^2$ is a dimensionless quantity in natural units and given by $\approx 1/137$.

4.3.1 Quantizing the free scalar field

The quantization condition (4.42) for a real scalar field φ is enforced by using a wave-packet solution to the equations of motion. In the Heisenberg picture the field is given the Fourier or *mode decomposition*

$$\hat{\varphi}(\boldsymbol{x}, t) = \int \frac{\mathrm{d}^3p}{(2\pi)^{3/2}\sqrt{2\varepsilon_p}}\left(\hat{a}_p\, e^{-ip\cdot x} + \hat{a}_p^\dagger\, e^{ip\cdot x}\right), \tag{4.43}$$

where $p_0 = \varepsilon_p = \sqrt{m^2 + \boldsymbol{p}^2}$ and the relativistic shorthand $p \cdot x = p_\mu x^\mu$ is in use. Because of the *dispersion relation* $\varepsilon_p^2 = m^2 + \boldsymbol{p}^2$, it follows that $(\Box + m^2)\hat{\varphi} = 0$. Although it is not obvious, the combination $\mathrm{d}^3p/\sqrt{\varepsilon_p}$ is a Lorentz scalar since it can be obtained from the Lorentz invariant quantity $\mathrm{d}^4p\ \delta(p_0^2 - \boldsymbol{p}^2 - m^2)\,\theta(p_0)$ by performing the integration over p_0 in conjunction with the property (1.19) of Dirac functions. Using $\pi = \dot{\varphi}$ gives

$$\hat{\pi}(\boldsymbol{x}, t) = -i\int \frac{\mathrm{d}^3p}{(2\pi)^{3/2}}\sqrt{\frac{\varepsilon_p}{2}}\left(\hat{a}_p\, e^{-ipx} - \hat{a}_p^\dagger\, e^{ipx}\right). \tag{4.44}$$

Using equations (4.43), (4.44), and (1.26) shows that the commutation relations

$$[\hat{a}_p, \hat{a}_k^\dagger] = \delta^3(\boldsymbol{p} - \boldsymbol{k}), \quad [\hat{a}_p, \hat{a}_k] = 0, \quad [\hat{a}_p^\dagger, \hat{a}_k^\dagger] = 0, \tag{4.45}$$

yield the correct canonical quantization condition for scalar fields,

$$[\hat{\varphi}(\boldsymbol{x}, t), \hat{\pi}(\boldsymbol{y}, t)] = \frac{i}{2}\int \frac{d^3p}{(2\pi)^3}(e^{ip(x-y)} + e^{-ip(x-y)}) = i\,\delta^3(\boldsymbol{x} - \boldsymbol{y}). \tag{4.46}$$

It also follows that using equations (4.43) and (4.44) in the scalar Hamiltonian density (4.27) yields, after considerable algebra, the Hamiltonian operator for the scalar field,

$$\hat{H} = \int d^3x\left(\frac{1}{2}\hat{\pi}^2 + \frac{1}{2}\nabla\hat{\varphi}\cdot\nabla\hat{\varphi} + \frac{1}{2}m^2\hat{\varphi}^2\right) = \int d^3p\ \varepsilon_p\left(\hat{a}_p^\dagger\,\hat{a}_p + \frac{1}{2}\right). \tag{4.47}$$

For each value of $\boldsymbol{p}$ the quantization condition (4.45) is identical to the quantization condition (2.72) for the simple harmonic oscillator creation and annihilation operators. Like the harmonic oscillator, the states associated with the scalar field operator are defined by introducing the ground state or *vacuum*, denoted $|0\rangle$, which satisfies $\hat{a}_p|0\rangle = 0$ for all $\boldsymbol{p}$. The vacuum can be understood as a tensor product of ground states for each of the individual $a_{\boldsymbol{p}}$. The multiparticle states of the system are given by $|n_{p_1}, \ldots, n_{p_N}\rangle = (n_{p_1}!)^{-1/2}(\hat{a}_{p_1}^\dagger)^{n_{p_1}}\cdots(n_{p_N}!)^{-1/2}(\hat{a}_{p_N}^\dagger)^{n_{p_N}}|0\rangle$, where n_{p_j} is the number of particles with the momentum $\boldsymbol{p}_j$ in the state and is referred to as the *occupation number* for the momentum $\boldsymbol{p}_j$. Clearly, there is no restriction of the value of the occupation numbers and so the particles associated with the scalar field are said to obey *Bose–Einstein statistics* and the scalar field is referred to as *bosonic*. The energy of the state is found by noting that $\hat{N}_p = \hat{a}_p^\dagger\hat{a}_p$ functions identically to the number operator $\hat{N} = a^\dagger a$ in the harmonic oscillator Hamiltonian (2.74). It follows that multiparticle state has the energy eigenvalue $\hat{H}|n_{p_1}, \ldots, n_{p_N}\rangle = (n_{p_1}\varepsilon_{p_1} + \ldots + n_{p_N}\varepsilon_{p_N})|n_{p_1}, \ldots, n_{p_N}\rangle$. Similarly to the harmonic states, the multiparticle states are orthonormal in the continuum sense, so that $\langle n_{p_1}|n_{p_2}\rangle = \delta_{n_{p_1}, n_{p_2}}(\delta^3(\boldsymbol{p}_1 - \boldsymbol{p}_2))^{n_{p_1}}$. The extension of the harmonic oscillator Hilbert space to multiparticle tensor product states is referred to as a *Fock space*, and it forms the natural particle basis for quantized field theories.

4.3.2 Quantizing the free bispinor field

The quantized bispinor or Dirac field is given the mode expansion

$$\hat{\psi}(\boldsymbol{x}, t) = \int \frac{d^3p}{(2\pi)^{3/2}}\sqrt{\frac{m}{\epsilon_p}}\left(\hat{b}_p^s u_s(p)e^{-ip\cdot x} + \hat{d}_p^{s\,\dagger} v_s(p)e^{ip\cdot x}\right), \tag{4.48}$$

where $p_0 = \varepsilon_p$ and there is an implicit sum over the repeated index s from 1 to 2. The four functions of momentum $u_s(p)$ and $v_s(p)$ are chosen so that equation (4.48) satisfies the Dirac equation (4.30), $(i\gamma^\mu\partial_\mu - m)\psi = 0$. It follows that this requires

$$\begin{aligned}(p_\mu\gamma^\mu - m)u_s(p) &\equiv (\not{p} - m)u_s(p) = 0,\\ (p_\mu\gamma^\mu + m)v_s(p) &\equiv (\not{p} + m)v_s(p) = 0.\end{aligned} \tag{4.49}$$

The Dirac representation solutions to equation (4.49) are built from the following solutions for $\boldsymbol{p} = 0$, where $p_\mu\gamma^\mu - m \to m\gamma^0 - m$ and $p_\mu\gamma^\mu + m \to m\gamma^0 + m$,

$$u_1(0) = \begin{pmatrix} 1 \\ 0 \\ 0 \\ 0 \end{pmatrix}, \quad u_2(0) = \begin{pmatrix} 0 \\ 1 \\ 0 \\ 0 \end{pmatrix}, \quad v_1(0) = \begin{pmatrix} 0 \\ 0 \\ 1 \\ 0 \end{pmatrix}, \quad v_2(0) = \begin{pmatrix} 0 \\ 0 \\ 0 \\ 1 \end{pmatrix}. \tag{4.50}$$

These solutions can be combined with the momentum version of relation (4.15), $(m + \not{p})(m - \not{p}) = m^2 - p^2 = 0$, to find the solutions to equation (4.49),

$$u_s(p) = N_p(m + \not{p})u_s(0), \quad v_s(p) = N_p(m - \not{p})v_s(0), \tag{4.51}$$

where the normalization factor is given by $N_p = 1/\sqrt{2m(m + \varepsilon_p)}$. The spinors of equation (4.51) satisfy the orthonormality relations $\bar{u}_s(p)u_{s'}(p) = -\bar{v}_s(p)v_{s'}(p) = \delta_{ss'}$, $\bar{u}_s(p)v_{s'}(p) = \bar{v}_s(p)u_{s'}(p) = 0$, $\bar{u}_s(\tilde{p})u_{s'}(p) = -\bar{v}_s(\tilde{p})v_{s'}(p) = (\varepsilon_p/m)\delta_{ss'}$, and $u_s^\dagger(p)u_{s'}(p) = v_s^\dagger(p)v_{s'}(p) = (\varepsilon_p/m)\delta_{ss'}$, where $\tilde{p} = (\varepsilon_p, -\boldsymbol{p})$. Using these in the Dirac Hamiltonian (4.32)gives, after considerable algebra,

$$\hat{H} = \int \mathrm{d}^3x\, \hat{\mathcal{H}}(\boldsymbol{x}, t) = \int \mathrm{d}^3\boldsymbol{p}\, \varepsilon_p\left(\hat{b}_{\boldsymbol{p}}^{s\dagger}\hat{b}_{\boldsymbol{p}}^{s} - \hat{d}_{\boldsymbol{p}}^{s}\hat{d}_{\boldsymbol{p}}^{s\dagger}\right), \tag{4.52}$$

where there is an implicit sum over s from 1 to 2 and care has been taken to preserve the order of the operators. While similar to the scalar field Hamiltonian (4.47), the Dirac Hamiltonian (4.52) cannot be quantized with a commutation relation like equation (4.45). The reason is that commutation relations would give $-\hat{d}_{\boldsymbol{p}}^{s}\hat{d}_{\boldsymbol{p}}^{s\dagger} = -\hat{d}_{\boldsymbol{p}}^{s\dagger}\hat{d}_{\boldsymbol{p}}^{s} + 2\,\delta^3(0)$, and the Hamiltonian would become unbounded from below when $\hat{d}_{\boldsymbol{p}}^{s\dagger}$ is treated as a creation operator. Like the Grassmann oscillator of equation (2.99), the Dirac field requires *anticommutation relations,*

$$\{\hat{b}_{\boldsymbol{p}}^{s\dagger}, \hat{b}_{\boldsymbol{k}}^{s'}\} = \{\hat{d}_{\boldsymbol{p}}^{s\dagger}, \hat{d}_{\boldsymbol{k}}^{s'}\} = \delta^3(\boldsymbol{p} - \boldsymbol{k})\delta_{ss'}, \tag{4.53}$$

with all other *anticommutator pairs* vanishing. The second term in equation (4.52) then becomes $-\hat{d}_{\boldsymbol{p}}^{s}\hat{d}_{\boldsymbol{p}}^{s\dagger} = \hat{d}_{\boldsymbol{p}}^{s\dagger}\hat{d}_{\boldsymbol{p}}^{s} - 2\,\delta^3(0)$. The operator $\hat{d}_{\boldsymbol{p}}^{s\dagger}$ then creates a positive energy excitation on top of a sea of negative energy given by $-2\int \mathrm{d}^3p\, \varepsilon_p\delta^3(0)$. Using the properties of the solutions (4.51) and the algebra (4.53) shows that the spinor components of the Dirac field ψ_a and its canonical momentum $\pi_b = \frac{1}{2}i\psi_b^\dagger$ satisfy the *equal time anticommutation relation*

$$\{\psi_a(\boldsymbol{x}, t), \pi_b(\boldsymbol{y}, t)\} = \frac{1}{2}i\{\psi_a(\boldsymbol{x}, t), \psi_b^\dagger(\boldsymbol{y}, t)\} = i\delta_{ab}\,\delta^3(\boldsymbol{x} - \boldsymbol{y}). \tag{4.54}$$

Using the definition of rotations it can be shown that s characterizes the two measurable *spin angular momenta,* $\pm\frac{1}{2}\hbar$, for the Dirac particle, but a full demonstration of this aspect lies outside the scope of this book.

The Fock space for quantized Dirac fields is identified similarly to the scalar field, but with the additional feature that the vacuum state is a direct product, $|0\rangle = |0\rangle_b \otimes |0\rangle_d$. The two vacuum states are annihilated by the respective annihilation operators, so that $b_p^s|0\rangle_b = 0$ and $d_p^s|0\rangle_d = 0$. The particle states are also direct product states, so that $b_p^{s\,\dagger}|0\rangle = |\boldsymbol{p}, s\rangle_b \otimes |0\rangle_d$, with a similar statement for $d_p^{s\,\dagger}$. However, because the creation operators $b_p^{s\,\dagger}$ and $d_p^{s\,\dagger}$ are both nilpotent, the occupation number for a momentum $\boldsymbol{p}$ and spin s state is either zero or one for *both types of particle*. This is the essence of the *Pauli exclusion principle*, and so Dirac particles therefore obey *Fermi–Dirac statistics* and are referred to as *fermions*. The state created by $b^{s\,\dagger}(\boldsymbol{p})$ is typically associated with the electron, while the creation operator $d_p^{s\,\dagger}$ is associated with its *antiparticle*, typically referred to as a *positron* because its electric charge is positive. The reader interested in more details is recommended to one of the texts on quantum field theory.

4.3.3 Quantizing the free gauge field

Quantizing the free gauge field will be done in the manifestly covariant Landau gauge discussed earlier by identifying the *physical subspace* where the Lorenz condition $\partial_\mu \hat{a}^\mu = 0$ holds. The gauge field is given an expansion consistent with its massless vector nature, so that

$$\hat{a}^\mu(x) = \int \frac{\mathrm{d}^3 p}{(2\pi)^{3/2}\sqrt{2\omega_p}}\left(\hat{\alpha}_p^\mu\, e^{-ip\cdot x} + \hat{\alpha}_p^{\mu\,\dagger}\, e^{ip\cdot x}\right), \tag{4.55}$$

where $p_0 = \omega_p$ and $\omega_p^2 = \boldsymbol{p}^2$, so that $p_\mu p^\mu = 0$. This relationship results in equation (4.55) satisfying the Landau gauge equation of motion $\Box \hat{a}^\mu = 0$. Using $\hat{\pi}^j = \partial^j \hat{a}^0 - \partial^0 \hat{a}^j$ and $\hat{\pi}^0 = -\partial_\mu \hat{a}^\mu$ gives

$$\begin{aligned}
\hat{\pi}^j(x) &= i\int \frac{\mathrm{d}^3 k}{(2\pi)^{3/2}}\sqrt{\frac{\omega_k}{2}} \times \left[\left(\hat{\alpha}_k^j - \frac{k^j}{\omega_k}\hat{\alpha}^0(\boldsymbol{k})\right)e^{-ik\cdot x} - \left(\hat{\alpha}_k^{j\,\dagger} - \frac{k^j}{\omega_k}\hat{\alpha}_k^{0\,\dagger}\right)e^{ik\cdot x}\right],\\
\hat{\pi}^0(x) &= i\int \frac{\mathrm{d}^3 k}{(2\pi)^{3/2}\sqrt{2\omega_k}}\left[k_\mu \hat{\alpha}_k^{\mu\,\dagger} e^{ik\cdot x} - k_\mu \hat{\alpha}_k^\mu e^{-ik\cdot x}\right].
\end{aligned} \tag{4.56}$$

The quantized field $\hat{a}^\mu(x)$ and its canonically conjugate momentum $\hat{\pi}^\nu(y)$ must satisfy the *covariant* equal time commutation relation

$$[\hat{a}^\mu(\boldsymbol{x}, t), \hat{\pi}^\nu(\boldsymbol{y}, t)] = i\,\eta^{\mu\nu}\,\delta^3(\boldsymbol{x} - \boldsymbol{y}). \tag{4.57}$$

Using equations (4.55), (4.56), and (1.26) it can be shown that relation (4.57) is satisfied if $[\alpha_k^\mu, \alpha_p^{\nu\dagger}] = -\eta^{\mu\nu}\delta^3(\boldsymbol{k} - \boldsymbol{p})$. The total Fock space for the gauge field is generated by the action of the creation operators $\hat{\alpha}_p^{\mu\,\dagger}$ on the vacuum state defined by $\hat{\alpha}_p^\mu|0\rangle = 0$. It is important to note that the commutator $[\hat{\alpha}_0, \hat{\alpha}_0^\dagger]$ has a negative sign while the commutator $[\hat{a}_j, \hat{a}_j^\dagger]$ has a positive sign due to the metric tensor. As a result, the *timelike* state $\hat{a}_{\boldsymbol{p}}^{0\,\dagger}|0\rangle$ has a negative norm since the commutator gives $\langle 0|\hat{a}_{\boldsymbol{k}}^0 \hat{a}_{\boldsymbol{p}}^{0\,\dagger}|0\rangle = \langle 0|\hat{a}_{\boldsymbol{p}}^{0\,\dagger}\hat{a}_{\boldsymbol{k}}^0|0\rangle - \delta^3(\boldsymbol{p} - \boldsymbol{k}) = -\delta^3(\boldsymbol{p} - \boldsymbol{k})$. The space of particle states in a gauge theory therefore has both positive and negative norm states, and for that reason it is referred to as *an indefinite metric space*. The presence of negative norm states threatens the probabilistic interpretation of the theory. Like the Hamiltonian (4.40) that is unbounded from below, this problem is resolved by implementing the gauge constraint.

The Landau gauge, or equivalently the Lorenz gauge condition, requires implementing the gauge constraint

$$\hat{\pi}^0(\boldsymbol{x}, t) = -\partial_\mu \hat{a}^\mu(\boldsymbol{x}, t) = 0. \tag{4.58}$$

Using $\partial_0^2 \hat{a}^0 = \nabla^2 a^0$ and $\hat{\pi}_j = \partial_j \hat{a}^0 + \partial_0 \hat{a}^j$, it follows that the secondary constraint is given by the charge free version of Gauss's law,

$$\partial_0 \hat{\pi}^0 = -\partial_\mu \partial_0 \hat{a}^\mu = -\partial_0^2 \hat{a}^0 - \partial_j \partial_0 \hat{a}^j = -\partial_j(\partial_j \hat{a}^0 + \partial_0 \hat{a}^j) = -\partial_j \hat{\pi}^j. \tag{4.59}$$

Implementing these two constraints is somewhat different than the quantum mechanical case of the gauge constraint, where the physical subspace that satisfied the primary and secondary operator constraints was identified and then implemented by using the projection operator (3.216). This is because both constraints contain a combination of annihilation and creation operators, and so identifying the members $|P\rangle$ of the physical subspace of the total Fock space by demanding the *strong implementation* $\hat{\pi}_0|P\rangle = 0$ leads to states with infinite norms Instead, Gupta and Bleuler identified the physical subspace by the *weak implementation*, $\langle P'|\hat{\pi}_0|P\rangle = 0$. This allows the physical subspace to be identified by the action of the annihilation operators in the constraint, denoted $\hat{\pi}_0^{(-)}$ since they have a negative frequency, so that $\hat{\pi}_0^{(-)}|P\rangle = 0$. For the two constraints of the Landau gauge, the physical subspace must be annihilated by two Fock space operators, $k_\mu \hat{a}_{\boldsymbol{k}}^\mu$ from $\hat{\pi}^0$ and $p_j(p^j \hat{a}_{\boldsymbol{p}}^0 - \omega_p \hat{a}_{\boldsymbol{p}}^j)$ from $\partial_j \hat{\pi}^j$. Implementing these two constraints begins by noting that $[k_\mu \alpha_{\boldsymbol{k}}^\mu, p_\nu \alpha_{\boldsymbol{p}}^{\nu\,\dagger}] = -k_\mu p_\nu \eta^{\mu\nu} \delta^3(\boldsymbol{p} - \boldsymbol{k}) = k_\mu p^\mu \delta^3(\boldsymbol{p} - \boldsymbol{k}) = p_\mu p^\mu \delta^3(\boldsymbol{p} - \boldsymbol{k}) = 0$, where the property $tt'\delta(t - t') = t^2\delta(t - t')$ was used. It follows that $k_\mu \hat{a}_{\boldsymbol{k}}^{\mu\dagger}$ creates a Fock space state with zero norm since it commutes with $p_\nu \hat{a}_{\boldsymbol{p}}^\nu$. Since it cannot be normalized, such a state has no probabilistic interpretation. It is referred to as a *ghost state*, and implementing the gauge constraint must decouple such ghost states from physical processes. A full explanation of this requires further developments.

The physical subspace is best identified by introducing a new set of operators $\hat{a}_{\boldsymbol{p}}^\lambda$ using the definition $\hat{a}_{\boldsymbol{p}}^\mu = \epsilon_\lambda^\mu(\boldsymbol{p})\hat{a}_{\boldsymbol{p}}^\lambda$, where $\epsilon_\lambda^\mu(\boldsymbol{p})$ is treated as the contravariant μ component of four real covariant *polarization vectors* $\epsilon_\lambda(\boldsymbol{p})$ indexed by λ. The new operators satisfy the familiar commutation relation, $[\hat{a}_{\boldsymbol{p}}^\lambda, \hat{a}_{\boldsymbol{k}}^{\lambda'\dagger}] = -\eta^{\lambda\lambda'}\delta^3(\boldsymbol{p} - \boldsymbol{k})$, and this yields the requirement

$$
\begin{aligned}
\eta^{\mu\nu}\delta^3(\boldsymbol{p}-\boldsymbol{k}) &= -[\hat{a}_{\boldsymbol{p}}^{\mu}, \hat{a}_{\boldsymbol{k}}^{\nu\dagger}] = -\epsilon_{\lambda}^{\mu}(\boldsymbol{p})\epsilon_{\lambda'}^{\nu}(\boldsymbol{k})[\hat{a}_{\boldsymbol{p}}^{\lambda}, \hat{a}_{\boldsymbol{k}}^{\lambda'\dagger}] \\
&= \epsilon_{\lambda}^{\mu}(\boldsymbol{p})\epsilon_{\lambda'}^{\nu}(\boldsymbol{k})\eta^{\lambda\lambda'}\,\delta^3(\boldsymbol{p}-\boldsymbol{k}) \\
&\Longrightarrow \epsilon_{\lambda}^{\mu}(\boldsymbol{p})\epsilon_{\lambda'}^{\nu}(\boldsymbol{p})\,\eta^{\lambda\lambda'} = \eta^{\mu\nu}.
\end{aligned}
\tag{4.60}
$$

The polarization vectors can be chosen so that $p_\mu\epsilon_0^\mu(\boldsymbol{p}) = p_\mu\epsilon_3^\mu(\boldsymbol{p}) = \omega_{\boldsymbol{p}}$. These two polarization vectors have the components $\epsilon_0(\boldsymbol{p}) = (1, \boldsymbol{0})$ and $\epsilon_3(\boldsymbol{p}) = (0, \boldsymbol{p}/\omega_{\boldsymbol{p}})$. The vector $\epsilon_3(\boldsymbol{p})$ is *longitudinal* to the direction of propagation given by $\boldsymbol{p}$, so that $p_j\epsilon_3^j(\boldsymbol{p}) = \omega_{\boldsymbol{p}}$. The $\epsilon_{(1,2)}(\boldsymbol{p})$ vectors lie along the two spatial directions *transverse* to the direction of propagation $\boldsymbol{p}$. They can be written $\epsilon_{(1,2)}(\boldsymbol{p}) = (0, \boldsymbol{p}_{(1,2)}^T)$, where $\boldsymbol{p}\cdot\boldsymbol{p}_{(1,2)}^T = 0$, $\boldsymbol{p}_1^T\cdot\boldsymbol{p}_1^T = \boldsymbol{p}_2^T\cdot\boldsymbol{p}_2^T = 1$, and $\boldsymbol{p}_1^T\cdot\boldsymbol{p}_2^T = 0$. As a result, $p_\mu\epsilon_{(1,2)}^\mu(\boldsymbol{p}) = 0$. By choosing a coordinate system where $\boldsymbol{p}$ lies along the z-axis, so that $\epsilon_{(1,2)}(\boldsymbol{p})$ are unit vectors along the x and y axis respectively, it is straightforward to show that these definitions satisfy the required relationship (4.60).

Using these polarization vectors the two constraint operators become $p_\mu\hat{a}_{\boldsymbol{p}}^\mu = \omega_{\boldsymbol{p}}(\hat{a}_{\boldsymbol{p}}^0 - \hat{a}_{\boldsymbol{p}}^3)$ and $p_j(p^j\hat{a}_{\boldsymbol{p}}^0 - \omega_{\boldsymbol{p}}\hat{a}_{\boldsymbol{p}}^j) = \omega_{\boldsymbol{p}}^2(\hat{a}_{\boldsymbol{p}}^0 - \hat{a}_{\boldsymbol{p}}^3)$. The commutation properties for the $\hat{a}^\lambda$ show that the operator $\hat{g}_{\boldsymbol{p}} \equiv (\hat{a}_{\boldsymbol{p}}^0 - \hat{a}_{\boldsymbol{p}}^3)/\sqrt{2}$ has the ghost property $[\hat{g}_{\boldsymbol{p}}, \hat{g}_{\boldsymbol{k}}^\dagger] = 0$. However, the noncovariant operator $\hat{b}_{\boldsymbol{p}} \equiv (\hat{a}_{\boldsymbol{p}}^0 + \hat{a}_{\boldsymbol{p}}^3)/\sqrt{2}$ has the property that $[\hat{b}_{\boldsymbol{p}}, \hat{g}_{\boldsymbol{k}}^\dagger] = -\delta^3(\boldsymbol{p}-\boldsymbol{k})$. The physical subspace $|P\rangle$ of the Fock space is therefore identified as a combination of transverse excitations created by $\hat{a}_{1,2}^\dagger(\boldsymbol{p})$ and the *good ghosts* created by $\hat{g}_{\boldsymbol{p}}^\dagger$,

$$
|P\rangle = \hat{a}_{\boldsymbol{p}1}^{(1,2)\dagger}\cdots\hat{a}_{\boldsymbol{p}N}^{(1,2)\dagger}\hat{a}_{\boldsymbol{k}1}^{g\dagger}\cdots\hat{a}_{\boldsymbol{k}n}^{g\dagger}|0\rangle.
\tag{4.61}
$$

Since such a physical state $|P\rangle$ has no *bad ghosts* associated with $\hat{\boldsymbol{p}}^\dagger$, it will satisfy both constraints, $\hat{\pi}^{0(-)}|P\rangle = \partial_j\hat{\pi}^{j(-)}|P\rangle = 0$. Using the expansions for $a^\mu(x)$ and $\pi^\mu(x)$ in the Hamiltonian (4.40) yields

$$
\hat{H} = \int d^3p\;\omega_p\left(\hat{g}_{\boldsymbol{p}}^\dagger\hat{g}_{\boldsymbol{p}} + \sum_{\lambda=1}^{2}\hat{a}_{\boldsymbol{p}}^{\lambda\dagger}\hat{a}_{\boldsymbol{p}}^{\lambda}\right).
\tag{4.62}
$$

It should be noted that the ghost number operator $\hat{g}_{\boldsymbol{p}}^\dagger\hat{g}_{\boldsymbol{p}}$ has only zero for an eigenvalue *as long as there are only good ghosts in the state.* It is only the transverse excitations that contribute to the energy of a physical state, with the ghosts contributing nothing to the energy of a physical state. In effect, the physical states have a positive-definite energy since they are required to have equal numbers of negative energy timelike and positive energy longitudinal particle excitations. In the case of quantum electrodynamics, the particle excitations are referred to as *photons*, and are transverse to the direction of propagation. However, those physical states $|P\rangle$ containing good ghosts will have zero norm, and it is critical that such ghost states have no effect on physical processes involving the transverse photons. This requires an understanding of interacting field theories before the path integral formulation of quantized field theory processes can be developed.

4.4 Interacting fields and particle processes

All the free quantized fields of the previous section are characterized by a *free Hamiltonian*, denoted $\hat{H}_0$, that is quadratic and diagonal in the Fock space of the theory. The Fock space Schrödinger picture states, denoted generically as $|n, t\rangle$, have a time evolution given by $|n, T\rangle = \exp\{-i\hat{H}_0 T\}|n, 0\rangle = \exp\{-iE_n T\}|n, 0\rangle$, where E_n is the total energy of the Fock state. This shows that $\langle n', 0|n, T\rangle = \delta_{nn'} \exp\{-iE_n T\}$, so that the states remain orthonormal and therefore unchanged as time passes. Since one of the important goals of quantum field theory is to describe quantum processes where the particle states change due to interactions, it is clear that the quadratic actions of the previous sections are not adequate for that purpose. However, the states of the free theories provide a useful representation of the particles involved in many experimentally observed processes.

The remedy for the absence of particle state changing processes is to introduce interaction terms that are cubic and higher in the fields. The guidelines for the choice of such terms requires relativistic invariance and the presence of *symmetries* that reflect experimental observations. The latter includes such experimental facts as electric charge conservation and the strength and type of the interactions being modeled. In what follows, *all fields will be considered quantized*, and so the operator symbol will be dispensed with. For example, the term $\mathcal{H}_I = g\Phi(x)\bar{\Psi}(x)\Psi(x)$ models the interaction of a scalar field $\Phi(x)$ with a Dirac field $\Psi(x)$, and was used in early models of the *Yukawa interaction* between Dirac nucleons and scalar pi mesons. Adding this term to the combination of the scalar and Dirac free Lagrangians (4.25) and (4.29) results in coupled or *nonlinear Euler–Lagrange equations*. Assuming both the scalar field and Dirac field are massive, the two equations of motion become nonlinear and are given by $(\Box + m^2)\Phi - g\bar{\Psi}\Psi = 0$ and $(\not{\partial} + im)\Psi - g\Phi\Psi = 0$. It is assumed that the interacting quantized fields also obey the commutation and anticommutation relations $[\Phi(\boldsymbol{x}, t), \dot{\Phi}(\boldsymbol{y}, t)] = i\delta^3(\boldsymbol{x} - \boldsymbol{y})$ and $\frac{1}{2}\{\Psi_a(\boldsymbol{x}, t), \Psi_b^\dagger(\boldsymbol{y}, t)\} = \delta_{ab}\delta^3(\boldsymbol{x} - \boldsymbol{y})$ used to quantize the free fields. This should hold since the $g \to 0$ limit of the interacting theory apparently reduces it back to the two free field Lagrangians.

Unfortunately, almost all interacting field theories are only marginally understood despite many decades of intense efforts. This is especially true for theories in 3 + 1 dimensions. It is often the case that interacting field theories must be given a perturbative treatment. This consists of relating the interacting field to its free field counterpart in the form of an expansion in terms of powers of the interaction, much in the same way that nonlinear quantum mechanical systems can be analyzed perturbatively via the generating functional approach of equation (3.125). In that regard, the path integral has played a significant role in defining and analyzing perturbation expansions in field theory. Unfortunately, perturbative approaches to field theory can be misleading and are often beset by divergent mathematical expressions. This requires a mathematical process referred to as *renormalization* in order to extract finite results that can be compared to experimental observations. A thorough presentation of perturbation expansions and renormalization in field theory lies outside the scope of this book. Rather, the goal is to understand how the path integral for interacting fields is defined and used in their analysis.

The *Lehmann–Symanzik–Zimmermann (LSZ) reduction formalism* begins by postulating that at *asymptotic times*, $t \to \pm\infty$, the matrix elements involving interacting fields in the Heisenberg picture become proportional to the matrix elements of free fields, referred to as in-fields and out-fields, which have a well understood particle spectrum. This assumption is based on the idea that particle processes result in a large separation of the participants at such asymptotic times, and therefore any influence the fields and their associated particles have on each other is suppressed in this limit. Although this is a commonly used assumption, there are counterexamples such as the electric and magnetic fields that are present whenever a charge is moving as well as the possible presence and formation of bound states in strongly interacting theories. It is important to note that this assumption applies only in the *weak limit*, which means that the convergence to free fields applies only to matrix elements of the interacting fields. Nevertheless, the benefit of this assumption is that it provides a Fock space of well defined particle states to be used as the incoming and outgoing states in a nonlinear particle process.

For example, the quantized interacting scalar field $\Phi(x)$ is assumed to obey the weak asymptotic condition $\lim_{t\to\pm\infty}(\Phi(x) - \sqrt{Z}\varphi(x)) = 0$, where $\varphi(x)$ is the free quantized scalar field (4.43) and Z is referred to as the *wave-function renormalization constant*. If there are interactions, the constant Z will be less than one. This is because $\sqrt{Z}$ is interpreted as the probability amplitude for the interacting field to connect states differing by a single particle. The interacting field may also have nonzero matrix elements between states differing by an arbitrary number of particles. Since the single particle operator φ does not connect all the states connected by Φ, it must be that $0 < Z \leqslant 1$. However, the benefit of this assumption is that it allows the Heisenberg picture particle in-states, denoted $|\alpha\rangle_{\text{in}}$, and out-states, denoted ${}_{\text{out}}\langle\beta|$, to be defined in terms of the well understood free spectrum. Combining the function $f_p(x) = e^{-ip\cdot x}/\sqrt{(2\pi)^3 2\varepsilon_p}$ with the weak asymptotic relationship gives the scalar field matrix element,

$$\begin{aligned} {}_{\text{out}}\langle\alpha|\boldsymbol{p}, \beta\rangle_{\text{in}} &= {}_{\text{out}}\langle\alpha|\left(\hat{a}^{\dagger}_{p,\text{in}}|\beta\rangle_{\text{in}}\right) \\ &= -\frac{i}{\sqrt{Z}}\, {}_{\text{out}}\langle\alpha| \lim_{t\to-\infty} \int \mathrm{d}^3x \left(\frac{\partial f_p(x)}{\partial t}\Phi(x) - f_p(x)\frac{\partial\Phi(x)}{\partial t}\right)|\beta\rangle_{\text{in}}, \end{aligned} \tag{4.63}$$

with a similar expression for the out-state operator defined at $t \to \infty$.

Using these assumptions allows the LSZ reduction formulas to be derived. For simplicity of notation attention will be restricted to a scalar field Φ and the Fock space particle states of the associated free field φ. The starting point for the reduction process is the generalized matrix element

$$G_{\alpha\beta}(\boldsymbol{p}, x_1, \ldots, x_n) = {}_{\text{out}}\langle\alpha|T\{\Phi(x_1)\cdots\Phi(x_n)\}|\boldsymbol{p}, \beta\rangle_{\text{in}}, \tag{4.64}$$

where T refers to the time-ordered product first introduced in (1.44) and given by $T\{\Phi(x_1)\Phi(x_2)\} = \theta(t_1 - t_2)\Phi(x_1)\Phi(x_2) + \theta(t_2 - t_1)\Phi(x_2)\Phi(x_1)$. By combining (4.63) with the integral representation of the limits,

$$\lim_{t\to\infty} g(x) - \lim_{t\to-\infty} g(x) = \int_{-\infty}^{\infty} \mathrm{d}t \, \frac{\partial g(x)}{\partial t}, \tag{4.65}$$

the function (4.64) can be written

$$\begin{aligned} G_{\alpha\beta}(\boldsymbol{p}, x_1, \ldots, x_n) &= {}_{\text{out}}\langle\alpha|T\left\{\hat{\Phi}(x_1)\cdots\Phi(x_n)\right\}\hat{a}^{\dagger}_{\boldsymbol{p},\text{in}}|\beta\rangle_{\text{in}} \\ &= {}_{\text{out}}\langle\alpha|\hat{a}^{\dagger}_{\boldsymbol{p},\text{out}}\, T\{\Phi(x_1)\cdots\Phi(x_n)\}|\beta\rangle_{\text{in}} \\ &\quad - \frac{i}{\sqrt{Z}} \int_{-\infty}^{\infty} \mathrm{d}t \, \frac{\partial}{\partial t} \int \mathrm{d}^3x \\ &\quad \times \left[\frac{\partial f_{\boldsymbol{p}}(x)}{\partial t} {}_{\text{out}}\langle\alpha|T\{\Phi(x)\Phi(x_1)\cdots\Phi(x_n)\}|\beta\rangle_{\text{in}} \right. \\ &\quad \left. - f_{\boldsymbol{p}}(\boldsymbol{x}, t)\frac{\partial}{\partial t}\left({}_{\text{out}}\langle\alpha|T\{\Phi(x)\Phi(x_1)\cdots\Phi(x_n)\}|\beta\rangle_{\text{in}}\right)\right]. \end{aligned} \tag{4.66}$$

The limits on the time integral in equation (4.66) allow $\Phi(x)$ to be included in the time-ordered product since $\Phi(x)$ lies to the left or right of the field products when $t \to \pm\infty$. The first term on the right-hand side of equation (4.66) will typically vanish unless ${}_{\text{out}}\langle\alpha|$ contains a particle of momentum $\boldsymbol{p}$, in which case the term corresponds to an unscattered particle, referred to as *forward scattering*. In what follows it will be assumed to vanish. Taking the time derivative and using $\partial_0^2 f_{\boldsymbol{p}} = (\nabla^2 - m^2)f_{\boldsymbol{p}}$, the spatial integral on the right-hand side of equation (4.66) can be integrated by parts to obtain

$$\begin{aligned} G_{\alpha\beta}(\boldsymbol{p}, x_1, \ldots, x_n) &= -\frac{i}{\sqrt{Z}} \int \mathrm{d}^4x \\ &\quad \times \left(\frac{\partial^2 f_{\boldsymbol{p}}(x)}{\partial t^2} - f_{\boldsymbol{p}}(x)\frac{\partial^2}{\partial t^2} \right) {}_{\text{out}}\langle\alpha|T\{\hat{\Phi}(x)\hat{\Phi}(x_1)\cdots\hat{\Phi}(x_n)\}|\beta\rangle_{\text{in}} \\ &= \frac{i}{\sqrt{Z}} \int \mathrm{d}^4x\, f_{\boldsymbol{p}}(x)(\Box + m^2)\left({}_{\text{out}}\langle\alpha|T\{\hat{\Phi}(x)\hat{\Phi}(x_1)\cdots\hat{\Phi}(x_n)\}|\beta\rangle_{\text{in}}\right). \end{aligned} \tag{4.67}$$

Result (4.67) can be repeated until all particles are reduced from the incoming and outgoing states. The interested reader is recommended to one of the texts where these formulas are developed. However, for the purposes of this monograph, this result shows that the interacting field theory processes are defined entirely by the expectation value of time-ordered products of the interacting fields. Most of the rigorous results for interacting field theories have come from analyzing the properties of these time-ordered products, and it is the path integral representation of these time-ordered products that will be developed in the remainder of this chapter.

4.5 The interaction picture in field theory

Before deriving the relationship between the field theory path integral and the operator and state formalism, it is necessary to derive the evolution operator for the *interaction picture* Fock space states. This assumes the existence of a time-dependent unitary transformation $U(t)$ between the interacting fields and momenta, $\Phi_\alpha(x)$ and $\Pi_\beta(x)$, and the associated free fields and momenta, $\varphi_\alpha(x)$ and $\pi_\beta(x)$,

$$\varphi_\alpha(\boldsymbol{x}, t) = U(t)\Phi_\alpha(\boldsymbol{x}, t)U^{-1}(t), \quad \pi_\alpha(\boldsymbol{x}, t) = U(t)\Pi_\alpha(\boldsymbol{x}, t)U^{-1}(t). \tag{4.68}$$

This assumption is consistent with the equal-time commutation relation for both free and interacting fields since

$$\begin{aligned} i\,\delta_{\alpha\beta}\,\delta^3(\boldsymbol{x}-\boldsymbol{y}) &= \left[\varphi_\alpha(\boldsymbol{x}, t), \pi_\beta(\boldsymbol{y}, t)\right] = U(t)\left[\Phi_\alpha(\boldsymbol{x}, t), \Pi_\beta(\boldsymbol{y}, t)\right]U^{-1}(t) \\ &= U(t) i\,\delta_{\alpha\beta}\,\delta^3(\boldsymbol{x}-\boldsymbol{y})U^{-1}(t) = i\,\delta_{\alpha\beta}\,\delta^3(\boldsymbol{x}-\boldsymbol{y}), \end{aligned} \tag{4.69}$$

with a similar result for the case of anticommutation relations. Taking the time derivative of equation (4.68) and using $\dot{U}U^{-1} + U\dot{U}^{-1} = \dot{1} = 0$ gives

$$\dot{\varphi}_\alpha = \dot{U}U^{-1}\varphi_\alpha - \varphi_\alpha\dot{U}U^{-1} + U\dot{\Phi}_\alpha U^{-1} = [\dot{U}U^{-1}, \varphi_\alpha] + U\dot{\Phi}_\alpha U^{-1}. \tag{4.70}$$

The time evolution of the respective fields gives $\dot{\varphi}_\alpha = i[H_0[\varphi, \pi], \varphi_\alpha]$ and $\dot{\Phi}_\alpha = i[H[\Phi, \Pi], \Phi_\alpha]$, where H is the Hamiltonian containing interaction terms and H_0 is the free field Hamiltonian, which is typically quadratic. It follows from equation (4.68) that $U\dot{\Phi}_\alpha U^{-1} = iU[H[\Phi, \Pi], \Phi_\alpha]U^{-1} = i[H[\varphi, \pi], \varphi_\alpha]$. Returning this to equation (4.70) gives

$$[\dot{U}U^{-1}, \varphi_\alpha] = -i[H[\varphi, \pi] - H_0[\varphi, \pi], \varphi_\alpha] \equiv -i[H_I[\varphi, \pi], \varphi_\alpha], \tag{4.71}$$

where $H_I[\varphi, \pi] = \int d^3x\, \mathcal{H}_I(\varphi(\boldsymbol{x}, t), \pi(\boldsymbol{x}, t))$. Result (4.71) is solved identically to equation (1.40), so that the evolution operator for the Fock space states is given by

$$E(t, t_0) \equiv U(t)U^{-1}(t_0) = T\left\{\exp\left(-i\int_{t_0}^{t} dt\, H_I[\varphi, \pi]\right)\right\}. \tag{4.72}$$

While the innocuous looking assumption (4.68) yields the representation (4.72) of the interacting field evolution operator in terms of the free field, Haag and Wightman demonstrated such an assumption is flawed and that there are many *unitarily inequivalent representations of the commutation relations.* Nevertheless, the assumption (4.68) sits at the core of defining the evolution operator and its perturbative expansion, and it is likely the source of the many problems that perturbative representations of interacting field theories possess.

The LSZ formalism shows that the time-ordered products of interacting fields are of critical importance in analyzing quantum field theory processes. After the incoming and outgoing particle states are completely reduced, the interactions are encapsulated in the *vacuum expectation value of time-ordered field products*, often referred to as *generalized Green's functions*. The reason for this name stems from the free field version of time-ordered products. For example, the vacuum expectation value of the time-ordered product of two free scalar fields is given by

$$\begin{aligned}\Delta_F(x_1, x_2) &= \theta(t_2 - t_1)\langle 0|\varphi(x_2)\varphi(x_1)|0\rangle + \theta(t_1 - t_2)\langle 0|\varphi(x_1)\varphi(x_2)|0\rangle \\ &= \langle 0|T\{\varphi(x_1)\varphi(x_2)\}|0\rangle.\end{aligned} \tag{4.73}$$

Applying the differential operator $\Box_{x_1} + m^2$ gives

$$\begin{aligned}(\Box_{x_1} + m^2)\Delta_F(x_1, x_2) &= \delta(t_1 - t_2)\langle 0|[\dot{\varphi}(\boldsymbol{x}_1, t_1), \varphi(\boldsymbol{x}_2, t_2)]|0\rangle \\ &\quad - \langle 0|T\left\{(\Box_{x_1} + m^2)\varphi(x_1)\varphi(x_2)\right\}|0\rangle \\ &= -i\,\delta^4(x_1 - x_2),\end{aligned} \tag{4.74}$$

where $\dot{\theta}(t - t') = \delta(t - t')$, $\ddot{\theta}(t - t')\varphi(x) = \dot{\delta}(t - t')\varphi(x) = -\delta(t - t')\dot{\varphi}(x)$, $\dot{\varphi} = \pi$, and $(\Box + m^2)\varphi = 0$ were used. Result (4.74) shows that $\Delta_F(x_1, x_2)$ satisfies the definition of a Green's function for the Klein–Gordon equation differential operator $\Box + m^2$. It is referred to as the *Feynman propagator* since it can be used to solve the Klein–Gordon wave equation,

$$(\Box + m^2)\varphi(x) = j(x) \implies \varphi(x) = f(x) + i\int d^4y\, \Delta_F(x, y)j(y), \tag{4.75}$$

where f is an arbitrary solution of $(\Box + m^2)f = 0$ that can be chosen to match boundary conditions.

These time-ordered products can be put into the interaction picture,

$$\begin{aligned}G_{\alpha\cdots\beta}(x_1, \ldots, x_n) &= \langle 0|T\left\{\Phi_\alpha(x_1)\cdots\Phi_\beta(x_n)\right\}|0\rangle \\ &= \langle 0|U^{-1}(t_+)T\{U(t_+)U^{-1}(t_1)\varphi_\alpha(x_1)U(t_1)\cdots \\ &\quad \times U^{-1}(t_n)\varphi_\beta(x_n)U(t_n)U^{-1}(t_-)\right\}U(t_-)|0\rangle \\ &= \langle 0|U^{-1}(t_+)T\left\{\varphi_\alpha(x_1)\cdots\varphi_\beta(x_n)E(t_+, t_-)\right\}U(t_-)|0\rangle,\end{aligned} \tag{4.76}$$

where $t_\pm$ stands for the limits $t \to \pm\infty$. Combining the weak asymptotic condition with equation (4.68) gives $\lim_{t\to t_\pm}(U^{-1}(t)\varphi_\alpha(x)U(t) - \sqrt{Z}\varphi_\alpha(x)) = 0$. Since the vacuum state is annihilated by $\hat{a}_{p,\text{in}}$, it follows that $\lim_{t\to t_-}U(t)|0\rangle = \lambda_-|0\rangle$, where λ_- is a constant in order for $\hat{a}_{p,\text{in}}$ to commute with it. Using $U(t)U^{-1}(t) = 1$ shows that $\lim_{t\to-\infty}U^{-1}(t)|0\rangle = |0\rangle/\lambda_-$. Similarly, it must hold that $\lim_{t\to t_+}\langle 0|U^{-1}(t) = \lambda_+\langle 0|$. The *vacuum persistence amplitude* is then given by

$$\langle 0|E(t_+, t_-)|0\rangle = \langle 0|U(t_+)U^{-1}(t_-)|0\rangle = \frac{1}{\lambda_+\lambda_-}, \tag{4.77}$$

where $\langle 0|0\rangle = 1$ was used. Using equation (4.77), the function defined in equation (4.76) becomes

$$\langle 0|T\{\Phi_\alpha(x_1)\cdots\Phi_\beta(x_n)\}|0\rangle = \frac{\langle 0|T\{\varphi_\alpha(x_1)\cdots\varphi_\beta(x_n)E(t_+, t_-)\}|0\rangle}{\langle 0|E(t_+, t_-)|0\rangle}. \tag{4.78}$$

It can be shown that the denominator of equation (4.78) serves to cancel *disconnected processes* in the numerator, which are those processes that begin and end in the vacuum and are not connected to the in and out states that were reduced. Since the free fields φ_α are quantized in terms of well defined Fock space operators, result (4.78) gives a *perturbative* representation of the generalized Green's functions used in the LSZ formalism. The exponential in the evolution operator can be expanded and the time-ordered products of the free fields can be evaluated using *Wick's theorem* to obtain perturbative corrections to the free field Green's functions. This operator approach is carried out in many texts and the reader is recommended to them.

Introducing an *external source function*, denoted $J_\alpha(x)$, the generalized Green's functions can be written in terms of functional derivatives,

$$\langle 0|T\{\Phi_\alpha(x_1)\cdots\Phi_\beta(x_n)\}|0\rangle = i^n \frac{\delta^n}{\delta J_\alpha(x_1)\cdots\delta J_\beta(x_n)} Z_0[J]\Bigg|_{J_\alpha=0}, \tag{4.79}$$

where $Z_0[J_\alpha]$ is the *Schwinger generating functional* for the vacuum expectation value of time-ordered field products. It is given explicitly by

$$Z_0[J] = \langle 0|T\left\{\exp\left(-i\int d^4x\, J_\alpha(x)\Phi_\alpha(x)\right)\right\}|0\rangle. \tag{4.80}$$

The sign of J_α is the opposite of the choice in the quantum mechanical case (3.125) and is a matter of convention. Choosing the opposite sign in equation (4.80) simply changes i^n to $(-i)^n$ in equation (4.79). Slicing the time interval appearing in equation (4.80) according to the usual prescription, $\epsilon = (t_+ - t_-)/N = T/N$, allows the argument of the exponential to be written

$$\int d^4x\, J_\alpha(x)\Phi_\alpha(x) = \sum_{j=0}^{N-1} \epsilon \int d^3x\, J_\alpha(\boldsymbol{x}, t_j)\Psi_\alpha(\boldsymbol{x}, t_j). \tag{4.81}$$

Using the field theory versions of equations (1.60) and (2.6) shows that the time-ordered exponential can be written

$$\begin{aligned} T\left\{\exp\left(-i\int d^4x\, J_\alpha(x)\Phi_\alpha(x)\right)\right\} &= \exp\left\{-i\epsilon\int d^3x\, J_\alpha(\boldsymbol{x}, t_{N-1})\Phi_\alpha(\boldsymbol{x}, t_{N-1})\right\}\cdots \\ &\quad\times \exp\left\{-i\epsilon\int d^3x\, J_\alpha(\boldsymbol{x}, t_0)\Phi_\alpha(\boldsymbol{x}, t_0)\right\} \\ &\equiv \mathcal{Z}[J_\alpha, \Phi, t_{N-1}]\cdots\mathcal{Z}[J_\alpha, \Phi, t_0]. \end{aligned} \tag{4.82}$$

Using equation (4.77) shows that the interaction picture version of equation (4.80) is given by

$$\begin{aligned} Z_0[J_\alpha] &= \langle 0|U^{-1}(t_+)T\big\{U(t_+)U^{-1}(t_{N-1})\mathcal{Z}(J,\phi,t_{N-1})U(t_{N-1})\cdots \\ &\quad \times U(t_1)U^{-1}(t_-)\mathcal{Z}(J,\phi,t_-)\big\}U(t_-)|0\rangle \\ &= \lambda_+\lambda_-\langle 0|T\{\mathcal{Z}(J,\phi,t_{N-1})\cdots\mathcal{Z}(J,\phi,t_-)E(t_+,t_-)\}|0\rangle \\ &= \frac{\langle 0|E_J(t_+,t_-)|0\rangle}{\langle 0|E(t_+,t_-)|0\rangle}. \end{aligned} \tag{4.83}$$

where the external source factors $\mathcal{Z}[J,\phi,t_j]$ have been combined with E to give E_J,

$$E_J(t_+,t_-) = T\left\{\exp\left[-i\int_{t_-}^{t_+} \mathrm{d}t \int \mathrm{d}^3x\, (\mathcal{H}_I(\varphi_\alpha,\pi_\alpha) + J_\alpha\varphi_\alpha)\right]\right\}. \tag{4.84}$$

The two versions of the generating functional, equations (4.80) and (4.83), are normalized, so that $Z_0[0] = 1$. In what follows, the path integral version of equation (4.80) will be derived by using the interaction picture version (4.84) and coherent states.

4.6 Coherent states and the scalar field path integral

The Fock space states in the interaction picture of field theory are similar to the states of the quantum mechanical harmonic oscillator. The particle states are excitations given by applying the creation operator to the vacuum state, much in the same way that the energy eigenstates of the harmonic oscillator are given by the creation operator applied to the ground state. The incoming and outgoing particle states are reduced to the vacuum or ground state by the LSZ reduction process, which eventually results in a time-ordered product of field operators. As a result, the field theory path integral is often defined as a functional representation of the Schwinger generating functional (4.83). By using coherent states in the interaction picture Slavnov time-sliced the field theory evolution operator (4.84) in a manner similar to the harmonic oscillator path integral (2.85) and derived the field theory path integral.

For the sake of simplified notation, coherent states will be developed initially for the real free scalar field φ of equation (4.43). Two real c-number functions, $\psi(x)$ and $\rho(x)$, are introduced and defined by

$$\psi(x) = \int \mathrm{d}^3p\Big(\psi_p(t)f_p^*(x) + \psi_p^*(t)f_p(x)\Big), \tag{4.85}$$

$$\rho(x) = -i\int \mathrm{d}^3p\ \varepsilon_p\Big(\psi_p(t)f_p^*(x) - \psi_p^*(t)f_p(x)\Big), \tag{4.86}$$

where $f_p(x) = e^{ip\cdot x}/\sqrt{(2\pi)^3 2\varepsilon_p}$ and $p_0 = \varepsilon_p = \sqrt{\boldsymbol{p}^2 + m^2}$. The function $f_p(x)$ satisfies the Klein–Gordon free field equation, $(\Box + m^2)f_p(x) = 0$. The coefficients of these expansions, denoted $\psi_p(t)$, are complex c-number quantities *indexed by the time t* and the momentum $\boldsymbol{p}$. As a result, neither equation (4.85) nor equation (4.86) satisfy the free field equation, allowing them to manifest the nonlinear behavior of interacting fields. The two c-number fields of equations (4.85) and (4.86) will play the role of the quantized field and its canonical momentum in the context of the path integral action for field theory. While $\psi_p(t)$ is written as a function of time, it is not necessarily differentiable, and so equations (4.85) and (4.86) should not be viewed as Fourier transforms It is these coefficients that will become the variables of integration in the field theory path integral, like the coefficients a_k in the earlier quantum mechanical case (3.43).

Using these coefficients, the field theory coherent states in the interaction picture are defined similarly to equation (2.75),

$$|\psi, t\rangle = \exp\left\{\int d^3p\left(\psi_p(t)\hat{a}_p^\dagger - \psi_p^*(t)\hat{a}_p\right)\right\}|0\rangle. \tag{4.87}$$

Combining equation (4.45) with equation (1.54) gives the field theory versions of equation (2.82),

$$\hat{a}_k|\psi, t\rangle = \psi_k(t)|\psi, t\rangle, \quad \langle\psi, t|\hat{a}_k^\dagger = \langle\psi, t|\psi_k^*(t). \tag{4.88}$$

In turn, this gives the matrix elements of the free scalar field $\varphi(x)$,

$$\langle\psi, t'|\varphi(x)|\psi, t\rangle = (\psi^{(-)}(x, t') + \psi^{(+)}(x, t))\langle\psi, t'|\psi, t\rangle, \tag{4.89}$$

where $x = (\boldsymbol{x}, t)$ and the two functions are given by

$$\psi^{(-)}(x, t') = \int d^3p\, \psi_p^*(t')f_p(x), \quad \psi^{(+)}(x, t) = \int d^3p\, \psi_p(t)f_p^*(x). \tag{4.90}$$

These two functions satisfy $\psi^{(-)}(x, t) + \psi^{(+)}(x, t) = \psi(x)$. A similar relationship holds between the canonical momentum $\pi(x)$ and the c-number function $\rho(x)$,

$$\langle\psi, t'|\pi(x)|\psi, t\rangle = (\rho^{(-)}(x, t') + \rho^{(+)}(x, t))\langle\psi, t'|\psi, t\rangle, \tag{4.91}$$

where again $x = (\boldsymbol{x}, t)$ and the two functions are given by

$$\rho^{(-)}(x, t') = i\int d^3p\, \varepsilon_p\, \psi_p^*(t')f_p(x), \quad \rho^{(+)}(x, t) = -i\int d^3p\, \varepsilon_p\, \psi_p(t)f_p^*(x). \tag{4.92}$$

Because φ and π contain both annihilation and creation operators, extending equations (4.89) and (4.91) to products of φ and π requires introducing the concept of *normal ordering*. The normal ordered product, denoted $:\varphi^n(x):$ consists of placing all annihilation operators in the product to the right and all creation operators in the product to the left. In this process all commutators are ignored. As an example, the normal ordered version of products of the number operator are given by $:(\hat{a}_p^\dagger\hat{a}_p)^n: = (\hat{a}_p^\dagger)^n(\hat{a}_p)^n$. This allows an unambiguous evaluation of matrix elements for normal ordered field products. It is straightforward to show that normal ordered products of the field have the matrix elements

$$\langle\psi, t'|:\varphi^n(x):|\psi, t\rangle = (\psi^{(-)}(x, t') + \psi^{(+)}(x, t))^n\langle\psi, t'|\psi, t\rangle, \tag{4.93}$$

with a similar expression for normal ordered products of π.

The BCH theorem (1.55) can be applied to the definition (4.87). Using the algebra (4.45) gives a result similar to the inner product (2.77),

$$\langle\psi,\, t'|\psi,\, t\rangle = \exp\left\{\frac{1}{2}\int \mathrm{d}^3p\left[2\psi_p^*(t')\psi_p(t) - \psi_p^*(t)\psi_p(t) - \psi_p^*(t')\psi_p(t')\right]\right\}. \quad (4.94)$$

The field theory coherent states are also complete. Adapting the harmonic oscillator result (2.78) gives

$$\int [\mathcal{D}^2\psi(t)]\, \langle\psi,\, t''|\psi,\, t\rangle\langle\psi,\, t|\psi,\, t'\rangle = \langle\psi,\, t''|\psi,\, t'\rangle, \quad (4.95)$$

where the measure consists of an integration over the complex plane for each coefficient, $\psi_p(t) = \psi_p^R + i\psi_p^I$,

$$[\mathcal{D}^2\psi(t)] = \prod_p \frac{\mathrm{d}^3p}{\pi}\, \mathrm{d}\psi_p^R(t)\, \mathrm{d}\psi_p^I(t) = \prod_p \frac{\mathrm{d}^3p}{2\pi i}\, \mathrm{d}\psi_p^*(t)\, \mathrm{d}\psi_p(t). \quad (4.96)$$

The form of the measure (4.96) exhibits a mild complication for coherent states in field theory. There is a continuum of momentum states and the product over a continuum is a poorly defined quantity and is worsened by the presence of the infinitesimal factor d^3p. This can be remedied by placing the system in a cubic spatial box of width L. This results in discrete momenta, $p_n = 2\pi \boldsymbol{n}/L$, where $\boldsymbol{n} = (n_x, n_y, n_z)$ is a vector in the lattice of integers. The coefficients become $\psi_p(t) \to \psi_n(t)$, a discrete set that is indexed by $\boldsymbol{n}$. The integral in equation (4.94) becomes a sum according to $\int \mathrm{d}^3p \to \sum_n (2\pi/L)^3$, with the limit $L \to \infty$ understood. The product in equation (4.96) becomes $\prod_p \to \prod_n$, and is a product over all possible *negative and positive integers* n_x, n_y, and n_z. Using this approach the measure of equation (4.96) can be made well defined.

The scalar field theory path integral can now be defined by using the completeness (4.95) and the same time-slicing method employed in the quantum mechanical case (2.9). The transition element between two coherent states in the presence of a source term J is given by

$$\begin{aligned} Z_{\psi_\pm}[J] &= \langle\psi,\, t_+|E_J(t_+,\, t_-)|\psi,\, t_-\rangle \\ &= \int [\mathcal{D}^2\psi] \prod_{j=0}^{N-1} \langle\psi,\, t_{j+1}|E_J(t_{j+1},\, t_j)|\psi,\, t_j\rangle, \end{aligned} \quad (4.97)$$

where $t_j = t_- + j\epsilon$ and $\epsilon = (t_+ - t_-)/N \equiv T/N$, so that $t_N = t_+$ and $t_0 = t_-$. The measure in equation (4.97) is the product of $N - 1$ copies of the measure (4.96),

$$[\mathcal{D}^2\psi] = \prod_{j=1}^{N-1} [\mathcal{D}^2\psi(t_j)] = \prod_{j=1}^{N-1}\prod_p \frac{\mathrm{d}^3p}{\pi} \mathrm{d}\psi_p^R(t_j)\, \mathrm{d}\psi_p^I(t_j). \quad (4.98)$$

Since the limit $N \to \infty$ is understood, the time interval ϵ is infinitesimal. For the case that $t_{j+1} - t_j = \epsilon$ is infinitesimal, the evolution operator (4.84) becomes

$$E_J(t_{j+1}, t_j) = \exp\left\{-i\epsilon \int d^3x\, (:\mathcal{H}_I(\varphi, \pi):+J\varphi)\right\}, \tag{4.99}$$

where the interaction picture operators have been normal ordered and the fields and source are functions of $x_j = (\boldsymbol{x}, t_j)$. The coherent state matrix element for equation (4.99) can now be evaluated using equation (4.93) and the similar momentum expression. The method (3.54) for normalizing the quantum mechanical path integral using functional determinants will now be adapted to the field theory case. Like the quantum mechanical case this approach *assumes* that the coefficients $\psi_p(t)$ are *smooth or differentiable functions of time.* This means that $\psi_p(t_j + \epsilon) \approx \psi_p(t_j) + \epsilon\dot{\psi}_p(t_j)$. The sum of the functions in equation (4.90) becomes

$$\begin{aligned}\psi^{(-)}(x_j, t_{j+1}) + \psi^{(+)}(x_j, t_j) &= \psi^{(-)}(x_j, t_j) + \psi^{(+)}(x_i, t_j) + O(\epsilon)\\ &= \psi(x_j) + O(\epsilon),\end{aligned} \tag{4.100}$$

with a similar result for the sum $\rho^{(-)}(x_j, t_{j+1}) + \rho^{(+)}(x_j, t_j) = \rho(x_j) + O(\epsilon)$. Due to the overall factor of ϵ in the exponential of equation (4.99), the $O(\epsilon)$ term can be discarded since it becomes $O(\epsilon^2)$. As a result, the matrix element of the infinitesimal evolution operator is given by

$$\begin{aligned}&\langle\psi, t_{j+1}|E_J(t_{j+1}, t_j)|\psi, t_j\rangle\\ &= \langle\psi, t_{j+1}|\exp\left\{-i\epsilon \int d^3x\, (:\mathcal{H}_I(\varphi, \pi):+J\varphi)\right\}|\psi, t_j\rangle\\ &= \langle\psi, t_{j+1}|\psi, t_j\rangle \exp\left\{-i\epsilon \int d^3x\, (\mathcal{H}_I(\psi, \rho) + J\psi)\right\},\end{aligned} \tag{4.101}$$

where the fields and source are evaluated at x_j and the spatial integral is over the entire system.

The next step is to express the inner product appearing in equation (4.101) in terms of the functions (4.85) and (4.86). Using the *assumption of differentiability* gives $\psi_p(t_{j+1}) \approx \psi_p(t_j) + \epsilon\dot{\psi}_p(t_j)$. Discarding terms of $O(\epsilon^2)$, the infinitesimal version of the inner product (4.94) becomes

$$\begin{aligned}&\langle\psi, t_{j+1}|\psi, t_j\rangle\\ &= \exp\left\{-\frac{1}{2}\int d^3p\left[\psi_p^*(t_j)\psi_p(t_j) + \psi_p^*(t_{j+1})\psi_p(t_{j+1}) - 2\psi_p^*(t_{j+1})\psi_p(t_j)\right]\right\}\\ &= \exp\left\{\frac{1}{2}\epsilon\int d^3p\left[\dot{\psi}_p^*(t_j)\psi_p(t_j) - \psi_p^*(t_j)\dot{\psi}_p(t_j)\right]\right\}.\end{aligned} \tag{4.102}$$

The assumption of differentiability for the $\psi_p(t)$ as well as the results $\dot{f}_p(x) = i\varepsilon_p f_p(x)$ and $(\nabla^2 - m^2)f_p(x) = -\varepsilon_p^2 f_p(x)$ give the time derivatives of the two functions (4.85) and (4.86),

$$\dot{\psi}(x) = \rho(x) + \int d^3p\left(\dot{\psi}_p(t)f_p^*(x) + \dot{\psi}_p^*(t)f_p(x)\right), \tag{4.103}$$

$$\dot{\rho}(x) = (\nabla^2 - m^2)\psi(x) - i\int d^3p\, \varepsilon_p\Big(\psi_p(t)f_p^*(x) - \dot{\psi}_p^*(t)f_p(x)\Big). \tag{4.104}$$

Using equations (4.103) and (4.104), the infinitesimal inner product (4.102) can be written

$$\begin{aligned}
&\langle\psi, t_{j+1}|\psi, t_j\rangle \\
&= \exp\left\{\frac{1}{2}\epsilon\int d^3p\left[\dot{\psi}_p^*(t_j)\psi_p(t_j) - \psi_p^*(t_j)\dot{\psi}_p(t_j)\right]\right\} \\
&= \exp\left\{i\epsilon\int d^3x\left(\frac{1}{2}\rho\dot{\psi} - \frac{1}{2}\dot{\rho}\psi - \frac{1}{2}\rho^2 - \frac{1}{2}\nabla\psi\cdot\nabla\psi - \frac{1}{2}m^2\psi^2\right)\right\} \\
&= \exp\left\{i\epsilon\int d^3x\left(\frac{1}{2}\rho\dot{\psi} - \frac{1}{2}\dot{\rho}\psi - \mathcal{H}_0(\psi, \rho)\right)\right\},
\end{aligned} \tag{4.105}$$

where $\mathcal{H}_0$ is the free scalar Hamiltonian density defined in equation (4.27), while the ψ and ρ appearing in equation (4.105) are functions of $x_j \equiv (\boldsymbol{x}, t_j)$. The key to demonstrating equation (4.105) is using the two identities

$$\begin{aligned}
\int d^3x\, f_p^*(\boldsymbol{x}, t_j)f_k(\boldsymbol{x}, t_j) &= \varepsilon_p^{-1}\delta^3(\boldsymbol{p} - \boldsymbol{k}), \\
\int d^3x\, f_p(\boldsymbol{x}, t_j)f_k(\boldsymbol{x}, t_j) &= \varepsilon_p^{-1}e^{2i\varepsilon_p t_j}\delta^3(\boldsymbol{p} + \boldsymbol{k}),
\end{aligned} \tag{4.106}$$

along with their complex conjugates. Integrating the two terms $\frac{1}{2}\rho\dot{\psi} - \frac{1}{2}\dot{\rho}\psi$ using equation (4.106) and the expansions (4.85), (4.86), (4.103), and (4.104) demonstrates the equality between the first and second lines of equation (4.105).

Result (4.105) can now be combined with equation (4.101) and returned to the transition element of (4.97). Denoting $\rho_j = \rho(\boldsymbol{x}, t_j)$, $\psi_j = \psi(\boldsymbol{x}, t_j)$, and $J_j = J(\boldsymbol{x}, t_j)$ gives the path integral for the scalar field coherent state transition element,

$$\begin{aligned}
Z_\pm[J] &\equiv \langle\psi, t_+|E_J(t_+, t_-)|\psi, t_-\rangle \\
&= \int_{\psi_-}^{\psi_+}[\mathcal{D}^2\psi]\exp\left\{i\sum_{j=0}^{N-1}\epsilon\int d^3x\left(\frac{1}{2}\rho_j\dot{\psi}_j - \frac{1}{2}\dot{\rho}_j\psi_j - \mathcal{H}(\rho_j, \psi_j) - J_j\psi_j\right)\right\} \\
&= \int_{\psi_-}^{\psi_+}[\mathcal{D}^2\psi]\exp\left\{i\int_{t_-}^{t_+}dt\int d^3x\left(\frac{1}{2}\rho\dot{\psi} - \frac{1}{2}\dot{\rho}\psi - \mathcal{H}(\rho, \psi) - J\psi\right)\right\},
\end{aligned} \tag{4.107}$$

where $\mathcal{H} = \mathcal{H}_0 + \mathcal{H}_I$ is the full interacting Hamiltonian density and the measure is given by equation (4.98). The limits on the path integral, $\psi_- = \psi(\boldsymbol{x}, t_-)$ and $\psi_+ = \psi(\boldsymbol{x}, t_+)$, are those defined by the choice of incoming and outgoing coherent state. The spacetime integral and the path integral action are relativistically invariant. Up to a time integration by parts, the continuum form of the path integral action displayed in the last step of equation (4.107) is the Hamiltonian form of the Lagrangian density, $\mathcal{L}_J(\rho, \psi) = \rho\dot{\psi} - \mathcal{H}(\rho, \psi) - J\psi$, in the presence of an external source. As a result, equation (4.107) is the field theory counterpart to the quantum mechanical path integral (2.11). It is important to note that an integration by parts on the action term $-\frac{1}{2}\dot{\rho}\psi$ generates the surface term factor $\mathcal{S}$, where

$$\mathcal{S} = \exp\left\{-\frac{1}{2}i\int \mathrm{d}^3x\,(\rho(\boldsymbol{x},\,t_+)\psi(\boldsymbol{x},\,t_+) - \rho(\boldsymbol{x},\,t_-)\psi(\boldsymbol{x},\,t_-))\right\}. \tag{4.108}$$

The surface term phase factor $\mathcal{S}$ is unity if the initial and final coherent states of the transition element are identical, which corresponds to the choice $\psi_{\boldsymbol{p}}(t_+) = \psi_{\boldsymbol{p}}(t_-)$. The case of the vacuum transition element corresponds to $\psi_{\boldsymbol{p}}(t_\pm) = 0$, and for that case the functional $Z_\pm[J]$ of equation (4.107) becomes the path integral version of the Schwinger generating functional $Z_0[J]$ of equation (4.80). However, there are important physical processes, such as the spontaneous breakdown of symmetry, where the choice $\psi_\pm \neq 0$ plays an important role. This is discussed in the next chapter.

As it stands, the path integral of equation (4.107) is often viewed intuitively as a *sum over paths through the space of field configurations* that begin with the initial coherent state configuration and end with the final coherent state configuration. The intermediate paths are weighted by the phase of the classical field action $\exp(iS_c)$ for the fluctuation. In this sense, the field theory path integral (4.107) is a generalization of the quantum mechanical path integral (2.11) to *function space.*

The field theory version of the Lagrangian path integral (2.13) can be obtained by a change of variables that allows $\rho(x)$ to be integrated. Denoting $R = \frac{1}{2}N - 1$, the large N and T limit for the change of variables is given by the Fourier transform (1.24),

$$\frac{e^{-i\varepsilon_p t_j}}{\sqrt{2\varepsilon_p}}\psi_{\boldsymbol{p}}(t_j) = \lim_{N\to\infty}\sum_{n=-R}^{R}\frac{\pi}{T}\phi_{(\omega_n,\boldsymbol{p})}\,e^{-i\omega_n t_j} = \int_{-\infty}^{\infty}\mathrm{d}\omega\;\phi_p\,e^{-i\omega t_j}, \tag{4.109}$$

where $\omega_n = n\pi/T$. The range of the sum allows the new fluctuation variables $\phi_{(\omega_n,\boldsymbol{p})}$ to match the number of the previous fluctuation variables $\psi_{\boldsymbol{p}}(t_j)$. In the large N and T limit ω_n becomes the continuous variable ω, so that $p = (\omega,\,\boldsymbol{p})$ becomes an arbitrary four-momentum where $\omega = p_0$ is no longer identified as $\varepsilon_{\boldsymbol{p}}$. The Fourier coefficients ϕ_p are relativistic scalars that are indexed by p. Under this change of variables the path integral measure becomes

$$[\mathcal{D}^2\psi] = \prod_{n=-R}^{R}\prod_{\boldsymbol{p}}\mathcal{N}(\omega_n,\,\boldsymbol{p})\,\mathrm{d}\phi^*_{(\omega_n,\,\boldsymbol{p})}\,\mathrm{d}\phi_{(\omega_n,\boldsymbol{p})} \rightarrow \prod_p \mathcal{N}(p)\mathrm{d}\phi^*_p\,\mathrm{d}\phi_p, \tag{4.110}$$

so that it is a product over both of the continuous variables ω and $\boldsymbol{p}$. The factor $\mathcal{N}(p)$ appearing in equation (4.110) is an amalgamation of the Jacobian from the change of variables (4.109) and the factors originally in the measure (4.98), as well as the factor $\mathcal{N}_S J$ from equation (3.49) necessary to compensate for the assumption that $\psi_{\boldsymbol{p}}(t)$ is a differentiable function of time. All these factors combine to give $\mathcal{N}(p) = \mathcal{N}_S^2 J^2\varepsilon_p\,\mathrm{d}^3p\;\mathrm{d}^2\omega/2\pi^2 i$. However, this explicit form will not be necessary since the normalization relation $Z_0[0] = 1$ can be used to infer the total normalization coefficient. As a result, the specific form for $\mathcal{N}$ is not necessary.

Because the path integral measure (4.110) is a product over all p it can be written in several alternative forms. The first is that

$$\begin{aligned} \prod_p \mathcal{N}(p)\, \mathrm{d}\phi_p^*\, \mathrm{d}\phi_p &= \left(\prod_p \mathcal{N}(p)\, \mathrm{d}\phi_p^*\right)\left(\prod_p \mathrm{d}\phi_p\right) \\ &= \left(\prod_p \mathcal{N}(p)\, \mathrm{d}\phi_p^*\right)\left(\prod_p \mathrm{d}\phi_{-p}\right) \\ &= \prod_p \mathcal{N}(p)\, \mathrm{d}\phi_p^*\, \mathrm{d}\phi_{-p}, \end{aligned} \tag{4.111}$$

while the second results from separating the positive and negative ω contributions,

$$\begin{aligned} \prod_p \mathcal{N}(p)\, \mathrm{d}\phi_p &= \prod_{\omega>0} \prod_{\boldsymbol{p}} \left(\mathcal{N}(\omega, \boldsymbol{p})\, \mathrm{d}\phi_{(\omega,\boldsymbol{p})}\, \mathcal{N}(-\omega, \boldsymbol{p})\, \mathrm{d}\phi_{(-\omega,\boldsymbol{p})}\right) \\ &= \left(\prod_{\omega>0} \prod_{\boldsymbol{p}} \mathcal{N}(\omega, \boldsymbol{p})\, \mathrm{d}\phi_{(\omega,\boldsymbol{p})}\right)\left(\prod_{\omega>0} \prod_{\boldsymbol{p}} \mathcal{N}(-\omega, \boldsymbol{p})\, \mathrm{d}\phi_{(-\omega,\boldsymbol{p})}\right) \\ &= \left(\prod_{\omega>0} \prod_{\boldsymbol{p}} \mathcal{N}(\omega, \boldsymbol{p})\, \mathrm{d}\phi_{(\omega,\boldsymbol{p})}\right)\left(\prod_{\omega>0} \prod_{\boldsymbol{p}} \mathcal{N}(-\omega, -\boldsymbol{p})\, \mathrm{d}\phi_{(-\omega,-\boldsymbol{p})}\right) \\ &= \prod_{\omega>0} \prod_{\boldsymbol{p}} \mathcal{N}(p)\, \mathcal{N}(-p)\, \mathrm{d}\phi_p\, \mathrm{d}\phi_{-p}. \end{aligned} \tag{4.112}$$

The change of variables (4.109) allows equations (4.85) and (4.86) to be written

$$\psi(x) = \int \frac{\mathrm{d}^4p}{(2\pi)^2}\left(\phi_p^* + \phi_{-p}\right)e^{ip\cdot x} \equiv \int \frac{\mathrm{d}^4p}{(2\pi)^2}\, \psi_p\, e^{ip\cdot x}, \tag{4.113}$$

$$\rho(x) = i\int \frac{\mathrm{d}^4p}{(2\pi)^2}\, \varepsilon_p\left(\phi_p^* - \phi_{-p}\right)e^{ip\cdot x} \equiv i\int \frac{\mathrm{d}^4p}{(2\pi)^2}\, \varepsilon_p\, \rho_p\, e^{ip\cdot x}. \tag{4.114}$$

A key element of equation (4.114) is the relation $\varepsilon_{-p} = \varepsilon_p$. The new variables, $\psi_p = \phi_p^* + \phi_{-p}$ and $\rho_p = \phi_p^* - \phi_{-p}$, have the necessary properties that $\psi_{-p} = \psi_p^*$ and $\rho_{-p} = -\rho_p^*$ so that $\psi(x)$ and $\rho(x)$ are real. Combining identity (4.111) with the change of variables $\mathrm{d}\phi_p^*\, \mathrm{d}\phi_{-p} = \frac{1}{2}\mathrm{d}\psi_p\, \mathrm{d}\rho_p$ shows that the measure (4.110) becomes

$$[\mathcal{D}^2\psi] = \prod_p \frac{1}{2}\mathcal{N}(p)\, \mathrm{d}\psi_p\, \mathrm{d}\rho_p. \tag{4.115}$$

The final step is to use identity (4.112) for both factors in equation (4.115) as well as the relations $d\rho_{-p} = -d\rho_p^*$ and $d\psi_{-p} = d\psi_p^*$ to obtain

$$
\begin{aligned}
[\mathcal{D}^2\psi] &= \prod_{\omega>0}\prod_{p} \frac{1}{4}\mathcal{N}(p)\mathcal{N}(-p)\, d\rho_p\, d\rho_{-p}\, d\psi_p\, d\psi_{-p} \\
&= \prod_{\omega>0}\prod_{p} \left(-\frac{1}{4}\mathcal{N}(p)\mathcal{N}(-p)\, d\rho_p^*\, d\rho_p\, d\psi_p^*\, d\psi_p\right) \\
&= \prod_{\omega>0}\prod_{p} \mathcal{N}(p)\mathcal{N}(-p)\, d\rho_p^R\, d\rho_p^I\, d\psi_p^R\, d\psi_p^I,
\end{aligned}
\tag{4.116}
$$

which shows that the field ψ and the momentum ρ are independent degrees of freedom in the path integral.

For the case that the interaction $\mathcal{H}_I(\psi, \rho)$ depends solely on ψ, the ρ variables can be integrated from the path integral. After the time integration by parts that generates equation (4.108), the relevant piece of the action can be evaluated using the general property of integrating a function $f(p)$ over the entirety of phase space,

$$
\begin{aligned}
\int_{-\infty}^{\infty} d\omega \int d^3p\, f(p) &= \int_0^{\infty} d\omega \int d^3p(f(\omega, \mathbf{p}) + f(-\omega, \mathbf{p})) \\
&= \int_0^{\infty} d\omega \int d^3p(f(p) + f(-p)).
\end{aligned}
\tag{4.117}
$$

Combining equation (4.117) with the expansions (4.113) and (4.114), the Dirac delta integral representation (1.26), and the properties $\rho_{-p} = -\rho_p^*$ and $\psi_{-p} = \psi_p^*$, the part of the action that involves ρ, denoted S_ρ, becomes

$$
\begin{aligned}
S_\rho = i\int d^4x\left(\rho\dot{\psi} - \frac{1}{2}\rho^2\right) &= i\int_{-\infty}^{\infty} d\omega \int d^3p\left(\frac{1}{2}\varepsilon_p^2\rho_p\rho_{-p} - \omega\varepsilon_p\rho_p\psi_{-p}\right) \\
&= -i\int_0^{\infty} d\omega \int d^3p\left(\frac{1}{2}\varepsilon_p^2(\rho_p^*\rho_p + \rho_{-p}^*\rho_{-p}) + \omega\varepsilon_p(\rho_p\psi_p^* - \rho_{-p}\psi_{-p}^*)\right) \\
&= -i\int_0^{\infty} d\omega \int d^3p\left(\varepsilon_p^2\, \rho_p^*\rho_p + \omega\varepsilon_p(\rho_p\psi_p^* + \rho_p^*\psi_p)\right) \\
&= -i\int_0^{\infty} d\omega \int d^3p\left(\varepsilon_p^2\left(\rho_p^* + \frac{\omega}{\varepsilon_p}\psi_p^*\right)\left(\rho_p + \frac{\omega}{\varepsilon_p}\psi_p\right) - \omega^2\psi_p^*\psi_p\right) \\
&= -i\int_0^{\infty} d\omega \int d^3p\left(\varepsilon_p^2\pi_p^*\pi_p - \omega^2\psi_p^*\psi_p\right),
\end{aligned}
\tag{4.118}
$$

where $\pi_p = \rho_p + \omega\psi_p/\varepsilon_p$ and $\pi_p^*\pi_p = \pi_p^{R2} + \pi_p^{I2}$. The limits on the integral in the last line of equation (4.118) are extremely important since the measures for ρ_p and ψ_p involve *only positive values of* ω. The measure for ρ_p is translationally invariant, and so $d\rho_p \to d\pi_p$ leaves the path integral invariant.

Result (4.118) shows that the integrations over π_p are Fresnel integrals familiar from equation (1.101). However, field theory path integrals are more commonly evaluated in the *Euclidean formulation* based around the Wick rotation $t \overset{\text{W}}{\to} -it$. The origin of the term *Euclidean* is revealed by noting that $x_\mu x^\mu \overset{\text{W}}{\to} -t^2 - \boldsymbol{x}^2 < 0$ corresponds to a four-dimensional metric that places space and time on an equal footing. Since frequencies correspond to an inverse time, the Wick rotation for ω, k_0, or ε_p is the opposite sign from t, and are given by $\omega,\ k_0,\ \varepsilon_p \overset{\text{W}}{\to} i\omega,\ ik_0,\ i\varepsilon_p$. The Euclidean form then gives $\omega t \overset{\text{W}}{\to} \omega t$, so that the inner product remains $k \cdot x = \omega t - \boldsymbol{k} \cdot \boldsymbol{x} \overset{\text{W}}{\to} \omega t - \boldsymbol{k} \cdot \boldsymbol{x}$. This allows waves to remain traveling in the same direction after the Wick rotation. Using this recipe for the integration over π_p in the action (4.118) then yields two copies of the well defined Gaussian integral (1.99) for each value of ω and $\boldsymbol{p}$,

$$\int\left(\prod_{\omega>0,\boldsymbol{p}} \mathrm{d}\pi_p^R \mathrm{d}\pi_p^I\right)\exp\left\{-\int_{\omega>0} \mathrm{d}^4p_E\ \varepsilon_p^2\left(\pi_p^{R\,2} + \pi_p^{I\,2}\right)\right\} = \prod_{\omega>0,\boldsymbol{p}} \frac{\pi}{\varepsilon_p^2\, \mathrm{d}^4p_E}. \tag{4.119}$$

The inverse Wick rotation, $\mathrm{d}^4p_E \overset{\text{M}}{\to} -i\mathrm{d}^4p$ and $\varepsilon_p^2 \overset{\text{M}}{\to} -\varepsilon_p^2$, gives $\varepsilon_p^2\, \mathrm{d}^4p_E \overset{\text{M}}{\to} i\varepsilon_p^2\, \mathrm{d}^4p$. Result (4.119) becomes identical to the result that would have been obtained using the Fresnel formula (1.101) to integrate the original complex Gaussian integrals. The product (4.119) is combined with the factor $\mathcal{N}(p)$ in the measure (4.116) by writing $\mathrm{d}^4p = \mathrm{d}^3p\ \pi/T$ and using $\prod_{\omega>0} = \prod_{n=1}^{R}$, where $R = \frac{1}{2}N - 1$. The final factor is denoted $\tilde{\mathcal{N}}(p)$.

Using equation (4.117), $-\omega\omega'\,\delta^4(p + p') = \omega^2\,\delta^4(p + p')$, and $\psi_{-p} = \psi_p^*$ shows that the remaining integral in equation (4.118) can be written

$$i\int_0^\infty \mathrm{d}\omega \int \mathrm{d}^3p\ \omega^2\psi_p^*\psi_p = i\int \mathrm{d}^4p\ \frac{1}{2}\omega^2\psi_p^*\psi_p = i\int \mathrm{d}^4x\ \frac{1}{2}\dot{\psi}^2(x). \tag{4.120}$$

After integrating the momentum the scalar field path integral (4.107) becomes

$$Z_\pm[J] = \mathcal{S}\int [\tilde{\mathcal{D}}\psi] \exp\left\{i\int \mathrm{d}^4x\left(\frac{1}{2}\partial_\mu\psi\ \partial^\mu\psi - \frac{1}{2}m^2\psi^2 - \mathcal{H}_I(\psi) - J\psi\right)\right\}, \tag{4.121}$$

where $\mathcal{S}$ is the surface term given by equation (4.108) and the measure is

$$[\tilde{\mathcal{D}}\psi] = \prod_p \tilde{\mathcal{N}}(p)\ \mathrm{d}\psi_p^*\ \mathrm{d}\psi_p = \prod_p 2i\tilde{\mathcal{N}}(p)\ \mathrm{d}\psi_p^R\ \mathrm{d}\psi_p^I. \tag{4.122}$$

Results (4.121) and (4.122) constitute the Lagrangian version of the scalar field path integral. Combining it with the expansion (4.113) served as the starting point for Feynman's path integral version of quantum field theory.

Using equations (4.113) and (4.114) allows a *configuration space measure* to be defined by relating ψ_p to $\psi(x)$ and $\rho(x)$. Using the Dirac delta (1.15), it can be shown that

$$\psi_p = \frac{1}{2}\int \frac{\mathrm{d}^4x}{(2\pi)^2}\, e^{ip\cdot x}\left(\psi(x) + \frac{i}{\varepsilon_p}\rho(x)\right), \tag{4.123}$$

with a similar expression for ψ_p^* found by taking the complex conjugate. It then follows that the path integral measure (4.110) can be written

$$[\mathcal{D}^2\psi] = [\mathcal{D}\psi\ \mathcal{D}\rho] = \prod_x N(x)\, \mathrm{d}\psi(x)\, \mathrm{d}\rho(x), \tag{4.124}$$

where the product over x can be regularized by treating spacetime as a lattice. The factor $N(x)$ is found from the Jacobian of the change of variables given by equation (4.123) and the original measure. However, the details of this Jacobian will not be presented since the path integral normalization can be accomplished by easier methods. A more thorough discussion of the configuration space measure (4.124) is given in a later section.

4.7 Coherent states and the Dirac bispinor path integral

Deriving the path integral for the Dirac spinor field shares many similarities to the derivation of the scalar field path integral in the previous section. The major difference lies in the need to accommodate the anticommuting creation and annihilation operators (4.53). This will be done by extending the approach used to derive the coherent state path integral (2.111) for the Grassmann harmonic oscillator to the case of Grassmann fields.

Like the c-number scalar field expansions (4.85) and (4.86), the starting point is to introduce the Grassmann expansion

$$\eta(x) = \int \frac{\mathrm{d}^3p}{(2\pi)^{3/2}}\sqrt{\frac{m}{\varepsilon_p}}\sum_{s=1}^{2}\left(\beta_p^s(t)u_s(p)e^{-ip\cdot x} + \alpha_p^{s*}(t)\, v_s(p)e^{ip\cdot x}\right), \tag{4.125}$$

where $p_0 = \varepsilon_p = \sqrt{\boldsymbol{p}^2 + m^2}$, $u^s(p)$ and $v^s(p)$ are the solutions of equation (4.49) given by (4.51). The coefficients of the expansions are complex time-dependent Grassmann variables, so that $\{\beta_p^s(t), \beta_{p'}^{s'}(t')\} = 0$, with similar anticommutators for $\alpha_p^s(t)$. The definition (4.125) gives the functional dependence between the coefficients and the field expansion (4.125),

$$\frac{\delta\eta_a(x)}{\delta\beta_p^s(t)} = \sqrt{\frac{m}{(2\pi)^3\varepsilon_p}}\, u_a^s(p)e^{-ip\cdot x}, \quad \frac{\delta\eta_a(x)}{\delta\alpha_p^{s*}(t)} = \sqrt{\frac{m}{(2\pi)^3\varepsilon_p}}\, v_a^s(p)e^{ip\cdot x}. \tag{4.126}$$

Using equation (4.125) gives the conjugate field expansion for $\bar{\eta}(x)$ since it is defined as $\bar{\eta}(x) = \eta^\dagger(x)\gamma_0$,

$$\bar{\eta}(x) = \int \frac{d^3p}{(2\pi)^{3/2}} \sqrt{\frac{m}{\varepsilon_p}} \sum_{s=1}^{2} \left(\beta_p^{s*}(t) \bar{u}_s(p) e^{-ip\cdot x} + \alpha_p^s(t)\, \bar{v}_s(p) e^{ip\cdot x} \right). \tag{4.127}$$

The definition (4.127) gives the functional dependence between the complex conjugate of the coefficients and $\bar{\eta}(x)$,

$$\frac{\delta \bar{\eta}_a(x)}{\delta \beta_p^{s*}(t)} = \sqrt{\frac{m}{(2\pi)^3 \varepsilon_p}}\, \bar{u}_a^s(p) e^{-ip\cdot x}, \quad \frac{\delta \bar{\eta}_a(x)}{\delta \alpha_p^s(t)} = \sqrt{\frac{m}{(2\pi)^3 \varepsilon_p}}\, \bar{v}_a^s(p) e^{ip\cdot x}. \tag{4.128}$$

The expansion η will play the role of the Dirac field in the path integral, while $\bar{\eta}$ will play the role of the conjugate momentum.

The Grassmann coherent states $|\eta, t\rangle$ are defined by using the coefficients of equation (4.125) similarly to equation (4.87),

$$|\eta, t\rangle = \exp\left\{ \int d^3p \left(\hat{b}_p^{s\dagger} \beta_p^s(t) - \beta_p^{s*}(t) \hat{b}_p^s + \hat{d}_p^{s\dagger} \alpha_p^s(t) - \alpha_p^{s*}(t) \hat{d}_p^s \right) \right\} |0\rangle, \tag{4.129}$$

where the order of the operators and Grassmann variables is important since they anticommute, so that $\{\hat{b}_k^{s'}, \beta_p^s\} = 0$. Like the Grassmann coherent state inner product (2.104), combining the BCH theorem (1.55) with the algebra of equation (4.53) gives the inner product of the states defined by equation (4.129),

$$\begin{aligned}\langle \eta, t' | \eta, t\rangle = &\exp\left\{ \frac{1}{2} \int d^3p \left[2\beta_p^{s*}(t')\beta_p^s(t) - \beta_p^{s*}(t)\beta_p^s(t) - \beta_p^{s*}(t')\beta_p^s(t') \right] \right\} \\ &\times \exp\left\{ \frac{1}{2} \int d^3p \left[2\alpha_p^{s*}(t')\alpha_p^s(t) - \alpha_p^{s*}(t)\alpha_p^s(t) - \alpha_p^{s*}(t')\alpha_p(t') \right] \right\}.\end{aligned} \tag{4.130}$$

Combining equation (4.53) with equation (1.54) and $(\hat{b}_p^{s\dagger} \beta_p^s(t))^\dagger = \beta_p^{s*}(t) \hat{b}_p^s$ gives the bispinor field versions of equation (2.82),

$$\hat{b}_k^s |\eta, t\rangle = \beta_k^s(t) |\eta, t\rangle, \quad \langle \eta, t| \hat{b}_k^{s\dagger} = \langle \eta, t| \beta_k^{s*}(t), \tag{4.131}$$

with similar expressions involving $\hat{d}_k^s$ and α_k^s. In turn, this gives the matrix elements of the normal ordered free Dirac field $\hat{\psi}(x)$ given by equation (4.48),

$$\langle \eta, t' | {:}\hat{\psi}^n(x){:}\, |\eta, t\rangle = (\eta^{(-)}(x, t') + \eta^{(+)}(x, t))^n \langle \eta, t' | \eta, t\rangle, \tag{4.132}$$

where $x = (\boldsymbol{x}, t)$ and the two functions are given by

$$\eta^{(-)}(x, t') = \int \frac{d^3p}{(2\pi)^{3/2}} \sqrt{\frac{m}{\varepsilon_p}}\, \beta_p^s(t') u_s(p) e^{-ip\cdot x}, \tag{4.133}$$

$$\eta^{(+)}(x, t) = \int \frac{d^3p}{(2\pi)^{3/2}} \sqrt{\frac{m}{\varepsilon_p}}\, \alpha_p^{s*}(t) v_s(p) e^{ip\cdot x}. \tag{4.134}$$

A similar expression in terms of $\bar{\eta}^{(\pm)}$ follows for normal ordered products of $\bar{\psi}(x)$. Like the scalar field states of equation (4.95), the coherent states (4.129) are complete,

$$\int [\mathcal{D}^2\eta(t)]\, \langle \eta,\, t''|\eta,\, t\rangle\langle \eta,\, t|\eta,\, t'\rangle = \langle \eta,\, t''|\eta,\, t'\rangle, \tag{4.135}$$

where the measure is given by

$$[\mathcal{D}^2\eta(t)] = \prod_{s,\boldsymbol{p}} (\mathrm{d}^3p)^2\, \mathrm{d}\beta_{\boldsymbol{p}}^{s*}(t)\, \mathrm{d}\beta_{\boldsymbol{p}}^{s}(t)\, \mathrm{d}\alpha_{\boldsymbol{p}}^{s*}(t)\, \mathrm{d}\alpha_{\boldsymbol{p}}^{s}(t). \tag{4.136}$$

The statement of completeness is demonstrated by combining the inner product (4.130) with the integral version (1.86) of the Grassmann Dirac delta (1.87).

All the necessary components are now in place to derive the Dirac spinor path integral. Like the scalar field theory definition (4.97), the path integral for the Dirac field is the time-sliced version of the transition element between two coherent states in the presence of an external bispinor Grassmann source K,

$$\begin{aligned} Z_{\eta_\pm}[K,\, \bar{K}] &= \langle \eta,\, t_+|E_K(t_+,\, t_-)|\eta,\, t_-\rangle \\ &= \int [\mathcal{D}^2\eta]\prod_{j=0}^{N-1} \langle \eta,\, t_{j+1}|E_K(t_{j+1},\, t_j)|\eta,\, t_j\rangle, \end{aligned} \tag{4.137}$$

where the total measure is given by a product of the measure of equation (4.136),

$$[\mathcal{D}^2\eta] = \prod_{j=1}^{N-1} [\mathcal{D}^2\eta(t_j)] = \prod_{j=1}^{N-1}\prod_{s,\boldsymbol{p}} (\mathrm{d}^3p)^2\, \mathrm{d}\beta_{\boldsymbol{p}}^{s*}(t_j)\, \mathrm{d}\beta_{\boldsymbol{p}}^{s}(t_j)\, \mathrm{d}\alpha_{\boldsymbol{p}}^{s*}(t_j)\, \mathrm{d}\alpha_{\boldsymbol{p}}^{s}(t_j). \tag{4.138}$$

As before $t_j = t_- + j\epsilon$, where $\epsilon = (t_+ - t_-)/N = T/N$ is infinitesimal. For the Dirac field case the infinitesimal evolution operator (4.84) is given by

$$E_K(t_{j+1},\, t_j) = \exp\left\{-i\epsilon\int \mathrm{d}^3x\, (:\mathcal{H}_I(\psi,\, \bar{\psi}): + \bar{\psi}K + \bar{K}\psi)\right\}, \tag{4.139}$$

where $\bar{K} = K^\dagger\gamma_0$ and the interaction picture operators ψ and $\bar{\psi}$ have been normal ordered in any interaction $\mathcal{H}_I$ present in the Hamiltonian density, The fields and source terms in equation (4.139) are all functions of x_j, where $x_j = (\boldsymbol{x},\, t_j)$. The coherent state matrix element for equation (4.139) can now be evaluated using equation (4.132) and the similar expression for $\bar{\psi}$. As with the scalar field, the assumption of differentiability for $\beta_{\boldsymbol{p}}^s(t)$ and $\alpha_{\boldsymbol{p}}^s(t)$ gives the Dirac field counterpart to equation (4.100), $\eta^{(-)}(\boldsymbol{x},\, t_{j+1}) + \eta^{(+)}(\boldsymbol{x},\, t_j) \approx \eta(\boldsymbol{x},\, t_j) + O(\epsilon)$. Suppressing the $O(\epsilon)$ contribution, the matrix element is given by

$$\begin{aligned} &\langle \eta,\, t_{j+1}|E_K(t_{j+1},\, t_j)|\eta,\, t_j\rangle \\ &= \exp\left\{-i\epsilon\int \mathrm{d}^3x\, (\mathcal{H}_I(\eta,\, \bar{\eta}) + \bar{\eta}K + \bar{K}\eta)\right\}\langle \eta,\, t_{j+1}|\eta,\, t_j\rangle, \end{aligned} \tag{4.140}$$

where the fields and source are now all functions of x_j.

As in the scalar field case (4.102), using equation (4.130) and the differentiability of the Grassmann coefficients $\beta_p^s(t)$ and $\alpha_p^s(t)$ allows the infinitesimal inner product in equation (4.140) to be written to $O(\epsilon)$ as

$$\begin{aligned}\langle \eta,\, t_{j+1}|\eta,\, t_j\rangle = \exp\left\{\frac{1}{2}\epsilon\int \mathrm{d}^3p\left[\dot{\beta}_p^{s*}(t_j)\beta_p^s(t_j) - \beta_p^{s*}(t_j)\dot{\beta}_p^s(t_j)\right.\right.\\ \left.\left. + \dot{\alpha}_p^{s*}(t_j)\alpha_p^s(t_j) - \alpha_p^{s*}(t_j)\dot{\alpha}_p^s(t_j)\right]\right\}.\end{aligned} \tag{4.141}$$

Both $u_s(p)e^{-ip\cdot x}$ and $v_s(p)e^{ip\cdot x}$ satisfy the Dirac equation (4.30) when combined with equation (4.49), and this gives

$$\begin{aligned}i\gamma^0\partial_0(u_s(p)e^{-ip\cdot x}) &= \left(\varepsilon_p\gamma^0 u_s(p)\right)e^{-ip\cdot x}\\ &= \left(\left(-p_k\gamma^k + m\right)u_s(p)\right)e^{-ip\cdot x}\\ &= (-i\gamma^k\partial_k + m)(u_s(p)e^{-ip\cdot x}),\end{aligned} \tag{4.142}$$

with a similar statement for $v_s(p)e^{ip\cdot x}$. This can be combined with the assumption of differentiability for $\beta_p^s(t)$ and $\alpha_p^s(t)$ to show that the time derivative of the function (4.125) is given by

$$\begin{aligned}\frac{1}{2}i\gamma^0\partial_0\eta(x) = \frac{1}{2}i\int\frac{\mathrm{d}^3p}{(2\pi)^{3/2}}\sqrt{\frac{m}{\varepsilon_p}}\left(\dot{\beta}_p^s(t)\gamma^0 u_s(p)e^{-ip\cdot x} + \dot{\alpha}_p^{s*}(t)\,\gamma^0 v_s(p)e^{ip\cdot x}\right)\\ + \frac{1}{2}(-i\gamma^k\partial_k\eta(x) + m)\eta(x).\end{aligned} \tag{4.143}$$

Combining equation (4.143) with $\gamma^{0\dagger} = \gamma^0$, $\gamma^{0\,2} = \mathbf{1}$, $\gamma^0\gamma^{k\dagger}\gamma^0 = \gamma^k$, and $\bar{\eta} = \eta^\dagger\gamma^0$ gives

$$\begin{aligned}-\frac{1}{2}i\partial_0\bar{\eta}(x)\gamma^0 = -\frac{1}{2}i\int\frac{\mathrm{d}^3p}{(2\pi)^{3/2}}\sqrt{\frac{m}{\varepsilon_p}}\left(\dot{\beta}_p^{s*}(t)u_s^\dagger(p)e^{ip\cdot x} + \dot{\alpha}_p^s(t)\,v_s^\dagger(p)e^{-ip\cdot x}\right)\\ + \frac{1}{2}i\partial_j\bar{\eta}(x)\gamma^j + \frac{1}{2}m\bar{\eta}(x),\end{aligned} \tag{4.144}$$

Combining equations (4.143) and (4.144) with the expansions (4.125) and (4.127) shows that the infinitesimal inner product (4.141) is given by

$$\langle \eta,\, t_{j+1}|\eta,\, t_j\rangle = \exp\left\{i\epsilon\int \mathrm{d}^3x\left(\frac{1}{2}i\bar{\eta}\gamma^0\partial_0\eta - \frac{1}{2}i(\partial_0\bar{\eta})\gamma^0\eta - \mathcal{H}_0(\eta,\, \bar{\eta})\right)\right\}, \tag{4.145}$$

where $\mathcal{H}_0$ is the free Dirac Hamiltonian density (4.32) written in terms of η and $\bar{\eta}$ and evaluated at $x_j = (\boldsymbol{x},\, t_j)$,

$$\mathcal{H}_0(x_j) = -\frac{1}{2}i\bar{\eta}(x_j)\gamma^k\partial_k\eta(x_j) + \frac{1}{2}i\partial_k\bar{\eta}(x_j)\gamma^k\eta(x_j) + m\bar{\eta}(x_j)\eta(x_j). \tag{4.146}$$

The key to demonstrating equation (4.145) is the bispinor identity

$$u_s^b(p)u_s^{\dagger a}(p) + v_s^b(\tilde{p})v_s^{\dagger a}(\tilde{p}) = \delta^{ab}, \tag{4.147}$$

where there is a sum over the index s, $\tilde{p} = (\varepsilon_p, -\boldsymbol{p})$, and a and b are the spinor indices. The proof of equation (4.147) follows from the form of the bispinor solutions (4.51) and the properties of the Dirac matrices (4.14). The argument of the exponential is then given by

$$\frac{1}{2}i\bar{\eta}\gamma^0\partial_0\eta - \frac{1}{2}i(\partial_0\bar{\eta})\gamma^0\eta - \mathcal{H}_0(\eta, \bar{\eta}) = \frac{1}{2}i\bar{\eta}\gamma^\mu\partial_\mu\eta - \frac{1}{2}i(\partial_\mu\bar{\eta})\gamma^\mu\eta - m\bar{\eta}\eta, \tag{4.148}$$

which is the Lagrangian (4.29) for the free Dirac field. Using equation (4.148) in equation (4.140) and returning the result to equation (4.137) gives the final form for the interacting Dirac field path integral,

$$\begin{aligned} Z_{\eta_\pm}[K, \bar{K}] &= \int_{\eta_-}^{\eta_+} [\mathcal{D}^2\eta] \\ &\times \exp\left\{i\int \mathrm{d}^4x \left(\frac{1}{2}i\bar{\eta}\gamma^0\partial_0\eta - \frac{1}{2}i(\partial_0\bar{\eta})\gamma^0\eta - \mathcal{H}(\eta, \bar{\eta}) - \bar{\eta}K - \bar{K}\eta\right)\right\}, \end{aligned} \tag{4.149}$$

where $\mathcal{H} = \mathcal{H}_0 + \mathcal{H}_I$ is the full Hamiltonian that includes interactions, while the measure is given by equation (4.138). Like the scalar field path integral The argument of the exponential is again the action for the interacting field written in Hamiltonian form. This follows from the previous identifications $\pi = \frac{1}{2}i\bar{\eta}\gamma^0$ and $\bar{\pi} = -\frac{1}{2}i\gamma^0\eta$, so that the action appearing equation (4.149) can be understood as $\mathcal{L} = \pi\dot{\eta} + \dot{\bar{\eta}}\bar{\pi} - \mathcal{H}(\eta, \bar{\eta}) - \bar{K}\eta - \bar{\eta}K$. This is consistent with the previous definition of the free bispinor Hamiltonian density (4.32) with the addition of external Grassmann sources. However, it is not possible to break the measure into separate integrations over π and η as it was with the scalarfield path integral.

4.8 Configuration space techniques and quadratic path integrals

It is possible to define the field theory path integral with a configuration space version of the measure. Such a measure is often combined *a priori* with the classical configuration space action in the path integral's exponential, and the fields appearing in the action are treated as *differentiable functions of spacetime*. It is a more intuitive prescription to define the path integral by treating the value of the field at a spacetime point as the fluctuation variable. For example, the constraints required to quantize the gauge field are more easily expressed using a configuration space measure. However, this approach lacks the direct connection to the annihilation and creation operators associated with the quantized fields in the interaction picture that is provided by complete sets of coherent states with well defined measures. As a result, the path integral's connection to particle processes becomes less obvious and the normalization process is handled in a much more ad hoc manner. Nevertheless, configuration space measure combined with the assumption of field differentiability has often served as the starting point for defining the path integral for quantum field theory. In what follows attention will be limited to four-dimensional spacetime, although the development can be presented in an arbitrary dimension.

The configuration approach begins by breaking spacetime into a lattice. Each cell in the lattice is characterized by a spacetime point x_j that is contained in the lattice cell. This allows spacetime integrals to be expressed as Riemann sums,

$$\int \mathrm{d}^4x\ \varphi(x) \to \sum_j \mathrm{d}^4x_j\ \varphi(x_j), \tag{4.150}$$

where d^4x_j is the volume of the lattice cell. The expression (4.150) can be made more rigorous using arguments like those discussed after equation (4.96), where the spacetime volume is initially finite, $V = L^3T$. This allows d^4x to be defined as $\mathrm{d}^4x = L^3T/N^4$, where the limit $N \to \infty$ is understood. Subsequent to passing to the $N \to \infty$ limit, the infinite volume $L,\ T \to \infty$ limits are then taken. In what follows, the sum over j and d^4x_j will stand as shorthand for these more rigorous lattice arguments. Care must be taken in this process if the underlying spacetime manifold has a non-trivial topology, in which case the covering of the manifold with surfaces or volumes requires a careful application of *homology theory*. The continuum limit process is assumed to be without complication, but this needs to be verified on a case by case basis. However, using Riemann sums allows the functional measure to be defined as an integration over the value of the field functions $\varphi(x_j)$ and canonical momentum $\pi(x_j)$ at each lattice cell. For the case of a real valued field $\varphi(x)$ this measure will be written as

$$[\mathcal{D}^2\varphi] = [\mathcal{D}\varphi][\mathcal{D}\pi] = \prod_j N_j\ \mathrm{d}\varphi(x_j) \prod_k \mathrm{d}\pi(x_k), \tag{4.151}$$

where the product is over all lattice cells. The N_j represent a collection of factors typically determined by normalizing the functional integral to a known result for an associated transition element, as was done with the quantum mechanical case (3.54). The approach of equation (4.151) allows continuum expressions in path integrals to be evaluated by using matrix results that have been already obtained.

In particular, the Gaussian integral of equation (1.105) was central to analyzing the quantum mechanical path integral. Based on the previous results (4.107) and (4.124) for the scalar field, it is *assumed* that the path integral for a continuous action counterpart is obtained for the configuration space path integral for the vacuum transition amplitude after integrating the canonical momentum,

$$Z_0[J] = \int_0^0 [\mathcal{D}^2\varphi]\exp\{iS_c(\varphi,\pi)\} = \int_0^0 [\mathcal{D}\varphi]\exp\{iS_c(\varphi)\}. \tag{4.152}$$

It will also be assumed that the action $S_c(\varphi)$ obtained by integrating the momentum contains terms that are quadratic and linear in the fields. In the case of a free field, only terms quadratic and linear in the fields are present. In the case of an interacting field there will also be higher order terms In what follows the higher order terms, if any, will be ignored, and the terms quadratic and linear in the fields will be used to define a *basis path integral* that can be used in a *perturbative expansion* as was done in the quantum mechanical case in expression (3.125). It is therefore the quadratic and linear terms that will undergo functional integration. In the case of a single scalar field φ with an external source J, the Gaussian functional integral is given by

$$Z_0[J] = \int [\mathcal{D}\varphi] \exp\left\{-\frac{1}{2}\int \mathrm{d}^4x\, \mathrm{d}^4y\, \varphi(x)M(x, y)\varphi(y) - i\int \mathrm{d}^4x\, J(x)\varphi(x)\right\}. \tag{4.153}$$

The scalar field φ has the units of inverse length and the source $J(x)$ has units of inverse length cubed, while the function M has the units inverse length to the sixth power. The units of M can be understood from the Klein–Gordon action, which can be written

$$\begin{aligned} iS &= i\int \mathrm{d}^4x\left(\frac{1}{2}\partial_\mu\varphi\, \partial^\mu\varphi - \frac{1}{2}m^2\varphi^2 - J\varphi\right) = -i\int \mathrm{d}^4x\left(\frac{1}{2}\varphi(\Box + m^2)\varphi + J\varphi\right) \\ &= -\frac{1}{2}\int \mathrm{d}^4x\, \mathrm{d}^4y\, \varphi(x)i(\Box + m^2)\delta^4(x-y)\varphi(y) - i\int \mathrm{d}^4x\, J(x)\varphi(x). \end{aligned} \tag{4.154}$$

The function $M(x, y)$ is therefore identified as $M(x, y) = i(\Box_x + m^2)\delta^4(x - y)$, where $\Box$ is the d'Alembertian. The function $M(x, y)$ acts as a matrix with an infinite continuum of elements. This allows the matrix methods used to evaluate equation (1.105) to be brought to bear in the analysis of equation (4.153).

This is made clearer by introducing a constant α with the units of length to define the dimensionless quantities

$$\tilde{\varphi}(x_j) = \alpha^{-1}\sqrt{\mathrm{d}^4x_j}\; \varphi(x_j), \tag{4.155}$$

$$\tilde{M}(x_j, x_k) = \sqrt{\mathrm{d}^4x_j\, \mathrm{d}^4x_k}\; \alpha^2 M(x_j, x_k), \tag{4.156}$$

$$\tilde{J}(x_j) = \sqrt{\mathrm{d}^4x_j}\; \alpha J(x_j), \tag{4.157}$$

and then replacing the integrals of equation (4.153) with Riemann sums over the lattice cell variables defined in equation (4.151). The constant α can be related to a coupling constant or a mass that occurs in the theory, but it can also be an arbitrary choice of length. This results in the continuous Euclidean functional of equation (4.153) becoming an infinite dimensional matrix functional,

$$
\begin{aligned}
Z_0[J] &\to \int \prod_{\ell} N'_\ell \, \mathrm{d}\tilde{\varphi}(x_\ell) \exp\left\{-\frac{1}{2}\sum_{j,k} \tilde{\varphi}(x_j)\tilde{M}(x_j, x_k)\tilde{\varphi}(x_k) - i\sum_{j} \tilde{J}(x_j)\tilde{\varphi}(x_j)\right\} \\
&= \frac{N'}{\sqrt{\det \tilde{M}}} \exp\left\{-\frac{1}{2}\sum_{j,k} \tilde{J}(x_j)\tilde{M}^{-1}(x_j, x_k)\tilde{J}(x_k)\right\} \\
&\to \frac{N}{\sqrt{\det(\alpha^2 M)}} \exp\left\{-\frac{1}{2}\int \mathrm{d}^4x \, \mathrm{d}^4y \, J(x)M^{-1}(x, y)J(y)\right\}.
\end{aligned}
\tag{4.158}
$$

Result (4.158) follows from the formal use of the *matrix result* (1.110) to evaluate the first line of equation (4.158). It is important to recall that result (1.110) requires that the real part of the eigenvalues of $\tilde{M}$ are positive-definite. The final step is to continue (4.153) back to real time. The details of this require a specific for the matrix M^{-1} and the analysis for the Klein–Gordon theory will be presented later. In these expressions N'_ℓ, N', and N represent an infinite collection of constants resulting from the change of variables and the subsequent Gaussian integrations. For example, there is a factor of $\sqrt{2\pi}$ associated with each lattice cell integration that has been absorbed into the definition of N'. The infinite product of d^4x factors associated with replacing $\tilde{M}(x_j, x_k)$ with $\sqrt{\mathrm{d}^4x_j \, \mathrm{d}^4x_k} \; \alpha^2 M(x_j, x_k)$ have also been absorbed into the definition of N in the last line of equation (4.158). The reason for doing so follows from the definition of the determinant of the continuous function $\det(\alpha^2 M)$ made later in equation (4.167). The continuum version of M^{-1} present in the last line of equation (4.158) also needs to be defined.

The clarification starts by defining the multiplication of two continuous matrices with the elements $A(x, y)$ and $B(x, y)$ as the matrix $C(x, y)$ with the elements

$$
C(x, y) = (A \cdot B)(x, y) = \int \mathrm{d}^4z \, A(x, z)B(z, y). \tag{4.159}
$$

Using the definition (4.159) allows the identification of the *continuum unit matrix* in four dimensions, $I(x, y) = \delta^4(x - y)$, since this definition and equation (4.159) show that $(I \cdot I)(x, y) = I(x, y)$. The continuum unit matrix allows the definition of the inverse $A^{-1}(x, y)$ from the requirement that

$$
(A \cdot A^{-1})(x, z) = \int \mathrm{d}^4y \, A(x, y)A^{-1}(y, z) = I(x, z) = \delta^4(x - z). \tag{4.160}
$$

In addition, it shows that the function $f(x)$ can be treated as a vector with the usual matrix result,

$$
(I \cdot f)(x) \to \int \mathrm{d}^4y \, I(x, y)f(y) = \int \mathrm{d}^4y \, \delta^4(x - y)f(y) = f(x). \tag{4.161}
$$

A continuous four-dimensional diagonal matrix then takes the form $D(x, y) = \lambda(x)\,\delta^4(x - y)$, which possesses the inverse $D^{-1}(x, y) = \lambda^{-1}(x)\,\delta^4(x - y)$ as long as the inverse of the function $\lambda(x)$ is well defined everywhere.

Many of the structures of finite linear algebra can be adapted to the continuum version by the use of Riemann sums. However, it is necessary to note that the scaling technique used earlier, $\tilde{M}(x_i, x_j) = \sqrt{\mathrm{d}^4x_i\,\mathrm{d}^4x_j}\;\alpha^2 M(x_i, x_j)$, has some counterintuitive results, one of which follows from

$$\begin{aligned}
&\sum_j \tilde{M}(x_i, x_j)\tilde{M}^{-1}(x_j, x_k) = \delta_{ik} \\
&\Rightarrow \sum_j \mathrm{d}^4x_j\; M(x_i, x_j)\frac{\alpha^2}{\sqrt{\mathrm{d}^4x_j\,\mathrm{d}^4x_k}}\tilde{M}^{-1}(x_j, x_k) = \frac{\delta_{ik}}{\sqrt{\mathrm{d}^4x_i\,\mathrm{d}^4x_k}} = \delta^4(x_i - x_k) \qquad (4.162) \\
&\Rightarrow \tilde{M}^{-1}(x_j, x_k) = \frac{1}{\alpha^2}\sqrt{\mathrm{d}^4x_j\,\mathrm{d}^4x_k}\; M^{-1}(x_j, x_k).
\end{aligned}$$

Using equations (4.162) and (4.157) verifies the exponential in the last step of equation (4.158),

$$\sum_{j,k} \tilde{J}(x_j)\tilde{M}^{-1}(x_j, x_k)\tilde{J}(x_k) \rightarrow \int \mathrm{d}^4x\,\mathrm{d}^4y\; J(x)M^{-1}(x, y)J(y). \qquad (4.163)$$

The next step is to clarify the meaning of the expression $\det(\alpha^2 M)$ appearing in equation (4.158). This is accomplished by adapting the identity for an arbitrary finite matrix $\mathbf{M}$, which states that $\det \mathbf{M} = \exp(\mathrm{Tr}\ln \mathbf{M})$. The proof of this finite matrix identity follows from the fact that both the left and right side of the equation yield the product of the *matrix eigenvalues*. The expression $\ln \mathbf{M}$ is understood as a power series involving the matrix $\mathbf{M}$ obtained from the Taylor series expansion,

$$\mathrm{Tr}\ln \mathbf{M} = \mathrm{Tr}\ln[\mathbf{I} + (\mathbf{M} - \mathbf{I})] = \sum_{n=1}^{\infty} \frac{(-1)^{n+1}}{n}\mathrm{Tr}(\mathbf{M} - \mathbf{I})^n, \qquad (4.164)$$

where convergence of the infinite series will be assumed. In order to adapt this expression to continuum matrices it is necessary to define the trace of the function $\alpha^2 M(x, y)$ as a dimensionless quantity. This is accomplished using integration,

$$\mathrm{Tr}(\alpha^2 M) = \int \mathrm{d}^4x\; \alpha^2 M(x, x). \qquad (4.165)$$

Setting $\alpha^2 M(x, y) = I(x, y) = \delta^4(x - y)$ shows that equation (4.165) gives the trace of the identity matrix as $\int \mathrm{d}^4x \int \mathrm{d}^4k/(2\pi)^4$, the dimensionless total volume of phase space. The trace of the power n of the function $\alpha^2 M(x, y)$ is then given by

$$\mathrm{Tr}(\alpha^2 M)^n = \alpha^{2n}\int \mathrm{d}^4x_1 \cdots \mathrm{d}^4x_n\; M(x_1, x_2)M(x_2, x_3)\cdots M(x_n, x_1), \qquad (4.166)$$

The quantity $(\alpha^2 M - I)(x, y) = \alpha^2 M(x, y) - \delta^4(x - y)$ is easily adapted to expression (4.164), giving the determinant of the continuum function $\alpha^2 M(x, y)$,

$$\det(\alpha^2 M) = \exp\left\{\sum_{n=1}^{\infty} \frac{(-1)^{n+1}}{n} \operatorname{Tr}(\alpha^2 M - I)^n\right\}. \tag{4.167}$$

Setting $\alpha^2 M = I$ in expression (4.167) shows that it yields the required result $\det I = 1$. For the general case that $M(x, y)$ is arbitrary, equation (4.167) defines the expression $\det(\alpha^2 M)$ appearing in result (4.158).

Although the emphasis has been on configuration space functions and measure, it is often useful to express the functions appearing in the path integral action in terms of a discrete set of *complete orthonormal functions*, denoted $\varphi_j(x)$. This process begins by assuming a large but finite spacetime manifold, so that the Dirac delta function can be represented as an infinite series,

$$\delta^4(x - y) = \sum_j \varphi_j(x)\varphi_j(y) = \sum_{j,k} \delta_{jk}\, \varphi_j(x)\varphi_k(y), \tag{4.168}$$

where the $\varphi_j(x)$ have the units of inverse length squared and are orthonormal,

$$\delta_{jk} = \int \mathrm{d}^4x\, \varphi_j(x)\varphi_k(x). \tag{4.169}$$

Since these functions obey the completeness relation (4.168) it is possible to expand the arbitrary functions appearing in the path integral action in terms of them, so that the real functions $\varphi(x)$ and $M(x, y)$ can be expressed as

$$\varphi(x) = \sum_n a_n \varphi_n(x), \qquad M(x, y) = \sum_{j,k} \tilde{M}_{jk}\varphi_j(x)\varphi_k(y). \tag{4.170}$$

The orthonormality of equation (4.169) shows that the trace of M is given by

$$\operatorname{Tr} M = \int \mathrm{d}^4x\, M(x, x) = \int \mathrm{d}^4x \sum_{jk} \tilde{M}_{jk}\varphi_j(x)\varphi_k(x) = \sum_j \tilde{M}_{jj} = \operatorname{Tr} \tilde{M}, \tag{4.171}$$

with a similar statement for powers of $\alpha^2 M$,

$$\operatorname{Tr}(\alpha^2 M)^n = \sum_{j_1, j_2, \ldots, j_n} \alpha^{2n} \tilde{M}_{j_1 j_2} \tilde{M}_{j_2 j_3} \cdots \tilde{M}_{j_n j_1} = \operatorname{Tr}(\alpha^2 \tilde{M})^n. \tag{4.172}$$

The determinant (4.167) can then be expressed in terms of the expansion coefficients $\alpha^2 \tilde{M}_{jk} - \delta_{jk} \equiv (\alpha^2 \tilde{M} - \tilde{I})_{jk}$, yielding

$$\det(\alpha^2 M) = \exp\left\{\sum_{n=1}^{\infty} \frac{(-1)^{n+1}}{n} \operatorname{Tr}(\alpha^2 \tilde{M} - \tilde{I})^n\right\}. \tag{4.173}$$

For the case that $\tilde{M}$ is diagonal, so that $\tilde{M}_{jk} = \lambda^{(j)}\delta_{jk}$, expression (4.173) takes a particularly simple form since $(\tilde{M} - \tilde{I})_{jk} = (\lambda^{(j)} - 1)\delta_{jk}$ is a diagonal matrix,

$$\sum_{n=1}^{\infty} \frac{(-1)^{n+1}}{n} \mathrm{Tr}(\alpha^2 \tilde{M} - \tilde{I})^n = \sum_j \sum_{n=1}^{\infty} \frac{(-1)^{n+1}}{n} (\alpha^2 \lambda^{(j)} - 1)^n = \sum_j \ln(\alpha^2 \lambda^{(j)})$$
$$\Longrightarrow \det(\alpha^2 M) = \exp\left\{\sum_j \ln(\alpha^2 \lambda^{(j)})\right\} = \prod_j \alpha^2 \lambda^{(j)}, \tag{4.174}$$

which is the product of the diagonal elements, as it is with a finite diagonal matrix.

These methods are extremely useful in evaluating a continuum field theory path integral of the form (4.158) when a linear differential operator $\hat{L}$ occurs in the quadratic part of the Euclidean path integral action. An example of this is the quadratic scalar field action (4.154) where $\hat{L} = \Box + m^2$ and $M(x, y) = \hat{L}\delta^4(x - y)$. Such a general differential operator $\hat{L}$ is assumed to satisfy linearity, so that $\hat{L}(af(x) + bg(x)) = a\hat{L}f(x) + b\hat{L}g(x)$ where a and b are constants. Linear differential operators of the Sturm–Liouville or self-adjoint type are common in physical applications in quantum mechanics and quantum field theory. They typically possess a complete set of orthonormal eigenfunctions satisfying $\hat{L}\varphi_j(x) = \lambda^{(j)}\varphi_j(x)$, where $\lambda^{(j)}$ is assumed to be positive-definite for the Euclidean theory. It should be noted that these eigenfunctions may also incorporate combinations of discrete and continuous indices as well as satisfying the boundary conditions created by the spatial manifolds on which they are defined. These subtleties will be ignored in the following general presentation.

For the case that $M(x, y) = \hat{L}\delta^4(x - y)$ appears in the quadratic part of the Euclidean path integral action, it is easiest to employ the eigenfunctions of $\hat{L}$ to expand the Dirac delta. Restricting attention to the scalar field case, it is assumed that a complete set of orthonormal eigenfunctions $\varphi_j(x)$ can be found that satisfy $\hat{L}\varphi_j(x) = \lambda^{(j)}\varphi_j(x)$, where $\lambda^{(j)}$ is the positive-definite eigenvalue associated with the eigenfunction $\varphi_j(x)$. Using these eigenfunctions in equation (4.168) allows the function $M(x, y)$ to be written

$$M(x, y) = \hat{L}\delta^4(x - y) = \sum_j \lambda^{(j)}\varphi_j(x)\varphi_j(y) = \sum_j \lambda^{(j)}\delta_{jk}\, \varphi_j(x)\varphi_k(y), \tag{4.175}$$

so that $\tilde{M}_{jk} = \lambda^{(j)}\delta_{jk}$ is a diagonal matrix identical to the form used to obtain equation (4.174). The determinant in the prefactor of the quadratic field theory path integral (4.158) is simply the product of the eigenvalues of the differential operator present in the action. For such a case the determinant is often written

$$\det(\alpha^2 M) = \prod_j \alpha^2 \lambda^{(j)} = \det(\alpha^2 \hat{L}). \tag{4.176}$$

Expression (4.175) is also easy to invert, since $M^{-1}(x, y)$ is given formally by

$$M^{-1}(x, y) = \sum_j \frac{1}{\lambda^{(j)}}\varphi_j(x)\varphi_j(y) = \sum_j \frac{1}{\lambda^{(j)}}\delta_{jk}\, \varphi_j(x)\varphi_k(y). \tag{4.177}$$

Expression (4.177) assumes that there are *no zero eigenvalues*, often referred to as *zero modes*. Zero modes must be excluded from the inversion process of equation (4.177) since they render M^{-1} undefined.

For example, in the case of the Klein–Gordon field operator $\hat{L} = i(\Box + m^2)$ has a complete set of orthonormal eigenfunctions, $\varphi_k(x) = e^{-ik\cdot x}$, which can be used to define the Dirac delta using equation (1.26). The matrix M is therefore given by

$$\begin{aligned} M(x, y) &= \hat{L}\delta^4(x - y) = i\int \frac{\mathrm{d}^4k}{(2\pi)^4}(-k^2 + m^2)e^{-ik\cdot(x-y)} \\ \Longrightarrow M^{-1}(x, y) &\equiv \Delta_F(x - y) = \lim_{\epsilon\to 0}\int \frac{\mathrm{d}^4k}{(2\pi)^4}\frac{ie^{-ik\cdot(x-y)}}{k^2 - m^2 + i\epsilon}. \end{aligned} \tag{4.178}$$

The term $i\epsilon$ is present in the second step to avoid the infinite set of zero modes corresponding to the case that $k^2 = m^2$. These zero modes are referred to as being *on mass-shell* since a classical scalar particle of mass m satisfies the relativistic energy condition $k_0 = \varepsilon_{\boldsymbol{k}} = +\sqrt{\boldsymbol{k}^2 + m^2}$. The inversion of the Klein–Gordon operator given in equation (4.178) avoids the zero mode problem by moving the on-shell poles off the real axis through an ϵ prescription. The function $M^{-1}(x, y)$ is identified as the Feynman propagator $\Delta_F(x - y)$ since it satisfies the real time Klein–Gordon Green's function equation, $(\Box + m^2)\Delta_F(x - y) = -i\,\delta^4(x - y)$, in the limit that $\epsilon \to 0$. This is identical to the time-ordered product of equation (4.73). For this particular ϵ prescription, Cauchy's residue theorem shows that $\Delta_F(x - y)$ is indeed identical to the Feynman propagator defined by the time-ordered product of equation (4.73), as demonstrated in equation (4.74). This follows from the poles in the integrand of equation (4.178) occurring at $\omega = \pm\sqrt{\boldsymbol{k}^2 + m^2 - i\epsilon} \approx \pm(\varepsilon_{\boldsymbol{k}} - \frac{1}{2}i\epsilon)$. The details of how this reproduces the time-ordered product (4.73) are available in Peskin and Schroeder. The final result of evaluating the free Klein–Gordon scalar field path integral is

$$Z_0[J] = \frac{N}{\sqrt{\det(\alpha^2 M)}}\exp\left\{-\frac{1}{2}\int \mathrm{d}^4x\,\mathrm{d}^4y\,J(x)\Delta_F(x - y)J(y)\right\}. \tag{4.179}$$

The functional derivatives of equation (4.179) give the time-ordered products of the free Klein–Gordon scalar field when evaluated at $J = 0$,

$$\left.\frac{1}{Z_0[J]}\frac{\delta^2 Z_0[J]}{\delta J(x)\,\delta J(y)}\right|_{J=0} = -\langle 0|T\{\varphi(x)\varphi(y)\}|0\rangle = -\Delta_F(x - y). \tag{4.180}$$

As a result, $Z_0[J]$ is *the generating functional for time-ordered products.*

Evaluating a field theory path integral most often proceeds by first continuing it to imaginary time, $t \overset{\mathrm{W}}{\to} -it$. As an example, the free Klein–Gordon path integral becomes

$$iS[\varphi, J] \overset{\mathrm{W}}{\to} -\frac{1}{2}\int \mathrm{d}^4x\,\varphi(x)(\Box_E + m^2)\varphi(x) - \int \mathrm{d}^4x\,J(x)\varphi(x), \tag{4.181}$$

where the Euclidean version of the d'Alembertian is given by $\square_E = -\partial_0^2 - \nabla^2$. The eigenfunctions of $\square_E$ are the Wick rotated version of $e^{-ip\cdot x}$, which is obtained from the dual recipe $k_0 \overset{W}{\to} ik_0$ and $t \overset{W}{\to} -it$ so that $k_0 t \overset{W}{\to} k_0 t$. The advantage of the Wick rotation is that $\square_E + m^2$ has positive definite eigenvalues given by $\lambda^{(k)} = -k_E^2 + m^2 = k_0^2 + \boldsymbol{k}^2 + m^2$. In the general case the Wick rotated differential operator $\hat{L}_E$ is assumed to have a complete set of orthonormal eigenfunctions given by $\varphi_j(x)$ with positive definite eigenvalues $\lambda^{(j)}$. The field variables and the source are expanded in terms of these eigenfunctions using equation (4.170). In effect, this is a change of variables in the original functional (4.158). For example, in the scalar field case using the orthonormal eigenfunctions $\varphi_j(x)$ of $\hat{L}$ in (4.170),

$$\varphi(x) = \sum_j a_j\varphi_j(x), \quad J(x) = \sum_j J_j\varphi_j(x), \tag{4.182}$$

allows the Euclidean quadratic scalar action to be written as

$$iS \overset{W}{\to} -\int d^4x\left(\frac{1}{2}\varphi\hat{L}\varphi + J\varphi\right) = \sum_j\left(-\frac{1}{2}\lambda^{(j)}a_j^2 - J_j a_j\right). \tag{4.183}$$

The change of variables causes the original measure (4.151) to become

$$[\mathcal{D}\varphi] \to \prod_j N_j'\, da_j = [\mathcal{D}a], \tag{4.184}$$

where the Jacobian of the change of variables, det $\delta\varphi/\delta a$, is independent of the a_j and has therefore been absorbed into the normalization factor N_j'. Given this form for the action the quadratic functional integral (4.158) becomes

$$Z_0[J] = \int [\mathcal{D}a] \exp\left\{-\sum_j\left(\frac{1}{2}\lambda^{(j)}a_j^2 + J_j a_j\right)\right\}. \tag{4.185}$$

The path integral of equation (4.185) is simply an infinite product of Gaussian integrations with the familiar form (1.99). Assuming $\lambda^{(j)}$ is positive for all j, the result is

$$Z_0[J] = \prod_j N_j'\sqrt{\frac{\pi}{\lambda^{(j)}}} \exp\left\{\frac{1}{2}\sum_k \frac{1}{\lambda^{(k)}}J_k^2\right\}. \tag{4.186}$$

Using the inverse expansion for J_k,

$$J_k = \int d^4x\, J(x)\varphi_k(x), \tag{4.187}$$

allows the exponential to be written in terms of the Euclidean inverse M_E^{-1},

$$\exp\left\{\frac{1}{2}\sum_k \frac{1}{\lambda^{(k)}} J_k^2\right\} = \exp\left\{\frac{1}{2}\int \mathrm{d}^4x\, \mathrm{d}^4y\, J(x) M_E^{-1}(x, y) J(y)\right\}, \qquad (4.188)$$

where $M_E^{-1}(x, y)$ is given by the previous result (4.177) using the Euclidean eigenvalues. Choosing $N_j' = N/\sqrt{\alpha^2\pi}$ matches the prefactor in equation (4.186) to the prefactor $N/(\det(\alpha^2\hat{L}_E))^{1/2}$ in equation (4.158). The inverse Wick rotation gives $\mathrm{d}^4x\, \mathrm{d}^4y \overset{\mathrm{M}}{\rightarrow} -\mathrm{d}^4x\, \mathrm{d}^4y$, and the exponential factor of equation (4.188) then matches that of equation (4.158), while the prefactor becomes $N/\det(\alpha^2\hat{L})^{1/2}$, identical to the prefactor (4.176).

Another commonly used configuration space method for analyzing quadratic path integrals uses a change of configuration space variables in the context of a spacetime lattice. For the scalar field action (4.153) the field variables are changed from $\varphi(x)$ to the variable $\tilde{\varphi}(x)$ by the continuum version of an orthogonal transformation. Using the length scale α the new variables are defined as

$$\varphi(x) = \alpha \int \mathrm{d}^4y\, H(x, y)\phi(y). \qquad (4.189)$$

The new variable ϕ has units of inverse length squared, while the function $H(x, y)$ has units of inverse length to the fourth and is chosen so that

$$\alpha^2 \int \mathrm{d}^4x\, \mathrm{d}^4y\, H(x, z) M(x, y) H(y, w) = \delta^4(z - w). \qquad (4.190)$$

The units of these functions allow the Riemann sums to be written as matrix equations by scaling to the dimensionless variables of equations (4.156) and (4.157), as well as $\tilde{H}(x_j, x_k) = \sqrt{\mathrm{d}^4x_j\, \mathrm{d}^4x_k}\; H(x_j, x_k)$ and $\tilde{\phi}(x_k) = \sqrt{\mathrm{d}^4x_k}\; \phi(x_k)$. Relations (4.189) and (4.190) become the continuum matrix equations

$$\begin{aligned} \tilde{\varphi}(x_j) &= \sum_k \tilde{H}(x_j, x_k)\tilde{\phi}(x_k), \\ \delta_{jk} &= \sum_{i,n} \tilde{H}(x_i, x_j)\tilde{M}(x_i, x_n)\tilde{H}(x_n, x_k). \end{aligned} \qquad (4.191)$$

Relation (4.191) combines with equations (4.162) and (4.156) to give

$$\begin{aligned} &\sum_i \tilde{H}(x_j, x_i)\tilde{H}(x_k, x_i) = \tilde{M}^{-1}(x_j, x_k) \\ &\rightarrow \int \mathrm{d}^4z\, H(x, z) H(y, z) = \frac{1}{\alpha^2} M^{-1}(x, y). \end{aligned} \qquad (4.192)$$

In effect, the continuous matrix $H(x, y)$ is the square root of the continuous matrix $M^{-1}(x, y)/\alpha^2$. Using equation (4.190) shows that the scalar field action (4.153) becomes

$$
\begin{aligned}
&- \int \mathrm{d}^4x\, \mathrm{d}^4y\, \frac{1}{2}\varphi(x)M(x,\, y)\varphi(y) - i\int \mathrm{d}^4x\, J(x)\varphi(x) \\
&\to \sum_j \left(-\frac{1}{2}\tilde{\phi}^2(x_j) - i\sum_k \tilde{J}(x_j)\tilde{H}(x_j,\, x_k)\tilde{\phi}(x_j) \right),
\end{aligned}
\tag{4.193}
$$

where $\tilde{J}(x_j)$ is given by equation (4.157).

Using equation (4.193), the action functional (4.153) becomes

$$
Z_0[J] = \int_0^0 [\mathcal{D}\tilde{\phi}]\exp\left\{ -\sum_j \left(\frac{1}{2}\tilde{\phi}^2(x_j) + i\sum_k \tilde{J}(x_k)\tilde{H}(x_k,\, x_j)\tilde{\phi}(x_j) \right) \right\}, \tag{4.194}
$$

where the measure is now given by

$$
[\mathcal{D}\tilde{\phi}] = \det(\tilde{H})\prod_j N_j\, \mathrm{d}\tilde{\phi}(x_j). \tag{4.195}
$$

The Jacobian associated with the change of variables (4.189) is $\det \tilde{H}$, and combining equation (4.192) with the properties of the determinant give

$$
(\det \tilde{H})^2 = \det(\tilde{H}^2) = \det(\tilde{M}^{-1}) = \frac{1}{\det \tilde{M}} \implies \det \tilde{H} = \frac{1}{\sqrt{\det \tilde{M}}}. \tag{4.196}
$$

In the continuum limit this yields the same prefactor that appears in equation (4.158). The remaining integrations in equation (4.194) are a product of Gaussian integrals,

$$
Z_0[J] = \frac{N''}{\sqrt{\det \tilde{M}}}\prod_j \mathrm{d}\tilde{\phi}(x_j)\exp\left\{ -\frac{1}{2}\tilde{\phi}^2(x_j) - i\sum_k \left(\tilde{J}(x_k)\tilde{H}(x_k,\, x_j)\tilde{\phi}(x_j) \right) \right\}. \tag{4.197}
$$

Using equation (1.99) and relation (4.192) gives

$$
\begin{aligned}
Z_0[J] &= \frac{N'}{\sqrt{\det \tilde{M}}}\exp\left\{ -\frac{1}{2}\sum_{j,k} \tilde{J}(x_j)\left(\sum_i \tilde{H}(x_j,\, x_i)\tilde{H}(x_k,\, x_i) \right)\tilde{J}(x_k) \right\} \\
&= \frac{N'}{\sqrt{\det \tilde{M}}}\exp\left\{ -\frac{1}{2}\sum_{j,k} \tilde{J}(x_j)\tilde{M}^{-1}(x_j,\, x_k)\tilde{J}(x_k) \right\} \\
&\to \frac{N}{\sqrt{\det(\alpha^2 M)}}\exp\left\{ -\frac{1}{2}\int \mathrm{d}^4x\, \mathrm{d}^4y\, J(x)M^{-1}(x,\, y)J(y) \right\},
\end{aligned}
\tag{4.198}
$$

which is identical to result (4.158).

The method of equation (4.189) can be adapted to evaluate a general Euclidean quadratic bispinor path integral defined by the Grassmann functional integral

$$Z_0[K, \bar{K}] = \int_0^0 [\mathcal{D}^2\eta] \exp\left\{-\int d^4x\, d^4y\, \bar{\eta}(x)\cdot \mathbf{M}(x, y)\cdot \eta(y) - i\int d^4x\, (\bar{\eta}(x)\cdot K(x) + \bar{K}(x)\cdot \eta(x))\right\}. \tag{4.199}$$

The matrix $\mathbf{M}(x, y)$ has the bispinor components $M_{ab}(x, y) = -i\hat{D}_{ab}\delta^4(x - y)$, where $\hat{\mathbf{D}}$ is a spinor differential operator. The matrix $\mathbf{M}$ will have the units of inverse length to the fifth power. The configuration space version of the bispinor measure in lattice notation is written

$$[\mathcal{D}^2\eta] = \prod_j N_j d\bar{\eta}(x_j)\cdot d\eta(x_j) = \prod_j \prod_{a=1}^{4} N_j d\bar{\eta}_a(x_j)\, d\eta_a(x_j). \tag{4.200}$$

This can be demonstrated by using the transformation matrix to configuration measure given by the functional derivatives (4.126) and (4.128). Using these to define the Grassmann Jacobian J_G, given by equation (1.116), results in the measure

$$\prod_{s,\mathbf{p},t} (d^3p)^2\, d\beta_{\mathbf{p}}^{s*}(t)\, d\beta_{\mathbf{p}}^{s}(t)\, d\alpha_{\mathbf{p}}^{s*}(t)\, d\alpha_{\mathbf{p}}^{s}(t) = \prod_{\mathbf{x},t} J_G\, d\bar{\eta}(\mathbf{x}, t)\cdot d\eta(\mathbf{x}, t). \tag{4.201}$$

As a result, the measure is often written $[\mathcal{D}^2\eta] = [\mathcal{D}\bar{\eta}\cdot \mathcal{D}\eta]$ to signify its relativistic invariance under unitary transformations of the form $\psi' = \mathbf{U}\cdot\psi$ where $\det \mathbf{U} = 1$. At first glance, it appears that there are twice as many integration variables in the previous measure (4.138). However, the product over the spin s in equation (4.138) is from 1 to 2 while the product over the spinor index a in equation (4.200) is from 1 to 4, so that the number of integration variables match.

Similarly to equation (4.189), two new bispinor Grassmann variables $\zeta(x)$ and $\bar{\zeta}(x)$ are introduced,

$$\eta(x) = \sqrt{\alpha}\int d^4y\, \mathbf{H}(x, y)\cdot \zeta(y), \quad \bar{\eta}(x) = \sqrt{\alpha}\int d^4y\, \bar{\zeta}(y)\cdot \mathbf{H}(x, y), \tag{4.202}$$

where $\mathbf{H}(x, y)$ is spinor valued matrix and is chosen so that

$$\alpha\int d^4x\, d^4y\, \mathbf{H}(x, z)\cdot \mathbf{M}(x, y)\cdot \mathbf{H}(y, w) = \mathbf{I}\,\delta^4(z - w), \tag{4.203}$$

where $\mathbf{I}$ is the unit spinor matrix. The original variables η and $\bar{\eta}$ have the units of inverse length to the 3/2 power. As a result of the factor $\sqrt{\alpha}$, the units of $\mathbf{H}$ are inverse length to fourth, while ζ and $\bar{\zeta}$ have the units of inverse length squared. This allows the Riemann sums that replace the integrals in equation (4.199) to be written as matrix equations with the dimensionless variables $\tilde{\mathbf{M}}(x_j, x_k) = \sqrt{d^4x_j\, d^4x_k}\, \alpha\mathbf{M}(x_j, x_k)$,

$\tilde{\mathbf{H}}(x_j, x_k) = \sqrt{d^4x_j\, d^4x_k}\; \mathbf{H}(x_j, x_k)$, $\quad \tilde{\zeta}(x_k) = \sqrt{d^4x_k}\; \zeta(x_k)$, $\quad \tilde{\bar{\zeta}}(x_k) = \sqrt{d^4x_k}\; \bar{\zeta}(x_k)$, $\tilde{K}(x_k) = \sqrt{d^4x_k}\; K(x_k)/\sqrt{\alpha}$, and $\tilde{\bar{K}}(x_k) = \sqrt{d^4x_k}\; \bar{K}(x_k)/\sqrt{\alpha}$. Relations (4.202) and (4.203) become the dimensionless matrix equations

$$\begin{aligned}
\tilde{\eta}(x_j) &= \sqrt{\frac{d^4x_j}{\alpha}}\; \eta(x_j) = \sum_k \tilde{\mathbf{H}}(x_j, x_k) \cdot \tilde{\zeta}(x_k), \\
\tilde{\bar{\eta}}(x_j) &= \sqrt{\frac{d^4x_j}{\alpha}}\; \bar{\eta}(x_j) = \sum_k \tilde{\bar{\zeta}}(x_k) \cdot \tilde{\mathbf{H}}(x_j, x_k), \\
\mathbf{I}\, \delta_{jk} &= \sum_{i,n} \tilde{\mathbf{H}}(x_i, x_j) \cdot \tilde{\mathbf{M}}(x_i, x_n) \cdot \tilde{\mathbf{H}}(x_n, x_k).
\end{aligned} \tag{4.204}$$

Relation (4.204) and $\tilde{\mathbf{M}}^{-1}(x_j, x_k) = \sqrt{d^4x_j\, d^4x_k}\; \mathbf{M}^{-1}(x_j, x_k)/\alpha$ gives

$$\begin{aligned}
&\sum_i \tilde{\mathbf{H}}(x_j, x_i) \cdot \tilde{\mathbf{H}}(x_k, x_i) = \tilde{\mathbf{M}}^{-1}(x_j, x_k) \\
&\to \int d^4z\; \mathbf{H}(x, z) \cdot \mathbf{H}(y, z) = \frac{1}{\alpha} \mathbf{M}^{-1}(x, y).
\end{aligned} \tag{4.205}$$

The Grassmann Jacobian rule (1.116) shows that the measure (4.200) becomes

$$\begin{aligned}
[\mathcal{D}^2\eta] &= \frac{1}{(\det \tilde{\mathbf{H}})^2} \prod_{j,a} N_j'\, d\tilde{\bar{\zeta}}_a(x_j)\, d\tilde{\zeta}_a(x_j) \\
&= \frac{1}{\det(\tilde{\mathbf{M}}^{-1})} \prod_{j,a} N_j'\, d\tilde{\bar{\zeta}}_a(x_j)\, d\tilde{\zeta}_a(x_j) \\
&= \det(\tilde{\mathbf{M}}) \prod_{j,a} N_j'\, d\tilde{\bar{\zeta}}_a(x_j)\, d\tilde{\zeta}_a(x_j).
\end{aligned} \tag{4.206}$$

The continuum action appearing in equation (4.199) becomes the lattice sum

$$\begin{aligned}
S[\tilde{\zeta}, \tilde{\bar{\zeta}}, \tilde{K}, \tilde{\bar{K}}] \;=\; &\sum_j \Big(-\tilde{\bar{\zeta}}(x_j) \cdot \tilde{\zeta}(x_j) - i \sum_k \Big(\tilde{\bar{K}}(x_k) \cdot \tilde{\mathbf{H}}(x_j, x_k) \cdot \tilde{\zeta}(x_j) \\
&- \tilde{\bar{\zeta}}(x_j) \cdot \tilde{\mathbf{H}}(x_j, x_k) \cdot \tilde{K}(x_k) \Big) \Big)
\end{aligned} \tag{4.207}$$

The Grassmann integrations in equation (4.199) can now be performed using the action (4.207), the measure (4.206), and the result (1.87), which show that the integrations over the $\tilde{\bar{\zeta}}(x_j)$ yield the Grassmann Dirac delta $\delta(\tilde{\zeta}(x_j) + i \sum_k \mathbf{H}(x_j, x_k) \cdot \tilde{K}(x_k))$ at every spacetime lattice point. The integrations over $\tilde{\zeta}(x_j)$ are now trivial. Replacing the scaled functions turns the Riemann sums into integrals, with the result that

$$Z_0[K, \bar{K}] = N' \det(\tilde{\mathbf{M}}) \exp\left\{-\sum_{j,k,n} \tilde{\bar{K}}(x_k) \cdot \tilde{\mathbf{H}}(x_j, x_k) \cdot \tilde{\mathbf{H}}(x_n, x_k) \cdot \tilde{K}(x_n)\right\}$$
$$\rightarrow N \det(\alpha \mathbf{M}) \exp\left\{-\int \frac{d^4x\, d^4y}{\alpha^2} \bar{K}(x) \cdot \mathbf{M}^{-1}(x, y) \cdot K(y)\right\}. \tag{4.208}$$

The Dirac equation operator $-i\hat{\mathbf{D}} = -i(i\not{\partial} - m\mathbf{1})$ is an example of a spinor valued linear differential operator that occurs in the quadratic part of the spinor action in equation (4.199). Using equation (1.26) for the Dirac delta gives the matrix $\mathbf{M}(x, y)$,

$$\mathbf{M}(x, y) = -i \int \frac{d^4p}{(2\pi)^4} (\not{p} - m\mathbf{1})\, e^{-ip\cdot(x-y)}. \tag{4.209}$$

The inverse matrix is formally written

$$\mathbf{M}^{-1}(x, y) = i \int \frac{d^4p}{(2\pi)^4} (\not{p} - m\mathbf{1})^{-1} e^{-ip\cdot(x-y)}. \tag{4.210}$$

Like the Klein–Gordon operator, the inversion process requires an ϵ prescription to avoid the on shell singularities at $p^2 = m^2$. This is given by

$$(\not{p} - m\mathbf{1})^{-1} = \lim_{\epsilon\to 0} \frac{\not{p} + m\mathbf{1}}{p^2 - m^2 + i\epsilon}, \tag{4.211}$$

so that

$$\mathbf{S}_F(x - y) \equiv \mathbf{M}^{-1}(x, y) = \lim_{\epsilon\to 0} \int \frac{d^4p}{(2\pi)^4} \frac{i(\not{p} + m\mathbf{1})}{p^2 - m^2 + i\epsilon} e^{-ip\cdot(x-y)}. \tag{4.212}$$

As with the Klein–Gordon equation, this results in the Green's function $\mathbf{S}_F(x - y)$ or *Feynman propagator for bispinor fields* since

$$(i\not{\partial} - m\mathbf{1}) \cdot \mathbf{S}_F(x - y) = i\delta^4(x - y)\, \mathbf{1}. \tag{4.213}$$

It can be shown that the elements of the propagator (4.212) are given by the time-ordered product of the free anticommuting Dirac fields $\hat{\psi}_a(x)$ given by (4.48),

$$S_{Fab}(x - y) = \langle 0| T\{\hat{\psi}_a(x)\, \hat{\bar{\psi}}_b(y)\} |0\rangle. \tag{4.214}$$

The details are given in Peskin and Schroeder. The final result for the free Dirac bispinor path integral is

$$Z_0[k, \bar{K}] = N \det(\alpha \mathbf{M}) \exp\left\{-\int \frac{d^4x\, d^4y}{\alpha^2} \bar{K}(x) \cdot \mathbf{S}_F(x, y) \cdot K(y)\right\}. \tag{4.215}$$

Another result that will be useful in the next section is the change of variables of integrals involving differential operators. Sometimes a function of one or more of the fields, denoted $G(\varphi(x))$, is imposed as a constraint, $G(\varphi(x)) = 0$, at all spacetime points. It is assumed that a unique solution $\varphi_c(x)$ can be found to this equation, so that $G(\varphi_c(x)) = 0$. In configuration space the measure of the constrained path integral then takes the form

$$[\mathcal{D}\varphi]_c = \prod_x N_x \,\mathrm{d}\varphi(x)\, \delta(\varphi(x) - \varphi_c(x)). \tag{4.216}$$

It is often useful to express the constraint in terms of the constraint equation rather than the solution. This is accompanied by the determinant of the change in the delta function argument,

$$[\mathcal{D}\varphi]_c = \det\left(\frac{\delta G}{\delta\varphi}\right) \prod_x N_x \,\mathrm{d}\varphi(x)\, \delta(G(\varphi(x))), \tag{4.217}$$

where the matrix appearing in the determinant has the continuum form $M(x, y) = \delta G(\varphi(x))/\delta\varphi(y)$. If $G(\varphi(x))$ is a *local function of* φ, which means that it contains only references to $\varphi(x)$ and no integrals or *global references*, then M is a diagonal matrix of the form $M(x, y) = \delta^4(x - y)\, \partial G/\partial\varphi(x)$.

4.9 The gauge field path integral

In this section the Landau gauge path integral for gauge fields will be evaluated using configuration space measure. The path integral for the gauge field can be derived using coherent states consistent with the vector extension of the scalar field variables,

$$A_\mu(x) = \int \mathrm{d}^3p \Big(\alpha_\mu(\boldsymbol{p}, t) f_p^*(x) + \alpha_\mu^*(\boldsymbol{p}, t) f_p(x)\Big), \tag{4.218}$$

$$\pi_\mu(x) = -i \int \mathrm{d}^3p\, \varepsilon_p \Big(\alpha_\mu(\boldsymbol{p}, t) f_p^*(x) - \alpha_\mu^*(\boldsymbol{p}, t) f_p(x)\Big), \tag{4.219}$$

where $f_p(x) = e^{ip\cdot x}/\sqrt{(2\pi)^3 2\varepsilon_p}$ and $p_0 = \varepsilon_p = |\boldsymbol{p}| = \sqrt{\boldsymbol{p}^2}$. The Landau gauge condition will be implemented by generalizing the quantum mechanical result (3.238) for ℓ gauge constraints. In this case, the ℓ quantum mechanical constraints will be increased to cover the entirety of lattice spacetime. Nevertheless, the general form of the constraint measure (3.239) will remain the same for such a case. The process used to derive the scalar field path integral is now applied to the four degrees of freedom represented by A_μ. The general form for the vacuum transition element derived for the scalar field occurs for all four degrees of freedom, so that the gauge field path integral is given by

$$Z_0[J] = \int_0^0 [\mathcal{D}^c\pi\; \mathcal{D}^c A] \exp\left\{i \int \mathrm{d}^4x \left(\pi_\mu \dot{A}^\mu - \mathcal{H}_G(A^\mu, \pi_\mu) - J_\mu A^\mu\right)\right\}, \tag{4.220}$$

where $[\mathcal{D}^c\pi\; \mathcal{D}^c A]$ is the *constrained measure*. Its definition requires further discussion of the constraints. However, it is important to note that the measure in configuration space includes all four components of A_μ in the product form $d^4A = dA_0\, dA_1\, dA_2\, dA_3$. The Lorentz invariance of this measure follows from

$$d^4A' = dA_0'\, dA_1'\, dA_2'\, dA_3' = \det(\Lambda)\, dA_0\, dA_1\, dA_2\, dA_3 = d^4A, \tag{4.221}$$

since it was shown in equation (4.5) that $\det \Lambda = 1$ for a proper Lorentz transformation. The current J_μ is a *gauge invariant conserved external source*. The source J^μ is used to generate time-ordered products of the gauge field through functional derivatives of $Z_0[J]$, and obeys the conservation law $\partial_\mu J^\mu = 0$. The property of conservation is necessary in order to allow the term $A_\mu J^\mu$ to be gauge invariant when added to the action, since an integration by parts gives

$$\begin{aligned} &\int d^4x\, A_\mu J^\mu \rightarrow \int d^4x\, (A_\mu J^\mu - \partial_\mu \Lambda\, J^\mu) \\ &= \int d^4x\, \left(A_\mu J^\mu - \partial_\mu(\Lambda J^\mu) + \Lambda\, \partial_\mu J^\mu\right) = \int d^4x\, A_\mu J^\mu, \end{aligned} \tag{4.222}$$

where it is assumed that the gauge function Λ vanishes at the boundaries of the spacetime manifold. The Hamiltonian density $\mathcal{H}_G$ appearing in equation (4.220) is given by the classical Landau gauge Hamiltonian (4.40). In natural units it is given by

$$\mathcal{H}_G(\pi_\mu, A^\mu) = \frac{1}{2}\left(\pi_i\pi_i - \pi_0^2 + \partial_j A^i \partial_j A^i - \partial_j A^i \partial_i A^j\right) + A_0 \partial_i \pi_i - \pi_0 \partial_i A^i. \tag{4.223}$$

The primary constraint is $\pi_0 = -\partial_\mu A^\mu = 0$. The secondary constraint was identified in equation (4.41) as Gauss's law, now modified by the external current J_μ,

$$G = \dot{\pi}_0 = -\partial_j \pi_j - J_0 = 0. \tag{4.224}$$

At the classical level, where $\pi_j = \partial_j A^0 + \dot{A}^j$, the secondary constraint is given by

$$-\partial_j \pi_j - J_0 = -\nabla^2 A_0 - \nabla \cdot \dot{A} - J_0 = \Box A_0 - \partial_\mu \dot{A}^\mu - J_0 = 0. \tag{4.225}$$

The solution to equation (4.225) is found using the Green's function $C(x, y)$ for ∇^2, which is the Coulomb potential $C(x, y) = -1/4\pi|\boldsymbol{x} - \boldsymbol{y}|$. This gives

$$A^0(\boldsymbol{x}, t) = \int \frac{d^3y}{4\pi} \frac{1}{|\boldsymbol{x} - \boldsymbol{y}|}\left(\nabla \cdot \dot{A}(\boldsymbol{y}, t) + J_0(\boldsymbol{y}, t)\right). \tag{4.226}$$

For the classical theory, it follows that $\nabla \cdot \ddot{A}(\boldsymbol{y}, t) = \nabla_y^2 \nabla \cdot A(\boldsymbol{y}, t) + \nabla \cdot J$. Using this result and $\partial_\mu J^\mu = 0$ in equation (4.226) with an integration by parts gives

$$\dot{A}^0(\boldsymbol{x}, t) = \int \frac{d^3y}{4\pi}\, \nabla \cdot A(\boldsymbol{y}, t) \nabla_y^2\left(\frac{1}{|\boldsymbol{x} - \boldsymbol{y}|}\right) = -\nabla \cdot A(\boldsymbol{x}, t) \implies \partial_\mu A^\mu = 0. \tag{4.227}$$

The solution of Gauss's law given by equation (4.226) is therefore consistent *classically* with the Landau gauge condition $\partial_\mu A^\mu = 0$.

The gauge theory path integral requires integrating over both π_μ and A^μ as independent degrees of freedom. However, the measure for the path integral must implement the gauge constraint dictated by the choice of the Landau gauge. The path integral measure can be written in two forms, which generalize the quantum mechanical result (3.239) to the case of a field theory,

$$\begin{aligned}[\mathcal{D}^c\pi\ \mathcal{D}^cA] &= \prod_{x,\mu} \mathrm{d}A_\mu(x)\mathrm{d}\pi_\mu(x)\prod_y \delta(\pi_0(y))\delta(A_0(y) - A_0^c(y)) \\ &= \det\left(\frac{\delta G}{\delta A_0}\right)\prod_{x,\mu} \mathrm{d}A_\mu(x)\mathrm{d}\pi_\mu(x)\prod_y \delta(\partial_\mu A^\mu(y))\,\delta(G(y)).\end{aligned} \tag{4.228}$$

In the first line A_0^c is the classical solution (4.226) for A_0 found from the secondary constraint of Gauss's law $G = -\partial_j\pi_j - J_0 = 0$ written in terms of A_μ. In the second version of the measure the primary constraint is written in terms of the fields A^μ as $\partial_\mu A^\mu = 0$, while the secondary constraint delta functions are written in terms of π_μ as $G = -\partial_j\pi_j - J_0 = 0$. The equality of the two measures follows from result (4.217), which requires including the determinant of the continuum matrix $\delta G(x)/\delta A_0(y)$. Since A_0 and π_j are variables of integration, the constraint G appears to have no dependence on A_0. This would cause the determinant to vanish in contradiction of the first version of the measure. It will be shown that the integration over π_j in the path integral induces a dependence for G on A_0, resulting in a nonzero determinant.

The process of analyzing the path integral (4.220) begins by writing out the Landau gauge action,

$$\begin{aligned}\mathcal{L}_G = &\frac{1}{2}\pi_0^2 + \pi_0\partial_\mu A^\mu - \frac{1}{2}\pi_j\pi_j + \pi_j\dot{A}^i - (\partial_j\pi_j + J_0)A^0 \\ &- \frac{1}{2}(\partial_j A^i\partial_j A^i - \partial_i A^i\partial_j A^j) - J_j A^j.\end{aligned} \tag{4.229}$$

The next step is to write the secondary constraint in exponential form (1.26),

$$\prod_x \delta(G(x)) = \int [\mathcal{D}\lambda]\exp\left\{i\int \mathrm{d}^4x\ \lambda(x)\big(\partial_j\pi_j(x) + J_0(x)\big)\right\}, \tag{4.230}$$

where the measure in equation (4.230) is given in configuration form by

$$[\mathcal{D}\lambda] = \prod_x \mathrm{d}^4x\ \mathrm{d}\lambda(x). \tag{4.231}$$

The exponentiated constraint in equation (4.230) becomes part of the action (4.229) in the path integral (4.220). Translating A^0 by λ, so that $A^0 \to A^0 + \lambda$, does not affect the measure term $[\mathcal{D}A^0]$. However, the primary constraint becomes

$$\prod_x \delta(\partial_\mu A^\mu) \to \prod_x \delta(\partial_\mu A^\mu + \dot{\lambda}), \tag{4.232}$$

while the action (4.229) becomes

$$\begin{aligned}\mathcal{L}_G \to &\frac{1}{2}\pi_0^2 + \pi_0(\partial_\mu A^\mu + \dot{\lambda}) - \frac{1}{2}\pi_j\pi_j + \pi_j\dot{A}^j - (\partial_j\pi_j + J_0)A^0 \\ &- \frac{1}{2}(\partial_j A^i\partial_j A^i - \partial_i A^i\partial_j A^j) - J_j A^j.\end{aligned} \tag{4.233}$$

The translation of A^0 has absorbed the secondary constraint into the primary constraint in both the measure (4.232) and the action (4.233).

The final step is to perform a gauge transformation, $\pi_\mu \to \pi_\mu$ and $A^\mu \to A^\mu - \partial^\mu\Lambda$. Because the action (4.233) is integrated over all spacetime, the terms generated by the gauge transformation may be integrated by parts. Combining $\pi_j\partial^j\dot{\Lambda} = -\pi_j\partial_j\dot{\Lambda}$ with the conserved current, $\partial_\mu J^\mu = 0$, it follows that three of these terms cancel,

$$\int \mathrm{d}^4x\left(-\pi_j\partial_j\dot{\Lambda} - (\partial_j\pi_j + J^0)\dot{\Lambda} - J^j\partial_j\Lambda\right) = \int \mathrm{d}^4x\ \Lambda\,\partial_\mu J^\mu = 0. \tag{4.234}$$

Subsequent to the gauge transformation, the action of equation (4.233) becomes

$$\begin{aligned}\mathcal{L}_G = &\frac{1}{2}\pi_0^2 + \pi_0(\partial_\mu A^\mu - \Box\Lambda + \dot{\lambda}) - \frac{1}{2}\pi_j\pi_j + \pi_j\dot{A}^i - (\partial_j\pi_j + J_0)A^0 \\ &- \frac{1}{2}(\partial_j A^i\partial_j A^i - \partial_i A^i\partial_j A^j) - J_j A^j,\end{aligned} \tag{4.235}$$

while the primary constraint (4.232) becomes

$$\prod_x \delta(\partial_\mu A^\mu + \dot{\lambda}) \to \prod_x \delta(\partial_\mu A^\mu - \Box\Lambda + \dot{\lambda}). \tag{4.236}$$

Although Λ must vanish at the boundaries of spacetime, it is otherwise arbitrary. As a result, it can be chosen to satisfy $\Box\Lambda - \dot{\lambda} = 0$. The action in equation (4.235) becomes

$$\begin{aligned}\mathcal{L}_G = &\frac{1}{2}\pi_0^2 + \pi_0\partial_\mu A^\mu - \frac{1}{2}\pi_j\pi_j + \pi_j\dot{A}^j - (\partial_j\pi_j + J_0)A^0 \\ &- \frac{1}{2}(\partial_j A^i\partial_j A^i - \partial_i A^i\partial_j A^j) - J_j A^j,\end{aligned} \tag{4.237}$$

while the primary constraint (4.236) becomes

$$\prod_x \delta(\partial_\mu A^\mu - \Box\Lambda + \dot{\lambda}) \to \prod_x \delta(\partial_\mu A^\mu). \tag{4.238}$$

The gauge field path integral is now given by

$$Z_0[J^\mu] = \int[\mathcal{D}\lambda]\int[\mathcal{D}\pi_\mu\mathcal{D}A^\mu]\det\left(\frac{\delta G}{\delta A^0}\right)\prod_x \delta(\partial_\mu A^\mu)\exp\left\{i\int \mathrm{d}^4x\ \mathcal{L}_G\right\}, \tag{4.239}$$

where the action is given by equation (4.237) and contains no reference to λ.

The volume associated with λ has been factored out of the path integral in the process of implementing the secondary constraint. The physical meaning of this factorization can be understood by using the momentum operator (4.56) from the canonical theory. It shows that the operator defined by

$$\hat{U}(t) = \exp\left\{i\int \mathrm{d}^3x\, \lambda(\boldsymbol{x}, t)\, \partial_j\hat{\pi}_j(\boldsymbol{x}, t)\right\} = \exp\left\{i\int \mathrm{d}^3x\, \lambda(\boldsymbol{x}, t)\, \dot{\hat{\pi}}_0(\boldsymbol{x}, t)\right\} \quad (4.240)$$

implements a gauge transformation on the operator version of $\hat{a}_\mu$ given by equation (4.55),

$$\hat{U}(t)\hat{a}_\mu(\boldsymbol{x}, t)\hat{U}^{-1}(t) = \hat{a}_\mu - \partial\lambda. \quad (4.241)$$

The particle operator content of $\hat{U}(t)$ consists entirely of the ghost operators $\hat{g}_p^\dagger$ that are allowed in the physical state (4.61). The exponentiation of the secondary constraint given by equation (4.230) is the path integral version of the generator of gauge transformations given by equation (4.240), so that the overall volume that has been factorized in equation (4.239) is often referred to as a *gauge volume*. It represents all possible gauge transformations generated by the ghost operators present in the particle spectrum by making all states generated by a gauge transformation on a positive norm transverse photon state equivalent and possessing *zero norm*. This means that, at least in the free theory, a state obtained by a gauge transformation adds nothing to the original energy of the state.

The next step is to integrate the momenta from the path integral. This is typically performed in the Euclidean version of the path integral (4.239), which is obtained from the following Wick rotation recipe for all the timelike components present in a gauge theory,

$$t \overset{\mathrm{W}}{\rightarrow} -it, \qquad \omega \overset{\mathrm{W}}{\rightarrow} i\omega, \qquad A^0 \overset{\mathrm{W}}{\rightarrow} iA^0, \qquad \pi_0 \overset{\mathrm{W}}{\rightarrow} -i\pi_0, \qquad J_0 \overset{\mathrm{W}}{\rightarrow} -iJ_0. \quad (4.242)$$

The Wick rotation gives $\partial_\mu A^\mu \overset{\mathrm{W}}{\rightarrow} -\dot{A}_0 - \partial_j A_j \equiv \partial_\mu A^\mu_W$. On the other hand, $\pi_j(\dot{A}_j - \partial_j A_0) = \pi_j F_{0j} \overset{\mathrm{W}}{\rightarrow} i\pi_j(\dot{A}_j - \partial_j A_0) = i\pi_j F_{0j}$. The Euclidean version of the action (4.237) can be written as

$$\begin{aligned}\mathcal{L}_G \overset{\mathrm{W}}{\rightarrow} &-\frac{1}{2}\left(\pi_0 + i\partial_\mu A^\mu_W\right)^2 - \frac{1}{2}\left(\pi_j + iF_{0j}\right)\left(\pi_j + iF_{0j}\right)\\ &-\frac{1}{4}F_{ij}F^{ij} - \frac{1}{2}F_{0j}F_{0j} - \frac{1}{2}(\partial_\mu A^\mu_W)^2 - J^E_\mu A^{\mu E},\end{aligned} \quad (4.243)$$

The momentum integration variables are translated according to $\pi_0 = \tilde{\pi}_0 - i\partial_\mu A^\mu_W$ and $\pi_j = \tilde{\pi}_j - iF_{0j}$. The Euclidean version of the path integral becomes

$$\begin{aligned}Z_0^E[J] = &\int [\mathcal{D}\lambda] \int [\mathcal{D}\tilde{\pi}_\mu]\exp\left\{-\frac{1}{2}\int \mathrm{d}^4x_E\, \tilde{\pi}_\mu\, \tilde{\pi}_\mu\right\}\\ &\times \int [\mathcal{D}A^\mu]\det\left(\frac{\delta G}{\delta A_0}\right)\prod_x \delta(\partial_\mu A^\mu)\exp\left\{\int \mathrm{d}^4x_E\, \tilde{\mathcal{L}}_G(A_\mu)\right\},\end{aligned} \quad (4.244)$$

where the Euclidean version of the gauge fixed Lagrangian (4.33) has appeared,

$$\tilde{\mathcal{L}}_G(A_\mu) = -\frac{1}{4}F_{ij}F^{ij} - \frac{1}{2}F_{0j}F_{0j} - \frac{1}{2}(\partial_\mu A_W^\mu)^2 - J_\mu^E A^{\mu E}. \tag{4.245}$$

The resulting momentum integrations in the path integral are standard Gaussians and irrelevant to the remaining path integral for A_μ.

However, in the process of translating the momenta in equation (4.244) the secondary constraint in the determinant of equation (4.239) becomes

$$G = -\partial_j \pi_j \rightarrow -\partial_j \tilde{\pi}_j + i\partial_j F_{0j}, \tag{4.246}$$

where $\partial_j F_{0j} = -\nabla^2 A_0 + \partial_j \dot{A}_j$. It is crucial to note that the secondary constraint can also be written $G = -\partial_j \tilde{\pi}_j + i\Box_E A_0 - i\partial_\mu \dot{A}_W^\mu$, where $\Box_E = -\partial_0^2 - \nabla^2$. In the Landau gauge this becomes $G = -\partial_j \tilde{\pi}_j - i\Box_E A_0$ since the primary constraint $\partial_\mu A_W^\mu = 0$ is enforced at all spacetime points, with the result that $\partial_\mu \dot{A}^\mu(x) = \lim_{\epsilon \to 0}(\partial_\mu A_W^\mu(\boldsymbol{x}, t + \epsilon) - \partial_\mu A_W^\mu(\boldsymbol{x}, t))/\epsilon = 0$. The argument of the determinant is the matrix whose elements are given by $i\Box_E \delta^4(x - y)$. Returning the delta function to Minkowski space uses

$$\delta(t) = \int \frac{d\omega}{2\pi} e^{i\omega t} \overset{\text{M}}{\rightarrow} -i \int \frac{d\omega}{2\pi} e^{i\omega t} = -i\delta(t). \tag{4.247}$$

As a result, $\delta^4(x - y) \overset{\text{M}}{\rightarrow} -i\delta^4(x - y)$ and $i\Box_E \delta^4(x - y) \overset{\text{M}}{\rightarrow} \Box\delta^4(x - y)$. The determinant in equation (4.244) is therefore $\det(\delta G/\delta A_0) = \det \Box$ in the notation of equation (4.176).

In order to complete the derivation of the gauge field path integral, it is necessary to write this determinant as a functional integral. This uses the field theory extension of the Grassmann integral (3.240) or the scalar version of equation (4.199),

$$\begin{aligned} \det\left(\frac{\delta G}{\delta A_0}\right) &= \det(\Box) = \int [\mathcal{D}\tilde{c}\ \mathcal{D}c] \exp\left\{\int d^4x\ \tilde{c}(x)\Box\, c(x)\right\} \\ &= \int [\mathcal{D}\tilde{c}\ \mathcal{D}c] \exp\left\{i\int d^4x\ \partial_\mu(i\tilde{c}(x))\partial^\mu c(x)\right\}, \end{aligned} \tag{4.248}$$

where $\tilde{c}$ and c are real scalar Grassmann variables and the measure is given by

$$[\mathcal{D}\tilde{c}\ \mathcal{D}c] = \prod_x N_x\, d\tilde{c}(x)\, dc(x). \tag{4.249}$$

It is important to note that the integrand in the second version of the functional integral (4.248) is real. This is because the two real Grassmann variables $\tilde{c}$ and c obey the complex conjugation rule of equation (1.75), so that $(\tilde{c}\ c)^* = c^*\tilde{c}^* = c\ \tilde{c} = -\tilde{c}\ c$. As a result, $(\partial_\mu(i\tilde{c})\ \partial^\mu c)^* = -\partial_\mu c\ \partial^\mu(i\tilde{c}) = \partial_\mu(i\tilde{c})\ \partial^\mu c$. The factor of i is often absorbed by defining $\bar{c} = i\tilde{c}$ with the understanding that $\bar{c}$ is therefore a pure imaginary Grassmann variable that obeys $\bar{c}^* = -\bar{c}$.

The Grassmann field theory defined by equation (4.248), with the Lagrangian given by

$$\mathcal{L}_c = \partial_\mu \bar{c}\, \partial^\mu c, \tag{4.250}$$

can be clarified by canonically quantizing the fields c and $\bar{c}$ as *anticommuting scalars*. This is a violation of the famous spin and statistics theorem, proven rigorously by Streater and Wightman, that requires integer spin fields to be quantized using commutation relations. The loophole in the theorem is the assumption of a positive definite metric space, which does not hold in the case of gauge fields. The two Grassmann fields satisfy the free field equation $\Box\bar{c} = \Box c = 0$, while the momentum canonically conjugate to $\bar{c}$ is given by $\dot{c}$. For such a case, the operator valued Grassmann fields can be given the expansions

$$c(\boldsymbol{x}, t) = \int \frac{d^3k}{(2\pi)^{3/2}} \frac{1}{\sqrt{2\omega_k}} \left(\hat{c}_k e^{-ik\cdot x} + \hat{c}_k^\dagger e^{ik\cdot x} \right), \tag{4.251}$$

$$\bar{c}(\boldsymbol{x}, t) = \int \frac{d^3p}{(2\pi)^{3/2}} \frac{1}{\sqrt{2\omega_p}} \left(\hat{\bar{c}}_p e^{-ip\cdot x} - \hat{\bar{c}}_p^\dagger e^{ip\cdot x} \right), \tag{4.252}$$

where $k_0 = |\boldsymbol{k}|$ insures the free field equation of motion is satisfied and the form for equation (4.252) insures $\bar{c}^* = -\bar{c}$. In order that the Grassmann properties $\{c(x), c(y)\} = \{\bar{c}(x), \bar{c}(y)\} = 0$ hold, it is necessary that

$$\left\{\hat{c}_p, \hat{c}_k\right\} = \left\{\hat{c}_p^\dagger, \hat{c}_k^\dagger\right\} = \left\{\hat{c}_p, \hat{c}_k^\dagger\right\} = \left\{\hat{\bar{c}}_p, \hat{\bar{c}}_k\right\} = \left\{\hat{\bar{c}}_p^\dagger, \hat{\bar{c}}_k^\dagger\right\} = \left\{\hat{\bar{c}}_p, \hat{\bar{c}}_k^\dagger\right\} = 0. \tag{4.253}$$

Satisfying the equal time anticommutation relation, $\{\bar{c}(\boldsymbol{x}, t), \bar{\pi}(\boldsymbol{y}, t)\} = i\delta^3(\boldsymbol{x} - \boldsymbol{y})$, requires the anticommutation relations

$$\left\{\hat{c}_p, \hat{\bar{c}}_k^\dagger\right\} = \left\{\hat{\bar{c}}_p, \hat{c}_k^\dagger\right\} = \delta^3(\boldsymbol{k} - \boldsymbol{p}), \quad \left\{\hat{c}_p, \hat{\bar{c}}_k\right\} = \left\{\hat{c}_p^\dagger, \hat{\bar{c}}_k^\dagger\right\} = 0. \tag{4.254}$$

This follows from using $\bar{\pi} = \dot{c}$, where the anticommutators of equation (4.254) give

$$\{\bar{c}(\boldsymbol{x}, t), \bar{\pi}(\boldsymbol{y}, t)\} = i\int \frac{d^3k}{(2\pi)^3} \cos(\boldsymbol{k}\cdot(\boldsymbol{x} - \boldsymbol{y})) = i\delta^3(\boldsymbol{x} - \boldsymbol{y}). \tag{4.255}$$

However, the algebra of equation (4.254) means that the states $c_p^\dagger|0\rangle$ and $\hat{\bar{c}}_p^\dagger|0\rangle$ have the inner products $\langle 0|\hat{c}_p\hat{c}_k^\dagger|0\rangle = \langle 0|\hat{\bar{c}}_p\hat{\bar{c}}_k^\dagger|0\rangle = 0$. Since they have zero norm they are ghost states, and are known as *Faddeev–Popov ghosts.*

The transformation of equation (4.244) back to Minkowski space gives $d^4x_E \overset{M}{\to} i\, d^4x$, $\partial_\mu A_W^\mu \overset{M}{\to} \partial_\mu A^\mu$, and $J_\mu^E A_E^\mu = J_\mu A^\mu$. Assuming the overall factor that results from integrating the momentum is canceled by normalization, the final form of the gauge field Lagrangian path integral is given by

$$Z_0[J] = \int [\mathcal{D}\lambda] \int [\mathcal{D}A_\mu\, \mathcal{D}\bar{c}\, \mathcal{D}c] \prod_x \delta(\partial_\mu A^\mu) \exp\left\{ i\int d^4x\, (\mathcal{L}_G + \mathcal{L}_c) \right\}, \tag{4.256}$$

where $\mathcal{L}_c$ is given by equation (4.250) and $\mathcal{L}_G$ is the gauge fixed Minkowski action,

$$\mathcal{L}_G = -\frac{1}{4}F_{\mu\nu}F^{\mu\nu} - \frac{1}{2}(\partial_\mu A^\mu)^2 - J_\mu A^\mu. \tag{4.257}$$

The Landau gauge path integral (4.256) possesses a residual form of gauge invariance under what is known as a *Becchi–Rouet–Stora–Tyupin (BRST) transformation*. The BRST transformation is given by

$$A_\mu(x) \to A_\mu(x) - \xi\, \partial_\mu c(x), \quad \bar{c}(x) \to \bar{c}(x) + \xi\, \partial_\mu A^\mu(x), \tag{4.258}$$

where ξ is a *Grassmann length* obeying $\xi^2 = 0$. The gauge invariant part of the Lagrangian $\mathcal{L}_G$ is left unchanged. The nilpotence of ξ causes the gauge fixing term to transform according to $-\frac{1}{2}(\partial_\mu A^\mu)^2 \to -\frac{1}{2}(\partial_\mu A^\mu)^2 + \xi\, \partial_\mu A^\mu \Box c$, while the ghost action becomes $\partial_\nu \bar{c}\, \partial^\nu c \to \partial_\nu \bar{c}\, \partial^\nu c + \xi(\partial_\nu \partial_\mu A^\mu)\partial^\nu c$. After an integration by parts on the ghost action contribution, the two terms proportional to ξ cancel. However, the measure has a delta function that becomes $\delta(\partial_\mu A^\mu - \xi\, \Box c)$ subsequent to the BRST transformation. Using the definition of the Grassmann delta function (1.87), the integration over $\bar{c}$ in the path integral gives

$$\int [\mathcal{D}\bar{c}] \exp\left\{-i\int \mathrm{d}^4x\, \bar{c}\, \Box c\right\} = \prod_x \delta(\Box c(x)), \tag{4.259}$$

As a result, the delta function constraint is *virtually invariant* under the BRST transformation since the integration results in $\delta(\partial_\mu A^\mu - \xi\, \Box c) = \delta(\partial_\mu A^\mu)$ because $\Box c = 0$. In effect, the BRST transformation reflects the invariance of equation (4.257) under gauge transformations that satisfy $\Box\Lambda = 0$, as well as the fact that the good ghost states of equation (4.61) are completely decoupled from the transverse excitations.

The final Lagrangian gauge field path integral (4.256) can be viewed as having factorized the infinite volume of gauge equivalent field configurations, leaving the remaining path integral uniquely gauge fixed. The method developed by Faddeev and Popov offers a more direct approach to this factorization procedure. The starting point for their method is the manifestly gauge invariant *Lagrangian version* of the path integral,

$$Z_0[J_\mu] = \int [\mathcal{D}A_\mu]\exp\left\{i\int \mathrm{d}^4x\left(-\frac{1}{4}F_{\mu\nu}F^{\mu\nu} - J_\mu A^\mu\right)\right\}. \tag{4.260}$$

The goal is to factorize the gauge equivalent field configurations consistent with an *arbitrary* gauge condition $\chi(A_\mu) = 0$. It is assumed that the gauge condition itself is *not* gauge invariant, and the gauge condition after a gauge transformation is denoted $\chi(A_\mu - \partial_\mu\Lambda) \equiv \chi^\Lambda$. For example, the Landau gauge condition becomes $\chi_F^\Lambda = \partial_\mu A^\mu - \Box\Lambda$. The next step defines a factor of unity,

$$1 = \int [\mathcal{D}\Lambda] \prod_x \delta(\Lambda(x)), \tag{4.261}$$

where the *gauge measure* is given by

$$[\mathcal{D}\Lambda] = \prod_x \mathrm{d}\Lambda(x). \tag{4.262}$$

This factor of unity can be inserted into the measure of the path integral (4.260) with no effect. The next step is to change the argument of the delta functions to the transformed gauge condition χ^Λ. Using result (4.217), this gives

$$\prod_x \delta(\Lambda(x)) = \det\left(\frac{\delta\chi^\Lambda}{\delta\Lambda}\right) \prod_x \delta(\chi^\Lambda(x)). \tag{4.263}$$

The argument of the determinant is the continuum matrix $\delta\chi^\Lambda(x)/\delta\Lambda(y)$. The path integral (4.260) becomes

$$Z_0[J_\mu] = \int [\mathcal{D}A_\mu][\mathcal{D}\Lambda]\det\left(\frac{\delta\chi^\Lambda}{\delta\Lambda}\right) \prod_x \delta(\chi^\Lambda)\exp\left\{i\int \mathrm{d}^4x\, \mathcal{L}(A_\mu, J_\mu)\right\}. \tag{4.264}$$

Because the Lagrangian $\mathcal{L}$ and the gauge field measure $[\mathcal{D}A_\mu]$ are both gauge invariant, it is possible to perform the inverse gauge transformation. However, it is important to note that $\Delta_\chi = \det(\delta\chi^\Lambda/\delta\Lambda)$ may or may not be affected by this inverse transformation since the functional derivative is taken *prior* to the inverse transformation. For a linear gauge condition, such as the Landau gauge condition $\chi_L = \partial_\mu A^\mu$ where $\delta\chi_L^\Lambda/\delta\Lambda = \Box$, it will not be affected by the inverse transformation. The final result for the Faddeev–Popov procedure is

$$Z_0[J_\mu] = \int [\mathcal{D}\Lambda] \int [\mathcal{D}A_\mu]\Delta_\chi \prod_x \delta(\chi)\exp\left\{i\int \mathrm{d}^4x\, \mathcal{L}(A_\mu, J_\mu)\right\}. \tag{4.265}$$

As a result of the delta functions $\prod_x \delta(\chi(x))$ present in the measure, a gauge fixing term involving χ, such as the Landau gauge fixing term $-\frac{1}{2}\chi_L^2$, can be added to the action. It is useful to note that the gauge fixing term need not be relativistically invariant, and there are useful choices of gauge that are not covariant, such as the *temporal gauge condition* $A_0 = 0$. The determinant can also be added to the action using the Grassmann ghost procedure,

$$\Delta_\chi = \int [\mathcal{D}\bar{c}\ \mathcal{D}c]\exp\left\{-i\int \mathrm{d}^4x\, \bar{c}\left(\frac{\delta\chi^\Lambda}{\delta\Lambda}\right)c\right\}. \tag{4.266}$$

If Δ_χ is independent of Λ, then the gauge volume has been successfully factorized. For the Landau gauge, $\chi^\Lambda = -\partial_\mu A^\mu + \Box\Lambda$ and $\Delta_\chi = \det\Box$. Therefore, for this specific case, the resulting Lagrangian path integral is *identical* to result (4.248). The method of factorizing the gauge volume in a manner consistent with the gauge condition will be extended to nonabelian gauge fields in the next chapter.

The last step in this section is to evaluate the path integral for the free quadratic gauge field path integral. The Faddeev–Popov ghosts will be ignored since they have no external source linked to them. The Landau gauge Lagrangian is

$$\mathcal{L}_G = -\frac{1}{2}\partial_\mu A^\nu \partial_\nu A^\mu - J_\mu A^\mu, \tag{4.267}$$

so that, after an integration by parts, the action can be written

$$\begin{aligned} iS_G &= i\int \mathrm{d}^4x\left(\frac{1}{2}A^\mu \Box A_\mu - J_\mu A^\mu\right) \\ &= -\frac{1}{2}\int \mathrm{d}^4x\, \mathrm{d}^4y\ A^\mu(x) g_{\mu\nu}(-i\Box)\delta^4(x-y)A^\nu(y) - i\int \mathrm{d}^4x J_\mu A^\mu \\ &= -\frac{1}{2}\int \mathrm{d}^4x\, \mathrm{d}^4y\ A^\mu(x) M_{\mu\nu}(x-y)A^\nu(y) - i\int \mathrm{d}^4x J_\mu A^\mu, \end{aligned} \tag{4.268}$$

where $M_{\mu\nu}(x-y) = -ig_{\mu\nu}\Box\delta^4(x-y)$.

The path integral can now be integrated by using a variant of equation (4.189), which introduces a change of variables given by

$$A^\mu(x) = \int \mathrm{d}^4y\ H^{\mu\nu}(x-y)\tilde{A}_\nu(y). \tag{4.269}$$

The gauge constraint $\delta(\partial_\mu A^\mu)$ in the measure requires that

$$\partial_\mu H^{\mu\nu}(x-y) = 0, \tag{4.270}$$

and, in order to match in number of degrees of freedom, the new variable $\tilde{A}_\mu$ must also satisfy the gauge condition $\partial_\mu \tilde{A}^\mu = 0$. The gauge field equivalent of the scalar field condition (4.190) is given by the requirement that

$$\int \mathrm{d}^4x\, \mathrm{d}^4y\ H^{\mu\rho}(z-x)M_{\rho\sigma}(x-y)H^{\sigma\nu}(y-w) = \delta^{T\mu\nu}(z-w). \tag{4.271}$$

For consistency with equation (4.270), the variant of the Dirac delta function $\delta^{T\mu\nu}(x-y)$ must satisfy $\partial_\mu \delta^{T\mu\nu}(x-y) = 0$, and so it is known as the *transverse delta function*. It can be written

$$\delta^T_{\mu\nu}(x-y) = g_{\mu\nu}\delta^4(x-y) - \partial_\mu\partial_\nu\Delta(x-y), \tag{4.272}$$

where the Green's function $\Delta(x-y)$ has the Fourier representation

$$\Delta(x-y) = -\int \frac{\mathrm{d}^4k}{(2\pi)^4}\frac{1}{k^2}e^{-ik\cdot(x-y)}. \tag{4.273}$$

The function defined by equation (4.273) satisfies $\Box\Delta(x-y) = \delta^4(x-y)$. Using equation (4.273) reveals the transverse delta function Fourier representation,

$$\delta^T_{\mu\nu}(x-y) = \int \frac{\mathrm{d}^4k}{(2\pi)^4}\left(g_{\mu\nu} - \frac{k_\mu k_\nu}{k^2}\right)e^{-ik\cdot(x-y)}. \tag{4.274}$$

The transverse delta function has the property that its continuum matrix product with a transverse function leaves that function unchanged. For example, this means

$$\int d^4y\, \delta^T_{\mu\nu}(x-y)\tilde{A}^\nu(y) = \tilde{A}_\mu(x), \tag{4.275}$$

Statement (4.275) follows by using equation (4.272) and $\partial_\mu \tilde{A}^\mu = 0$ after an integration by parts. The transverse delta also has the property that

$$\int d^4z\, \delta^T_{\mu\rho}(x-z)\delta^{T\rho}{}_\nu(z-y) = \delta^T_{\mu\nu}(x-y), \tag{4.276}$$

and so it is the unit continuum matrix for the *space of transverse functions.*

Using equations (4.275) and (4.276) the Landau gauge action of equation (4.268) becomes

$$iS_G = -\frac{1}{2}\int d^4x\, \tilde{A}_\mu(x)\tilde{A}^\mu(x) - i\int d^4x\, d^4y\, \tilde{A}_\mu(x)H^{\mu\nu}(x-y)J_\nu(y), \tag{4.277}$$

while result (4.275) allows requirement (4.271) to be written

$$\int d^4z\, H^{\mu\rho}(x-z)H_{\nu\rho}(z-y) = \int d^4z\, \delta^{\mu\rho}(x-z)M^{-1}_{\rho\nu}(z-y). \tag{4.278}$$

It is important to note that equation (4.278) holds even if $\partial^\mu M^{-1}_{\mu\nu}(x-y) \neq 0$. This is because the integral on the right-hand side of equation (4.278) projects $M^{-1}_{\mu\nu}(x-y)$ onto its transverse component. The gauge field path integral is now a set of standard Gaussians, and their integration gives

$$\begin{aligned} Z_0[J] &= N\exp\left\{-\frac{1}{2}\int d^4x\, d^4y\, d^4z\, J_\mu(x)H^{\mu\nu}(x-y)H_{\nu\rho}(y-z)J^\rho(z)\right\} \\ &= N\exp\left\{-\frac{1}{2}\int d^4x\, d^4y\, d^4z\, J_\mu(x)\delta^{T\mu\nu}(x-y)M^{-1}_{\nu\rho}(y-z)J^\rho(z)\right\} \\ &= N\exp\left\{-\frac{1}{2}\int d^4x\, d^4y\, J^\mu(x)\tilde{M}^{-1}_{\mu\nu}(x-y)J^\nu(y)\right\}, \end{aligned} \tag{4.279}$$

where

$$\tilde{M}^{-1}_{\mu\nu}(x-y) = \int d^4z\, \delta^T_{\mu\rho}(x-z)g^{\rho\sigma}M^{-1}_{\sigma\nu}(z-y). \tag{4.280}$$

Since $M_{\mu\nu}(x-y) = -ig_{\mu\nu}\Box\delta^4(x-y)$, it follows that

$$M^{-1}_{\mu\nu}(x-y) = \lim_{\epsilon\to 0}\int \frac{d^4k}{(2\pi)^4}\frac{-ig_{\mu\nu}}{k^2+i\epsilon}e^{-ik\cdot(x-y)}. \tag{4.281}$$

The epsilon prescription avoids the singularity at $k^2 = 0$ which would render the inversion undefined. Using equations (4.281) and (4.274) in equation (4.280) gives the Landau gauge photon propagator,

$$\tilde{M}^{-1}_{\mu\nu}(x - y) \equiv \Delta^{L}_{\mu\nu}(x - y) = \lim_{\epsilon \to 0} \int \frac{d^4k}{(2\pi)^4} \frac{-i}{k^2 + i\epsilon} \left(g_{\mu\nu} - \frac{k_\mu k_\nu}{k^2} \right) e^{-ik\cdot(x-y)}. \tag{4.282}$$

The normalization factor that occurs in the path integral (4.279) is complicated by the fact that the change of variables is associated with det $H \to 0$ and $\delta(\partial_\mu A^\mu) = \delta(0)$. As a result, it is understood that these terms must be *regulated*, i.e. made finite before the normalization is to be performed. The Landau propagator satisfies

$$\Box \Delta^{L}_{\mu\nu}(x - y) = i\delta^{T}_{\mu\nu}(x - y), \quad \partial^\mu \Delta^{L}_{\mu\nu}(x - y) = 0. \tag{4.283}$$

Using the canonically quantized gauge field (4.55) it can be shown that the epsilon prescription of equation (4.282) yields a result identical to the time-ordered product,

$$\Delta^{L}_{\mu\nu}(x - y) = -\frac{1}{Z_0[J]} \frac{\delta^2 Z_0[J]}{\delta J^\mu(x)\, \delta J^\nu(y)} \bigg|_{J=0} = \langle 0 | T\{\hat{a}_\mu(x) \hat{a}_\nu(y)\} | 0 \rangle. \tag{4.284}$$

The details of this are available in Peskin and Schroeder.

Further reading

Special relativity is presented in detail in

- P Tipler and R Llewellyn 2008 *Modern Physics* 4th edn (San Francisco, CA: W.H. Freeman)
- J Bernstein, P Fishbane and S Gasiorowicz 2000 *Modern Physics* (Englewood Cliffs, NJ: Prentice-Hall)

The relativistic formulation of electrodynamics is presented in

- W Greiner 1998 *Classical Electrodynamics* (Berlin: Springer)
- J D Jackson 1998 *Classical Electrodynamics* 3rd edn (New York: Wiley)
- J Westgard 1997 *Electrodynamics: A Concise Introduction* (Berlin: Springer)

Classical field theory is the subject matter of

- L D Landau and E M Lifshitz 1975 *The Classical Theory of Fields* vol 2, 4th edn (Amsterdam: Elsevier)
- D Soper 2008 *Classical Field Theory* (New York: Dover)
- M S Swanson 2015 *Classical Field Theory and the Stress-Energy Tensor* (Bristol: IOP Publishing)

Almost all modern texts on quantum field theory also include sections on path integrals and functional methods. These include

- S Weinberg 1995 *Quantum Theory of Fields* vol 1 (Cambridge: Cambridge University Press)
- M Peskin and D Schroeder 1995 *An Introduction to Quantum Field Theory* (Reading, MA: Addison-Wesley)
- L H Ryder 1996 *Quantum Field Theory* 2nd edn (Cambridge: Cambridge University Press)
- W Greiner and J Reinhardt 1996 *Field Quantization* (Berlin: Springer)
- M Srednicki 2007 *Quantum Field Theory* (Cambridge: Cambridge University Press)

The monograph by Kleinert includes numerous aspects of path integrals in field theory. There are a number of monographs that concentrate exclusively on path integrals in field theory. These include

- U Mosel 2004 *Path Integrals in Field Theory: An Introduction* (Berlin: Springer)
- K Fujikawa and H Suzuki 2004 *Path Integrals and Quantum Anomalies* (Oxford: Clarendon)
- J Zinn-Justin 2005 *Path Integrals in Quantum Mechanics* (Oxford: Oxford University Press)
- J R Klauder 2011 *A Modern Approach to Functional Integration* (Basel: Birkhäuser)
- A Das 2012 *Field Theory: A Path Integral Approach* 3rd edn (Singapore: World Scientific)
- R Shankar 2017 *Quantum Field Theory and Condensed Matter: An Introduction* (Cambridge: Cambridge University Press)

IOP Publishing

Path Integral Quantization

Mark S Swanson

Chapter 5

Basic quantum field theory applications

This chapter contains a very small subset of the basic applications of the path integral for quantum field theory. In particular, the generating functional approach to perturbation theory for interacting field theories is developed as a method for determining the Feynman rules for the field theory. The relationship of the perturbative Green's function to particle processes is briefly sketched. Unfortunately, the vast subject of renormalizing the perturbation series is beyond the scope of this text. The effective action is defined and its relationship to the connected Green's functions of the theory is presented. The subject of symmetry in interacting field theories is discussed, deriving the conservation laws and Ward–Takahashi identities essential to renormalization. This is placed in the context of the path integral for quantum electrodynamics and nonabelian gauge theories, with the latter assuming the reader has a passing knowledge of Lie group theory. The chapter concludes with the use of the effective potential in the analysis of spontaneously broken symmetry in terms of the functional determinants that occur in the analysis of the path integral.

5.1 Perturbation theory

The basic path integrals derived in the previous chapter included external sources which can be used to find time-ordered products of the fields using functional derivatives. For example, the Lagrangian form of the scalar field vacuum transition path integral $Z_0[J]$, given by equation (4.121) with $S = 1$, gives the time-ordered products,

$$\langle 0|T\{\psi(x_1)\cdots\psi(x_n)\}|0\rangle = i^n \frac{\delta^n Z_0[J]}{\delta J(x_1)\cdots\delta J(x_n)}\Bigg|_{J=0}, \tag{5.1}$$

where the external source is set to zero after the functional derivatives are taken. The demonstration of this for the free field case was given by equation (4.180). This holds

doi:10.1088/978-0-7503-3547-8ch5

true even in the presence of interactions represented by the Hamiltonian $\mathcal{H}_I$, although it is almost always impossible to evaluate the exact form for $Z_0[J]$ when nonlinear interactions are present.

However, the field theory path integral $Z_0[J]$ has an overall normalization factor, and the details of this factor were consistently ignored in the previous chapter. In the analysis of the periodic partition function in chapter 3, it was argued that continuum methods could be applied to a periodic path integral. Since the path integral representation of the vacuum transition element is also periodic, the quantum field theory path integral will be analyzed similarly. As a result, the lack of details regarding the normalization of the path integral is remedied by noting that disconnected processes, those that are unrelated to the incoming and outgoing particles and discussed in equation (4.78), are canceled by dividing equation (5.1) by $Z_0[J]$. The physically relevant Green's functions are therefore generated by

$$\frac{\langle 0|T\{\psi(x_1)\cdots\psi(x_n)\}|0\rangle}{\langle 0_+|0_-\rangle} = i^n \frac{1}{Z_0[J]}\frac{\delta^n Z_0[J]}{\delta J(x_1)\cdots\delta J(x_n)}\bigg|_{J=0}. \tag{5.2}$$

This has the effect of cancelling all factors unrelated to J, and this includes the prefactor determinants that result from evaluating quadratic path integrals such as equation (4.158). Therefore, the ignored normalization factors *need not be known in detail* since they will be cancelled in the process of calculating physical particle processes.

Like the perturbative expansion for the Green's functions of a quantum mechanical system given by equation (3.125), a nonlinear field theory can be given a similar functional approach to a perturbative solution. In the case of a nonlinear scalar field theory, this starts by writing the original path integral as

$$Z_0[J] = \exp\left\{-i\int \mathrm{d}^4x\, \mathcal{H}_I\left(i\frac{\delta}{\delta J(x)}\right)\right\} Z_0^B[J], \tag{5.3}$$

where the *basis path integral* $Z_0^B[J]$ is typically chosen to be an exactly solvable quadratic theory with the Lagrangian $\mathcal{L}_0(\psi) - J\psi$. The choice of minus sign for the source term $J\psi$ is opposite to that of the quantum mechanical version (3.125), and it is this choice that determines the sign of the functional derivatives in (5.3). Result (5.3) requires that $\mathcal{L}_0 - \mathcal{H}_I = \mathcal{L}$, the original Lagrangian of the system. The exponential defines the perturbation expansion for $Z_0[J]$ in powers of $\mathcal{H}_I$,

$$Z_0[J] = \sum_{n=0}^{\infty}\frac{1}{n!}\left(-i\int \mathrm{d}^4x\, \mathcal{H}_I\left(i\frac{\delta}{\delta J(x)}\right)\right)^n \int [\mathcal{D}\psi]\exp\left\{i\int \mathrm{d}^4x(\mathcal{L}_0(\psi) - J\psi)\right\}. \tag{5.4}$$

For example, using $\mathcal{L}_0(\psi) - J\psi = \frac{1}{2}(\partial_\mu\psi\partial^\mu\psi - m^2\psi^2) - J\psi$ with result (4.179) yields the perturbation series for the Klein–Gordon theory with an interaction $\mathcal{H}_I(\psi)$,

$$
\begin{aligned}
Z_0[J] = & \frac{N}{\sqrt{\det(\alpha^2(\Box + m^2))}} \sum_{n=0}^{\infty} \frac{1}{n!}\left(-i \int \mathrm{d}^4x\, \mathcal{H}_I\left(i\frac{\delta}{\delta J(x)}\right)\right)^n \\
& \times \exp\left\{-\frac{1}{2}\int \mathrm{d}^4x\, \mathrm{d}^4y\, J(x)\Delta_F(x-y)J(y)\right\},
\end{aligned} \tag{5.5}
$$

where $\Delta_F(x-y)$ is the Feynman propagator for the Klein–Gordon field. Throughout much of what follows the initial factors unrelated to J will simply be ignored. However, the important exception of spontaneously broken symmetry will require an analysis of the functional determinant present in the prefactor of equation (5.5).

Because the n-point functions of equation (5.2) appear in the LSZ reduction formulas (4.67) for scattering processes, understanding the effects of interaction is crucial to developing particle theories. Often the only recourse in doing this is perturbation theory, and so the perturbation series generated by forms such as equation (5.5) have been intensively studied. Early on, the perturbative expansion of the scalar functional (5.5), along with the similar expressions for spinor and gauge fields, gave rise to what is referred to as the *Feynman rules* for the field theory. The Feynman rules provide an extremely useful graphical method to determine the form of the integrals that occur at each order of the expansion of the exponential. The visualization consists of *lines*, which connect the two arguments of a propagator, and *vertices* where more than two lines join. The nature of the vertices are determined by the number of functional derivatives that occur in the interaction Hamiltonian, which in turn determines the number of propagators that can connect to the point that will be integrated. Each propagator attached to the point must connect to another point, either a field present in the LSZ reduction theorem or another vertex in the perturbative expansion, and so it can be visualized as a particle moving along the line whose endpoints are determined by the arguments of the propagator. This graphical visualization is referred to as a *Feynman diagram*. This gives an intuitive method to visualize particle production and destruction processes occurring at points as the particles propagate along the lines of the propagators. The details of this are contained in any standard introductory book on quantum field theory.

For example, the simplest particle process consists of one incoming particle and one outgoing particle. For a scalar particle the LSZ reduction formula Green's function associated with this process is given by the two-point function,

$$
\begin{aligned}
G(x_1, x_2) = \langle 0|T\{\Phi(x_1)\Phi(x_2)\}|0\rangle &= \frac{\langle 0|T\{\varphi(x_1)\varphi(x_2)E(t_+, t_-)\}|0\rangle}{\langle 0|E(t_+, t_-)|0\rangle} \\
&= -\frac{1}{Z_0[J]}\frac{\delta^2 Z_0[J]}{\delta J(x_1)\,\delta J(x_2)}\bigg|_{J=0}.
\end{aligned} \tag{5.6}
$$

The last two versions of the two-point function appearing in equation (5.6) are in the interaction picture, which is the picture where the path integral was derived. In the presence of interactions the two-point function is different from the free scalar propagator and reveals a wealth of information regarding the mass, stability, and range of interactions for the particles of the theory. The first order perturbative correction to the two-point function is given by

$$G(x_1, x_2) = -\frac{1}{Z_0[J]}\frac{\delta^2 Z_0[J]}{\delta J(x_1)\,\delta J(x_2)}\bigg|_{J=0} = -\frac{1}{Z_0[J]}\left(1 - i\int \mathrm{d}^4x\,\mathcal{H}_I\left(i\frac{\delta}{\delta J(x)}\right)\right) \times \frac{\delta^2}{\delta J(x_1)\,\delta J(x_2)}\exp\left\{-\frac{1}{2}\int \mathrm{d}^4y\,\mathrm{d}^4z\,J(y)\Delta_F(y-z)J(z)\right\}\bigg|_{J=0}. \tag{5.7}$$

Straightforward functional differentiation shows that $\mathcal{H}_I = \frac{1}{4!}\lambda\psi^4$ yields the first order correction when J is set to zero,

$$\begin{aligned} G(x_1, x_2) &= \Delta_F(x_1 - x_2) - \frac{1}{2}i\lambda\int \mathrm{d}^4x\,\Delta_F(x_1 - x)\Delta_F(x - x)\Delta_F(x - x_2) \\ &= \Delta_F(x_1 - x_2) - \frac{1}{2}i\lambda\Delta_F(0)\int \mathrm{d}^4x\,\Delta_F(x_1 - x)\Delta_F(x - x_2). \end{aligned} \tag{5.8}$$

The graphical visualization of the second term is a *bubble loop*, a particle line beginning and ending on the same point x. This visually represents the self-interaction described by $\mathcal{H}_I$ as the particle propagates from x_1 to x_2. The combinatorics of this interaction are entirely contained in the rules of functional differentiation.

However, even at this level, the problems with perturbative approaches to field theory processes manifest themselves. The overall factor of $\Delta_F(0)$ is given by

$$\Delta_F(0) = \int \frac{\mathrm{d}^4k}{(2\pi)^4}\frac{i}{k^2 - m^2 + i\epsilon}, \tag{5.9}$$

which is badly divergent. This follows from a simple scaling argument, $k_\mu \to \lambda k_\mu$, which shows that the integral of equation (5.9) is proportional to λ^2 as $\lambda \to \infty$. This divergence can be controlled and removed from the theory by redefining the mass appearing in the propagator as $m_0^2 = m^2 + \delta m^2$, where m_0 is known as the *bare mass* and δm^2 represents the divergent mass energy due to self-interactions. The systematic procedure to control and eliminate these perturbative divergences is complicated and goes by the general name of *renormalization*. One of the important outcomes of this process is an understanding of which interacting field theories are perturbatively renormalizable. In general, any theory with coupling constants that have units of positive powers of length is not renormalizable. For example, $\mathcal{H}_I = \frac{1}{4!}\lambda\psi^4$ is renormalizable since the coupling constant λ has no units. This does not necessarily mean that the perturbation series converges or yields physically meaningful results, just that the divergences can be systematically controlled and removed. Most of the important aspects of renormalization lie outside the scope of this monograph, and the reader is recommended to any of the many excellent texts where this procedure is detailed. However, a simple application of renormalization is presented in the last section of this chapter in the context of spontaneously broken symmetry.

In the process of performing the perturbative expansion, the types of particle processes are often characterized in terms of their *connectedness*. For example, the same ψ^4 scalar theory discussed earlier gives the four-point function to $O(\lambda)$,

$$\begin{aligned} G(x_1, x_2, x_3, x_4) &= \frac{1}{Z_0[J]} \frac{\delta^4 Z_0[J]}{\delta J(x_1)\, \delta J(x_2)\, \delta J(x_3)\, \delta J(x_4)} \bigg|_{J=0} \\ &= -i\lambda \int \mathrm{d}^4x\, \Delta_F(x_1 - x)\Delta_F(x_2 - x)\Delta_F(x_3 - x)\Delta_F(x_4 - x) \\ &\quad + \Delta_F(x_1 - x_2)\Delta_F(x_3 - x_4) + \Delta_F(x_1 - x_3)\Delta_F(x_2 - x_4) \\ &\quad + \Delta_F(x_1 - x_4)\Delta_F(x_2 - x_3). \end{aligned} \tag{5.10}$$

In the first term, the external points are all connected by lines to the vertex at x generated by $\mathcal{H}_I(i\delta/\delta J)$, while the last three terms are graphically represented by unconnected pairs of lines. It is useful for calculating particle processes to ignore processes that do not directly connect all the incoming and outgoing particles with lines. This requires isolating the *connected part of the Green*'s functions.

5.2 Generating functionals

In a scalar theory the generating functional for *connected Green*'s functions $W[J]$ is defined as $W[J] = i \ln Z_0[J]$ for the choice of negative sign for the external source term $J\psi$. The first functional derivative gives

$$G_c[J](x) = \frac{\delta W[J]}{\delta J(x)} = i\frac{1}{Z_0[J]}\frac{\delta Z_0[J]}{\delta J(x)} = \frac{\langle 0|\varphi(x)E_J(t_+, t_-)|0\rangle}{\langle 0|E_J(t_+, t_-)|0\rangle} \equiv \varphi_0(x). \tag{5.11}$$

It is important to note that the connected Green's functions are functionals of J. It is usually the case that the connected Green's functions are evaluated at $J = 0$, *but that need not be the case*. If the *vacuum expectation value* $\varphi_0(x)$ is nonzero, then the LSZ reduction formula shows that a particle in the in-state can disappear into the vacuum. This indicates either an unstable theory *or an unstable vacuum*. The latter can give rise to the phenomenon of *spontaneously broken symmetry*. Generating functionals derived from the path integral are an important tool in probing this phenomena and will be discussed in more detail.

The second derivative gives

$$\begin{aligned} i\frac{\delta^2 W[J]}{\delta J(x)\, \delta J(y)} &= -\frac{1}{Z_0[J]}\frac{\delta^2 Z_0[J]}{\delta J(x)\, \delta J(y)} + \frac{1}{Z_0^2[J]}\frac{\delta Z_0[J]}{\delta J(x)}\frac{\delta Z_0[J]}{\delta J(y)} \\ &= \frac{\langle 0|T\{\varphi(x)\, \varphi(y)\, E_J(t_+, t_-)\}|0\rangle}{\langle 0|E_J(t_+, t_-)|0\rangle} - \varphi_0(x)\, \varphi_0(y), \end{aligned} \tag{5.12}$$

which shows that the vacuum expectation values of equation (5.11) are subtracted from the propagator in the connected two-point function. For the case of the Klein–Gordon field with only a quartic nonlinear interaction, $\mathcal{H}_I = \frac{1}{4!}\lambda\psi^4$, it follows that

$$\left.\frac{\delta W[J]}{\delta J(x)}\right|_{J=0} = 0 \Longrightarrow \left.\frac{\delta^3 W[J]}{\delta J(x)\, \delta J(y)\, \delta J(z)}\right|_{J=0} = 0. \tag{5.13}$$

This follows by taking the third derivative of $W[J]$ and noting that

$$\left.\frac{1}{Z_0[J]}\frac{\delta^3 Z_0[J]}{\delta J(x)\, \delta J(y)\, \delta J(z)}\right|_{J=0} = 0 \tag{5.14}$$

since there are no vertices that are cubic in the field. For the case that equation (5.13) holds it follows that

$$\begin{aligned} G_c[J=0](x_1, x_2, x_3, x_4) = \Bigg(& \frac{i}{Z_0[J]}\frac{\delta^4 Z_0[J]}{\delta J(x_1)\, \delta J(x_2)\, \delta J(x_3)\, \delta J(x_4)} \\ & - \frac{i}{Z_0^2[J]}\frac{\delta^2 Z_0[J]}{\delta J(x_1)\, \delta J(x_2)}\frac{\delta^2 Z_0[J]}{\delta J(x_3)\, \delta J(x_4)} \\ & - \frac{i}{Z_0^2[J]}\frac{\delta^2 Z_0[J]}{\delta J(x_1)\, \delta J(x_3)}\frac{\delta^2 Z_0[J]}{\delta J(x_2)\, \delta J(x_4)} \\ & \left.- \frac{i}{Z_0^2[J]}\frac{\delta^2 Z_0[J]}{\delta J(x_1)\, \delta J(x_4)}\frac{\delta^2 Z_0[J]}{\delta J(x_2)\, \delta J(x_3)}\Bigg)\right|_{J=0}. \end{aligned} \tag{5.15}$$

The last three terms in equation (5.15) cancel the last three terms in equation (5.10), demonstrating that the connected part of the Green's functions is obtained from $W[J]$.

For a scalar field theory the counterpart to the quantum mechanical effective action (3.138) is given by a functional Legendre transformation,

$$\Gamma[\varphi_0] = W[J] - \int \mathrm{d}^4 y\, J(y)\varphi_0(y), \tag{5.16}$$

where φ_0 is given by equation (5.11) and is an implicit functional of the source J. The first derivative of the effective action is evaluated using the functional chain rule,

$$\frac{\delta\Gamma[\varphi_0]}{\delta\varphi_0(x)} = -J(x) + \int \mathrm{d}^4 y \left(\frac{\delta W[J]}{\delta J(y)}\frac{\delta J(y)}{\delta\varphi_0(x)} - \frac{\delta J(y)}{\delta\varphi_0(x)}\varphi_0(y)\right) = -J(x). \tag{5.17}$$

This gives a means to probe the vacuum expectation value of the scalar field. A value for $\varphi_0(x)$, denoted $\nu(x)$, such that

$$\left.\frac{\delta\Gamma[\varphi_0]}{\delta\varphi_0(x)}\right|_{\varphi(x)=\nu(x)} = 0, \tag{5.18}$$

indicates that a nonzero vacuum expectation value for φ may be present even when the external source vanishes. Equally as important, the second functional derivative of $\Gamma[\varphi_0]$ is related to the connected two-point function. This follows from

$$\delta^4(x_1 - x_2) = \frac{\delta\varphi_0(x_1)}{\delta\varphi_0(x_2)} = \frac{\delta^2 W[J]}{\delta\varphi_0(x_2)\,\delta J(x_1)} = \int \mathrm{d}^4x\, \frac{\delta^2 W[J]}{\delta J(x)\,\delta J(x_1)}\,\frac{\delta J(x)}{\delta\varphi_0(x_2)}$$
$$= -\int \mathrm{d}^4x\, G_c(x_1, x)\,\frac{\delta^2\Gamma[\varphi_0]}{\delta\varphi_0(x)\,\delta\varphi_0(x_2)}. \tag{5.19}$$

As a result, $\delta^2\Gamma[\varphi_0]/\delta\varphi_0(x)\,\delta\varphi_0(x_2)$ is the functional inverse of the connected propagator $G_c(x_1, x)$ even in the presence of interactions, so that it can be written

$$\frac{\delta^2\Gamma[\varphi_0]}{\delta\varphi_0(x)\,\delta\varphi_0(x_2)} = -G_c^{-1}(x, x_2). \tag{5.20}$$

If the system is translationally invariant, the Fourier representation of these two functions involves a single momentum,

$$G_c(x_1, x) = \int \frac{\mathrm{d}^4k}{(2\pi)^4}\,\tilde{G}(k)e^{-ik\cdot(x_1-x)},$$
$$\frac{\delta^2\Gamma[\varphi_0]}{\delta\varphi_0(x)\,\delta\varphi_0(x_2)} = \int \frac{\mathrm{d}^4p}{(2\pi)^4}\,\tilde{\Gamma}^{(2)}(p)e^{-ip\cdot(x-x_2)}. \tag{5.21}$$

Combining these with the Fourier representation (1.26) of the delta function and relation (5.20) gives

$$\tilde{G}(p)\tilde{\Gamma}^{(2)}(p) = -1. \tag{5.22}$$

As a result, the poles of the momentum space propagator are given by the zeros of the momentum space version of the second derivative for the effective action. Since the poles of the propagator correspond to the mass of the particle, it follows from equation (5.22) that the zeros of $\tilde{\Gamma}^{(2)}(p)$ must occur at the mass of the particle.

The higher order derivatives of $\Gamma[\varphi_0]$ yield the one-particle irreducible vertices of the interacting theory. The term *irreducible* means that the vertices cannot be decomposed into lower order vertices by severing a single line corresponding to a propagator. This can be seen by taking the functional derivative of equation (5.19) with respect to $J(y)$ and using the chain rule, which gives

$$0 = \int \mathrm{d}^4x\,\mathrm{d}^4z\, \frac{\delta^2 W[J]}{\delta J(x)\,\delta J(x_1)}\,\frac{\delta^3\Gamma[\varphi_0]}{\delta\varphi_0(x)\,\delta\varphi_0(z)\,\delta\varphi_0(x_2)}\,\frac{\delta^2 W[J]}{\delta J(y)\,\delta J(z)}$$
$$+ \int \mathrm{d}^4x\, \frac{\delta^3 W[J]}{\delta J(x)\,\delta J(y)\,\delta J(x_1)}\,\frac{\delta^2\Gamma[\varphi_0]}{\delta\varphi_0(x)\,\delta\varphi_0(x_2)}. \tag{5.23}$$

Integrating against $G_c(x, x_3)$ over d^4x, using equation (5.20), and invoking the definition of the three-point connected Green's function allows this to be written

$$G_c(x_1, x_2, x_3) = \int \mathrm{d}^4x\,\mathrm{d}^4y\,\mathrm{d}^4z\, G_c(x_1, x)G_c(x_2, y)G_c(x_3, z)V(x, y, z), \tag{5.24}$$

where the irreducible three-point vertex is given by

$$V(x, y, z) = \frac{\delta^3\Gamma[\varphi_0]}{\delta\varphi_0(x)\,\delta\varphi_0(y)\,\delta\varphi_0(z)}. \tag{5.25}$$

The connected three-point Green's function has been broken into a product of three connected propagators attached to a three-point vertex generated by the third functional derivative of $\Gamma[\varphi_0]$. This is the essence of decomposing particle processes into connected Green's functions and one-particle irreducible vertices. Although these results have been derived for the case of a single scalar field, the generalization to many fields as well as spinor and vector fields is straightforward.

The utility of the generating functionals is that the *physical masses and coupling constants* can be expressed in terms of them in the process of renormalization. It has already been pointed out that the mass of the interacting particle is determined by the value of $\tilde{\Gamma}^{(2)}(0)$, and this helps determine how renormalization must be instituted. In this process it is also necessary to renormalize the coupling constants by matching the value of the vertices at a chosen momentum. For example, it is possible to define the effective quartic coupling as the value of the quartic vertex when all participating particles have zero momentum, $\lambda_{\text{eff}} = \tilde{\Gamma}^{(4)}(0)$. This choice is not a unique method, and leads to what is known as the *renormalization group* description of how effective couplings scale with participating momentum.

5.3 Interaction symmetries and conservation laws

Interacting field theories are often constructed to possess a *symmetry*. In the context of a field theory a symmetry is an action on the components and argument of a field that leaves the action unchanged. In that regard, the Lorentz invariance of special relativity has been an essential *a priori* symmetry of the three free field theories developed so far for scalar, spinor, and gauge fields. This must continue to remain true even in the presence of interactions.

The scalar interaction, $\frac{1}{4!}\lambda\varphi^4$, considered in the previous section is relativistically invariant. In the case of a single bispinor field, ψ, interacting with a gauge field, A_μ, it is possible to add the relativistically invariant term, $e\bar{\psi}\gamma^\mu\psi A_\mu$, to the combined action of a free bispinor and a free gauge field. The resulting action with gauge fixing term is given by

$$\mathcal{L} = \bar{\psi}\gamma^\mu(i\partial_\mu - eA_\mu)\psi - m\bar{\psi}\psi - \frac{1}{4}F_{\mu\nu}F^{\mu\nu} - \frac{1}{2}(\partial_\mu A^\mu)^2, \tag{5.26}$$

and provides a description of the electromagnetic field interacting with a bispinor particle of charge e referred to as *quantum electrodynamics* or QED. Just as the free gauge field possessed the property of gauge invariance, this interacting theory is left invariant under an *abelian gauge transformation*,

$$\psi \to e^{ie\Lambda}\psi, \quad \bar{\psi} \to \bar{\psi}e^{-ie\Lambda}, \quad A_\mu \to A_\mu - \partial_\mu\Lambda, \tag{5.27}$$

as long as Λ satisfies $\Box\Lambda = 0$. The combination $D_\mu = \partial_\mu + ieA_\mu$ is referred to as the *gauge covariant derivative* since it has the property under a gauge transformation that $\gamma^\mu D_\mu \psi \to \gamma^\mu D_\mu' \psi' = \gamma^\mu D_\mu'(e^{ie\Lambda}\psi) = e^{ie\Lambda} D_\mu \psi$. This transformation is abelian since the result of two consecutive gauge transformations is simply the sum of the two gauge functions. The Landau gauge equations of motion for this system describe the gauge field A_μ interacting with a conserved charged matter current, $J_e^\mu = e\bar{\psi}\gamma^\mu\psi$, in the same manner, $\Box A^\mu = J_e^\mu$, as it did with an external source.

The presence of the symmetry, in this case the invariance under an abelian gauge transformation, is directly linked to the existence of a conserved matter current just as invariance under Lorentz transformations is directly linked to the conservation of relativistic energy and momentum. At the classical level these conservation laws are encapsulated in *Noether*'s theorem. If the classical Lagrangian $\mathcal{L}$ is invariant under a simultaneous change in the components of the field, $\psi_\alpha \to \psi_\alpha + \delta\psi_\alpha$, and the argument of the field, $x^\nu \to x^\nu + \delta x^\nu$, then the following current is conserved,

$$J^\mu = \left(g^\mu{}_\nu\, \mathcal{L} - \frac{\partial \mathcal{L}}{\partial(\partial_\mu \psi_\alpha)} \partial_\nu \psi_\alpha\right)\delta x^\nu + \frac{\partial \mathcal{L}}{\partial(\partial_\mu \psi_\alpha)} \delta\psi_\alpha, \tag{5.28}$$

as long as it is evaluated *using a classical solution* to the Euler–Lagrange equation for ψ_α.

For example, the gauge transformation (5.27) contains a subset of infinitesimal transformations that obey $\partial_\mu \Lambda = 0$. These are referred to as a *global transformation* because they are the same at every point. For such a case, the change in the bispinor field is $\delta\psi_a = ie\Lambda\psi$ and $\delta\bar{\psi}_a = -ie\Lambda\bar{\psi}_a$, while the gauge field change is $\delta A_\mu = 0$. The classical charged current associated with this transformation is given by

$$J_e^\mu = \frac{\partial \mathcal{L}}{\partial(\partial_\mu \psi_a)} \delta\psi_a + \frac{\partial \mathcal{L}}{\partial(\partial_\mu \bar{\psi}_a)} \delta\bar{\psi}_a = -\Lambda e\bar{\psi}\gamma^\mu\psi, \tag{5.29}$$

where the $-\Lambda$ can be dropped since it is an arbitrary parameter. As Noether's theorem predicts, this current is conserved, $\partial_\mu J_e^\mu = 0$, as long as the equations of motion, $i\gamma^\mu(\partial_\mu + iA_\mu)\psi - m\psi = 0$ and $-i(\partial_\mu - iA_\mu)\bar{\psi}\gamma^\mu - m\bar{\Psi} = 0$, both hold.

While classical currents are useful, it is important to understand how symmetries manifest themselves at the quantum level. In the case of the path integral version of quantum transition elements, it is assumed that the action appearing in the Lagrangian version that will be invariant under the same classical transformation. However, if the symmetry is to be present at the quantum level, it is also necessary for the measure in the path integral to be invariant. In the case of gauge fields and quantum electrodynamics, it is also necessary to perform gauge fixing to avoid the negative norm states present in such a theory. In the case of gauge transformations, the relevant configuration space measure before instituting the Faddeev–Popov procedure of equation (4.261) is given by

$$[\mathcal{D}\bar{\psi}\ \mathcal{D}\psi\ \mathcal{D}A] = \prod_x N_x \prod_{a,\mu} \mathrm{d}\bar{\psi}_a(x)\, \mathrm{d}\psi_a(x)\, \mathrm{d}A_\mu(x). \tag{5.30}$$

It is clear that $d\bar{\psi}_a(x)\, d\psi_a$ is invariant under the gauge transformation (5.27), while the measure $dA_\mu(x)$ is translationally invariant and therefore also unaffected by the gauge transformation. As a result, the symmetry of gauge invariance should remain in effect at the quantum level.

However, it is important to examine this subsequent to the gauge fixing procedure, which enforces the gauge constraints and adds the Faddeev–Popov ghosts to the measure and the action while factorizing the gauge volume $\mathcal{D}\Lambda$,

$$[\mathcal{D}\bar{\psi}\ \mathcal{D}\psi\ \mathcal{D}A\ \mathcal{D}\bar{c}\ \mathcal{D}c\ \delta(\partial_\mu A^\mu)] \\ = \prod_x N_x\, d\Lambda(x)\, d\bar{c}(x)\, dc(x)\, \delta(\partial_\mu A^\mu(x)) \prod_{a,\mu} d\bar{\psi}_a(x)\, d\psi_a(x)\, dA_\mu(x). \tag{5.31}$$

The path integral is now extended to the case of QED by attaching external source terms $\bar{K}\psi$, $\bar{\psi}K$, and $J_\mu A^\mu$ to the action, resulting in the vacuum transition element

$$Z_0[\bar{K}, K, J_\mu] = \int [\mathcal{D}\bar{\psi}\ \mathcal{D}\psi\ \mathcal{D}A\ \mathcal{D}\bar{c}\ \mathcal{D}c\ \delta(\partial_\mu A^\mu)] \\ \times \exp\left\{i \int d^4x(\mathcal{L} - \bar{K}\psi - \bar{\psi}K - J_\mu A^\mu)\right\}, \tag{5.32}$$

where the Lagrangian $\mathcal{L}$ is the sum of the gauge fixed QED Lagrangian (5.26) and the Faddeev–Popov ghost Lagrangian (4.250). Invariance of the Lagrangian $\mathcal{L}$ in equation (5.32) under the gauge transformation (5.27) requires that the external source J^μ is conserved, $\partial_\mu J^\mu = 0$. This means that if the Grassmann source terms are transformed according to

$$K \to e^{ie\Lambda}, \quad \bar{K} \to e^{-ie\Lambda}\bar{K}, \tag{5.33}$$

the terms $\bar{K}\psi$ and $\bar{\psi}K$ transform back to their original form.

Subsequent to the gauge fixing procedure the action in equation (5.32) is invariant under the abelian BRST transformation of equation (4.258) extended to include the bispinor fields,

$$\psi \to e^{i\xi c}\psi, \quad \bar{\psi} \to e^{-i\xi c}\psi, \quad A_\mu \to A_\mu - \xi\, \partial_\mu c, \quad \bar{c} \to \bar{c} - \xi\, \partial_\mu A^\mu, \tag{5.34}$$

where ξ is a Grassmann parameter such that $\xi^2 = 0$. While the bispinor field measure is clearly invariant under equation (5.34), it was shown that the constraint in the measure of the gauge field is also invariant under this transformation since the quantized ghost field satisfies $\Box c = 0$ via relation (4.259). The path integral (5.32) is therefore invariant under the gauge transformation of equation (5.27) or the related BRST transformation of equation (5.34).

The ramifications of this symmetry are best examined using the generating functionals for the connected Green's functions and the effective action in QED, defined by

$$W[\bar{K}, K, J_\mu] = i \ln Z_0[\bar{K}, K, J_\mu], \tag{5.35}$$

$$\Gamma[A_{0\mu}, \bar{\psi}_0, \psi_0] = W[\bar{K}, K, J_\mu] - \int \mathrm{d}^4x\, (\bar{K}\psi_0 + \bar{\psi}_0 K + J_\mu A_0^\mu). \tag{5.36}$$

The vacuum expectation value of A_μ in the presence of the source is given by

$$A_{0\mu}(x) = \frac{\delta W[\bar{K}, K, J_\mu]}{\delta J^\mu(x)}, \tag{5.37}$$

while those of the bispinor fields are given by

$$\psi_0(x) = \frac{\delta W[\bar{K}, K, J_\mu]}{\delta \bar{K}(x)}, \quad \bar{\psi}_0(x) = -\frac{\delta W[\bar{K}, K, J_\mu]}{\delta K(x)}. \tag{5.38}$$

The minus sign in the last definition is brought about by the Grassmann nature of both the source K and the variable $\bar{\psi}_0$. The counterparts to equation (5.17) are given by

$$\begin{aligned} \frac{\delta\Gamma[A_{0\mu}, \bar{\psi}_0, \psi_0]}{\delta A_{0\mu}(x)} &= -J^\mu(x), \\ \frac{\delta\Gamma[A_{0\mu}, \bar{\psi}_0, \psi_0]}{\delta \bar{\psi}_0^a(x)} &= -K^a(x), \\ \frac{\delta\Gamma[A_{0\mu}, \bar{\psi}_0, \psi_0]}{\delta \psi_0^b(x)} &= \bar{K}^b(x), \end{aligned} \tag{5.39}$$

with the last sign due to the Grassmann nature of the variables.

Because the variables of integration in the path integral can undergo gauge transformations or BRST transformations, both of the functionals (5.35) and (5.36) are invariant when the bispinor sources undergo the gauge transformation (5.33). This means that the generating functional $W[J_\mu, \bar{K}, K]$ satisfies the property

$$W[J_\mu, \bar{K}', K'] = W[J_\mu, e^{-ie\Lambda}\bar{K}, e^{ie\Lambda}K] = W[J_\mu, \bar{K}, K]. \tag{5.40}$$

In the case of an infinitesimal gauge transformation, where $\Lambda \approx 0$, it follows that $K' = (1 - ie\Lambda)K \equiv K - \delta K$ and $\bar{K}' = \bar{K} + \delta\bar{K}$. Expanding equation (5.40) to $O(\Lambda)$ using a functional Taylor series gives

$$\delta W = \int \mathrm{d}^4x\, ie\Lambda\left(\bar{K}^a(x)\frac{\delta W[J_\mu, \bar{K}, K]}{\delta\bar{K}^a(x)} - K^a(x)\frac{\delta W[J_\mu, \bar{K}, K]}{\delta K^a(x)}\right) = 0. \tag{5.41}$$

Because Λ is an arbitrary infinitesimal function, the integrand in equation (5.41) must vanish, and this gives an *exact local identity* that W must obey,

$$\bar{K}^a(x)\,\frac{\delta W[J_\mu, \bar{K}, K]}{\delta\bar{K}^a(x)} - K^a(x)\,\frac{\delta W[J_\mu, \bar{K}, K]}{\delta K^a(x)} = 0. \tag{5.42}$$

Since the QED vacuum transition element corresponds to the external sources vanishing, applying arbitrary functional derivatives to equation (5.42) and then setting J_μ, $\bar{K}$, and K to zero gives a set of identities for the exact QED Green's functions required by the gauge invariance of the theory. For example, applying $\delta/\delta K^b(y)$ gives

$$\delta^4(x-y)\frac{\delta W[J_\mu, \bar{K}, K]}{\delta K^a(x)}\Bigg|_{J=K=\bar{K}=0} = \delta^4(x-y)\,\psi_0^a(x)\big|_{J=K=\bar{K}=0} = 0. \tag{5.43}$$

Combining result (5.43) with the LSZ reduction formula (4.67) shows that a single charged spinor particle cannot disappear into the vacuum since $\psi_0^a(x) = 0$.

A similar analysis can be applied to the effective action (5.36), where gauge invariance gives

$$\Gamma[A_{0\mu} - \partial_\mu\Lambda, e^{ie\Lambda}\bar{\psi}_0, e^{-ie\Lambda}\psi_0] = \Gamma[A_{0\mu}, \bar{\psi}_0, \psi_0]. \tag{5.44}$$

Using the same method that derived equation (5.42) and integrating by parts shows that gauge invariance causes the effective action to satisfy the local identity

$$ie\frac{\delta\Gamma[A_{0\mu}, \bar{\psi}_0, \psi_0]}{\delta\psi_0^a(x)}\psi_0^a(x) - ie\frac{\delta\Gamma[A_{0\mu}, \bar{\psi}_0, \psi_0]}{\delta\bar{\psi}_0^a(x)}\bar{\psi}_0^a(x) + \partial_\mu\frac{\delta\Gamma[A_{0\mu}, \bar{\psi}_0, \psi_0]}{\delta A_{0\mu}(x)} = 0. \tag{5.45}$$

The identity given by equation (5.45) can be combined with functional derivatives to obtain relationships between the QED vertices and propagators. This starts by noting that the connected bispinor propagator is given by

$$S_{Fab}(x-y) = \frac{\delta^2 W[J_\mu, \bar{K}, K]}{\delta K^a(x)\,\delta\bar{K}^b(y)} = \frac{\delta\psi_0^b(y)}{\delta K^a(x)}. \tag{5.46}$$

This can be combined with the QED counterpart of the derivation (5.19), which gives the inverse of the bispinor propagator (5.46),

$$\begin{aligned}&\int d^4y\,\frac{\delta^2 W[J_\mu, \bar{K}, K]}{\delta K^a(x)\,\delta\bar{K}^b(y)}\frac{\delta^2\Gamma[A_{0\mu}, \bar{\psi}_0, \psi_0]}{\delta\bar{\psi}_0^b(y)\,\delta\psi_0^c(z)} = \delta_{ac}\,\delta^4(x-z),\\ &\Longrightarrow \frac{\delta^2\Gamma[A_{0\mu}, \bar{\psi}_0, \psi_0]}{\delta\bar{\psi}_0^b(y)\,\delta\psi_0^c(z)} = S_{Fbc}^{-1}(y-z).\end{aligned} \tag{5.47}$$

For example, applying $\delta^2/\delta\bar{\psi}_0^c(z)\,\delta\psi_0^b(y)$ to equation (5.45) and using equation (5.47) for the physical vacuum, where ψ_0 and $\bar{\psi}_0$ vanish, gives the relation

$$\begin{aligned}&ieS_{Fab}^{-1}(x-y)\,\delta^4(x-z) - ieS_{Fab}^{-1}(z-x)\,\delta^4(x-y)\\ &+\frac{\partial}{\partial x^\mu}\frac{\delta^3\Gamma[A_{0\mu}, \bar{\psi}_0, \psi_0]}{\delta A_{0\mu}(x)\,\delta\bar{\psi}_0^a(z)\,\delta\psi_0^b(y)} = 0.\end{aligned} \tag{5.48}$$

the gauge–spinor vertex to the inverse spinor propagator. Result (5.48) is an example of a *Ward–Takahashi identity*, which are of critical importance in implementing a renormalization procedure that respects the gauge invariance of the theory.

5.4 Yang–Mills gauge field theory

Gauge theories were generalized by Yang and Mills to include nonabelian Lie groups. A Lie group is characterized by a set of Hermitian generators, denoted τ^a, which obey a *Lie algebra*, $[\tau^a, \tau^b] = if^{abc}\tau^c$. The f^{abc} are referred to as the *structure constants* of the Lie algebra. Modeling physical phenomena typically requires choosing a particular *matrix representation* for the τ^a so that the members of the Lie group, $\mathbf{U}(\Lambda(x)) = \exp\{-ig\Lambda^a(x)\tau^a\}$, act on a *representation space* of bispinor matter fields $\Psi(x)$ as a local symmetry operation,

$$\Psi'(x) = \mathbf{U}(\Lambda(x))\Psi(x), \quad \bar{\Psi}'(x) = \bar{\Psi}(x)\mathbf{U}^{-1}(\Lambda(x)). \tag{5.49}$$

Yang–Mills gauge fields extend the abelian gauge field A_μ to the nonabelian gauge field A_μ^a by associating a gauge field with every generator. The index a is referred to as an *isospin index*. The nonabelian field $A_\mu(x) = A_\mu^a(x)\tau^a$ undergoes a nonabelian gauge transformation according to

$$A_\mu'(x) = \mathbf{U}(x)A_\mu(x)\mathbf{U}^{-1}(x) + \frac{i}{g}\mathbf{U}(x)\cdot\partial_\mu\mathbf{U}^{-1}(x), \tag{5.50}$$

where g is the generalization of electric charge. The gauge field action is obtained by defining the nonabelian extension of the abelian $F_{\mu\nu}$ as

$$F_{\mu\nu} = F_{\mu\nu}^a\tau^a = [D_\mu, D_\nu], \tag{5.51}$$

so that

$$F_{\mu\nu}^a = \partial_\mu A_\nu^a - \partial_\nu A_\mu^a + gf^{abc}A_\mu^b A_\nu^c. \tag{5.52}$$

Using the representation space matter fields, the generalization of quantum electrodynamics is given by the Lagrangian

$$\mathcal{L} = i\bar{\Psi}\gamma^\mu D_\mu\Psi - \frac{1}{4}F_{\mu\nu}^a F^{a\mu\nu}, \tag{5.53}$$

where the *covariant derivative* is defined as

$$D_\mu = \partial_\mu - igA_\mu. \tag{5.54}$$

The gauge invariance of the matter part of the action (5.53) follows from the identity $\partial_\mu(\mathbf{U}\Psi) = (\partial_\mu\mathbf{U}\cdot\mathbf{U}^{-1})\cdot\mathbf{U}\partial_\mu\Psi$, while the gauge field action invariance follows from the transformation property $F_{\mu\nu} \to \mathbf{U}F_{\mu\nu}\mathbf{U}^{-1}$.

The Lagrangian (5.53) is often used to represent the interaction of spinor quarks and vector gluons by choosing the generators from the unitary group SU(3), where it is typically referred to as *quantum chromodynamics (QCD)*. However, even without the bispinor matter fields, the Lagrangian (5.53) is a nonlinear field theory. This follows from the equation of motion for the gauge field, which is given by

$$\partial_\mu F^{a\mu\nu} + gf^{abc}A_\mu^b F^{c\mu\nu} = 0, \tag{5.55}$$

for the case the matter fields Ψ are not present. In the case of equation (5.55), the nonlinear terms proportional to g mean that the negative norm states present in the gauge fields *are not decoupled* from the positive norm transverse states, and this threatens the unitarity of the theory since such interactions will give rise to negative probabilities.

The method to avoid this problem is to invoke the Faddeev–Popov factorization procedure given by equation (4.265) for the abelian case of electrodynamics. For the case of an infinitesimal parameter, $\Lambda^a \approx 0$, the gauge field transformation of equation (5.50) becomes

$$A_\mu^{a'} = A_\mu^a - \partial_\mu \Lambda^a + g f^{abc} \Lambda^b A_\mu^c. \tag{5.56}$$

Picking the Lorenz gauge condition $\chi^a = -\partial_\mu A^{a\mu} = 0$, the differential operator appearing in the determinant of the Faddeev–Popov ghost action (4.266) is found from $\chi^{a\Lambda} = \chi^a + \Box\Lambda^a - g f^{abc} \partial_\mu \Lambda^b A^{c\mu}$, where the term proportional to $\partial_\mu A^{c\mu}$ has been dropped. This gives

$$\frac{\delta \chi^{a\Lambda}(x)}{\delta \Lambda^b(y)} = \delta^{ab} \Box \delta^4(x - y) - g f^{abc} A^{c\mu} \partial_\mu \delta^4(x - y). \tag{5.57}$$

The Faddeev–Popov ghost Lagrangian $\mathcal{L}_c$ for a Yang–Mills theory is obtained by substituting equation (5.57) into equation (4.266),

$$\mathcal{L}_c = \partial^\mu \bar{c}^a \partial_\mu c^a + g f^{abc} \partial^\mu \bar{c}^a A_\mu^b c^c. \tag{5.58}$$

For the case the matter field is absent, the path integral for the pure Yang–Mills field case is given by the nonabelian version of equation (4.265),

$$Z[J] = \int [\mathcal{D}\Lambda] \int \left[\mathcal{D}A_\mu \, \mathcal{D}\bar{c} \, \mathcal{D}c \, \delta(\chi^a)\right] \exp\left\{ i \int \mathrm{d}^4x \, (\mathcal{L}_G(A, \bar{c}, c) + J_\mu^a A^{a\mu}) \right\}, \tag{5.59}$$

where the Feynman gauge action for the Yang–Mills fields is given by

$$\mathcal{L}_G(A, \bar{c}, c) = -\frac{1}{4} F^{a\mu\nu} F_{\mu\nu}^a - \frac{1}{2} (\partial_\mu A^{a\mu})^2 + \partial^\mu \bar{c}^a \partial_\mu c^a + g f^{abc} \partial^\mu \bar{c}^a A_\mu^b c^c. \tag{5.60}$$

Result (5.60) shows that there is a nonlinear interaction between the Faddeev–Popov ghosts and the gauge fields. Whereas the Faddeev–Popov ghosts are uncoupled in the abelian theory (4.250), result (5.58) shows that the Faddeev–Popov ghosts will contribute nontrivially to the perturbation series for the case of Yang–Mills fields. In particular, because the Faddeev–Popov ghosts are Grassmann in nature, these perturbative terms will possess the correct signs to cancel the negative norm contributions from the gauge field nonlinearity and thereby maintain unitarity. Demonstrating this is far beyond the scope of this text, and the reader is referred to Peskin and Schroeder.

It is worth noting that the Faddeev–Popov ghost Lagrangian (5.58) is identical to the nonabelian version of equation (4.228), where the Faddeev–Popov determinant appears as a result of inverting the solution of Gauss's law $G^a = 0$,

$$\det\left(\frac{\delta G^a(x)}{\delta A_0^b(y)}\right) = \int [\mathcal{D}\bar{c}\ \mathcal{D}c] \exp\left\{i\int \mathrm{d}^4x\ \mathrm{d}^4y\ \bar{c}^a(x)\left(\frac{\delta G^a(x)}{\delta A_0^b(y)}\right)c^b(y)\right\}. \tag{5.61}$$

The Lagrangian form of the Yang–Mills path integral can be derived just as it was with electrodynamics. For the case where the matter field is ignored, the Landau gauge version of the Yang–Mills Lagrangian is given by

$$\mathcal{L}_G = -\frac{1}{4}F^a_{\mu\nu}F^{a\mu\mu} - \frac{1}{2}\partial_\mu A^{a\mu}\partial_\nu A^{a\nu}. \tag{5.62}$$

Deriving the Lagrangian version of the constrained Yang–Mills path integral starts from the Hamiltonian version of the gauge fixed Lagrangian that appears in the path integral. The canonical momenta are given by

$$\pi_i^a = \frac{\partial\mathcal{L}_G}{\partial\dot{A}_i^a} = -F^{a0i}, \quad \pi_0^a = \frac{\partial\mathcal{L}_G}{\partial\dot{A}_0^a} = -\partial_\mu A^{a\mu}, \tag{5.63}$$

and these relations give

$$\dot{A}_i^a = -\pi_i^a + \partial_i A_0^a - gf^{abc}A_0^bA_i^c, \quad \dot{A}_0^a = -\pi_0^a - \partial_j A^{aj}. \tag{5.64}$$

Using equation (5.64) the Landau gauge Hamiltonian density is given by

$$\mathcal{H}_G = -\frac{1}{2}\pi_j^a\pi_j^a - \frac{1}{2}\pi_0^a\pi_0^a - \pi_0^a\partial_j A^{aj} + \frac{1}{4}F_{ij}^aF^{aij} + A_0^a\left(\partial_j\pi^{aj} + gf^{abc}A_j^b\pi^{cj}\right). \tag{5.65}$$

It is straightforward to see that equation (5.90) is identical to the abelian Landau gauge case (4.223) in the case that $g = 0$.

For the case where the matter fields are ignored, the Gauss's law constraints for Yang–Mills theory in the Landau gauge are given by

$$G^a = -\frac{\partial\mathcal{H}_G}{\partial A_0^a} = \partial_i\pi_i^a + gf^{abc}A_i^b\pi_i^c = 0, \tag{5.66}$$

and these constraints can once again be exponentiated,

$$[\delta(G^a)] = \int [\mathcal{D}\Lambda] \exp\left\{-i\int \mathrm{d}^4x\ \Lambda^a G^a\right\}, \tag{5.67}$$

allowing them to be absorbed into the path integral action, resulting in the Lagrangian density

$$\begin{aligned}\mathcal{L}_G = \pi_\mu^a\dot{A}^{a\mu} - \mathcal{H}_G(A^\mu, \pi_\mu) &= \frac{1}{2}\pi_0^a\pi_0^a - \frac{1}{2}\pi_j^a\pi_j^a + \pi_0^a\partial_\mu A^{a\mu} \\ &+ \pi_j^a\dot{A}^{aj} - \frac{1}{4}F_{ij}^aF^{aij} - (\partial_j\pi_j^a + gf^{abc}A_j^b\pi_j^c)(A_0^a + \Lambda^a).\end{aligned} \tag{5.68}$$

The Gauss's law constraint is absorbed into the integration over A_0^a by the translation $A_0^a \to A_0^a - \Lambda^a$. As in the case of abelian electrodynamics in step

(4.246), the process of deriving the Lagrangian version of the path integral requires a translation of the momentum $\pi_j^a \to \pi_j^a + F_{0j}^a$, which gives

$$G^a \to G^a + \partial_\mu F^{a0\mu} = G^a - \Box A_0^a + \partial_\mu \dot{A}^{a\mu} + gf^{ade}\partial_\mu A_0^d A^{e\mu} + gf^{ade}A_0^d \partial_\mu A^{e\mu}, \quad (5.69)$$

where both terms proportional to $\partial_\mu A^{a\mu}$ can be dropped. This gives the argument of the constraint determinant of equation (5.61),

$$\begin{aligned} \left.\frac{\delta G^a(x)}{\delta A_0^b(y)}\right|_{\chi^a=0} &= \frac{\delta}{\delta A_0^b(y)}\left(-\Box A_0^a(x) + gf^{ade}\partial_\mu A_0^d(y)A^{e\mu}(y)\right) \\ &= -\delta^{ab}\Box\delta^4(x-y) + gf^{abc}A^{c\mu}(y)\partial_\mu\delta^4(x-y). \end{aligned} \quad (5.70)$$

Inserting the second form of equation (5.70) into equation (5.61) reproduces the Faddeev–Popov ghost Lagrangian of equation (5.58), leaving the gauge volume $[\mathcal{D}\Lambda]$ of (5.67) factorized. Like the case of electrodynamics, the constraint delta function $\delta(-\partial_\mu A^{a\mu})$ has become $\delta(-\partial_\mu A^{a\mu} + \dot{\Lambda}^a)$, just as the term in the action (5.68) becomes $\pi_0^a(\partial_\mu A^{a\mu} - \dot{\Lambda}^a)$. Like electrodynamics, the term $\dot{\Lambda}^a$ is removed using a gauge transformation on the A_μ^a, but in the case of Yang–Mills theory this is more complicated because of the nonlinear nature of the gauge transformation (5.50). The reader is recommended to the references where the details are presented.

Like QED, the combined Lagrangian of equation (5.60) possesses an invariance under the following BRST transformation,

$$\begin{aligned} A_\mu^a &\to A_\mu^a + \xi(\partial_\mu c^a - gf^{abc}A_\mu^b c^c), \\ c^a &\to c^a - \frac{1}{2}g\xi f^{abc}c^b c^c, \end{aligned} \quad (5.71)$$

where ξ is a Grassmann parameter. The BRST invariance (5.71) generates Ward–Takahashi identities for the nonabelian theory in the context of the effective action. These identities among the vertices of the theory are necessary to insure renormalizability of the theory.

In the absence of a mass term, the Lagrangian (5.53) possesses another symmetry. Using the Dirac representation (4.14) the γ^5 matrix is defined by

$$\gamma^5 = i\gamma^0\gamma^1\gamma^2\gamma^3 = \begin{pmatrix} 0 & I \\ I & 0 \end{pmatrix}. \quad (5.72)$$

This Hermitian matrix has the property that it anticommutes with the γ^μ and gives $\gamma^5 \cdot \gamma^5 = \mathbf{1}$. It is used to define a *chirality* or handedness projection operator for a bispinor. The bispinor ψ is written as the sum of a left- and right-handed piece, $\psi = \psi^{(L)} + \psi^{(R)}$, where $\psi^{(R)} = \frac{1}{2}(1 + \gamma^5)\psi$ and $\psi^{(L)} = \frac{1}{2}(1 - \gamma^5)\psi$. Both elements are eigenspinors of γ^5 since they satisfy $\gamma^5\psi^{(R)} = \psi^{(R)}$ and $\gamma^5\psi^{(L)} = -\psi^{(L)}$. Since $\{\gamma^0, \gamma^5\} = 0$, it follows that $\bar{\psi}^{(R)} = \frac{1}{2}\bar{\psi}(1 - \gamma^5)$ and $\bar{\psi}^{(L)} = \frac{1}{2}\bar{\psi}(1 + \gamma^5)$, so that $\bar{\psi}^{(R)}\psi^{(R)} = \frac{1}{4}\bar{\psi}(1 - \gamma^5)(1 + \gamma^5)\psi = 0$.

A *chiral transformation* on the bispinor ψ is defined as $\psi \to e^{i\alpha\gamma^5}\psi$. Because γ^5 anticommutes with the γ^μ, the term $\bar{\psi}\gamma^\mu\partial_\mu\psi$ is invariant under chiral transformations. However, the mass term $m\bar{\psi}\psi$ is not, instead transforming as $m\bar{\psi}\psi \to m\bar{\psi}e^{2i\alpha\gamma^5}\psi = m\bar{\psi}(e^{2i\alpha}\psi^{(R)} + e^{-2i\alpha}\psi^{(L)})$. As a result, invariance under a chiral transformation or *chiral symmetry* requires that the bispinor field is massless *at the classical level.* For $m = 0$ the classical theory (5.53) is chirally invariant. Noether's theorem (5.28) then states that the associated *axial current*, given by $J^{5\mu} = \bar{\psi}\gamma^\mu\gamma^5\psi$, will be conserved at the classical level, $\partial_\mu J^{5\mu} = 0$.

A problem with chiral symmetry at the quantum level is exposed by examining the configuration space measure (4.201) of the bispinor path integral, where it is clear that $\mathrm{d}\bar{\psi}(x)\,\mathrm{d}\psi$ is not invariant under a chiral transformation. In the context of the theory described by equation (5.53), a careful analysis of the bispinor measure under a chiral transformation reveals that the axial current is *no longer conserved.* In the case of a nonabelian gauge theory, it can be shown that the axial current obeys

$$\partial_\mu J^{5\mu} = \frac{g^2}{16\pi^2}\varepsilon^{\mu\nu\rho\sigma}F^a_{\mu\nu}F^a_{\rho\sigma}. \tag{5.73}$$

This failure is referred to as the *chiral anomaly* and is tied to topological properties of the gauge fields. The quantity on the right-hand side of equation (5.73) is known as the *Pontryagin density*, and its integral over all space yields an integer. The path integral analysis of the chiral anomaly and its properties are presented in Fujikawa and Suzuki.

5.5 Non-perturbative aspects of 1 + 1 Yang–Mills theory

Field theories become much simpler when they are considered in one spatial dimension. In particular, the SU(2) Yang–Mills theory in the absence of matter fields can be solved fairly easily. The Lagrangian for the gauge group SU(2) uses the nonabelian gauge field $A_\mu(x) = \frac{1}{2}A^a(x)\sigma^a$, where the σ^a are the three Pauli spin matrices (4.7). The structure constants for SU(2) correspond to the three-dimensional Levi–Civita symbol ε^{abc}. For SU(2) the gauge field action of equation (5.53) is given in the *Weyl or temporal gauge* by

$$\mathcal{L} = -\frac{1}{4}F^a_{\mu\nu}F^{a\mu\nu} + g\dot{A}^a_0 A^a_0, \tag{5.74}$$

where equation (5.52) becomes

$$F^a_{\mu\nu} = \partial_\mu A^a_\nu - \partial_\nu A^a_\mu + g\varepsilon^{abc}A^b_\mu A^c_\nu. \tag{5.75}$$

In one spatial dimension, the fields A^a_μ are dimensionless, while the coupling constant g has the units of inverse length. The only nonzero component for $F^a_{\mu\nu}$ is the electric field, denoted $E_a = F^a_{10} = -F^a_{01} = F^{a01}$, and the equation of motion for this electric field is given by equation (5.55) as

$$\partial_0 F^a_{01} + g\varepsilon^{abc}A^b_0 F^c_{01} = -\dot{E}_a - g\varepsilon^{abc}A^b_0 E_c = 0. \tag{5.76}$$

Using $F^a_{01} = \dot{A}^a_1 - \partial_1 A^a_0 + g\varepsilon^{abc} A^b_0 A^c_1 = -E_a$ gives the equation of motion

$$\dot{A}^a_1 = -E_a + \partial_1 A^a_0 - g\varepsilon^{abc} A^b_0 A^c_1 . \tag{5.77}$$

In the Weyl gauge the constraint $\Pi^a_0 = gA^a_0 = 0$ must also be enforced.

Although the temporal or Weyl gauge fixing term is not relativistically invariant, the Weyl gauge condition $\chi^a = gA^a_0 = 0$ has the advantage that the Faddeev–Popov ghosts are *decoupled* from the gauge fields. This follows from the infinitesimal gauge transformation on the Weyl gauge condition, which gives

$$\chi^a \to \chi^{a\Lambda} = A^a_0 - \dot{\Lambda}^a + g\varepsilon^{abc} \Lambda^b A^c_0 . \tag{5.78}$$

The last term can be ignored since it contains the gauge condition $A^a_0 = 0$, so that equation (5.57) becomes

$$\frac{\delta \chi^{a\Lambda}(x)}{\delta \Lambda^b(y)} = -\partial_0 \delta^2(x - y)\delta_{ab}. \tag{5.79}$$

Because equation (5.79) has no reference to the gauge fields, the Faddeev–Popov determinant can be ignored and there is no ghost–gauge field interaction in the Weyl gauge.

The action (5.74) is rewritten using the auxiliary field $\varphi^a_{\mu\nu}$, which satisfies $\varphi^a_{\mu\nu} = -\varphi^a_{\nu\mu}$, so that

$$\mathcal{L}_\varphi = \frac{1}{4}\varphi^a_{\mu\nu}\varphi^{a\mu\nu} - \frac{1}{2}\varphi^{a\mu\nu} F^a_{\mu\nu} + g\dot{A}^a_0 A^a_0 . \tag{5.80}$$

At the classical level, the Euler–Lagrange equation of motion for $\varphi^a_{\mu\nu}$ is simply $\varphi^a_{\mu\nu} = F^a_{\mu\nu}$, which gives equation (5.74) when substituted into equation (5.80). It will be shown that the two theories, $\mathcal{L}$ and $\mathcal{L}_\varphi$, are the same at the quantum level since the action (5.80) is quadratic in $\varphi^a_{\mu\nu}$, and the measure in the functional integral for $\varphi^a_{\mu\nu}$ can be chosen to yield the same result by completing the square,

$$\int \mathrm{d}^2x\, \mathcal{L}_\varphi = \int \mathrm{d}^2x\, \frac{1}{4}\left(\varphi^a_{\mu\nu} - F^a_{\mu\nu}\right)\left(\varphi^{a\mu\nu} - F^{a\mu\nu}\right) + \int \mathrm{d}^2x\, \mathcal{L}. \tag{5.81}$$

The action (5.80) can be rewritten by integrating by parts, so that it becomes

$$\begin{aligned}\mathcal{L}_\varphi = &\frac{1}{4}\varphi^{a\mu\nu}\varphi^a_{\mu\nu} - \frac{1}{2}\left(A^a_\mu \partial_\nu \varphi^{a\mu\nu} - A^a_\nu \partial_\mu \varphi^{a\mu\nu} + f^{abc}\varphi^{a\mu\nu} A^b_\mu A^c_\nu\right)\\ &+ g\dot{A}^a_0 A^a_0 .\end{aligned} \tag{5.82}$$

Analyzing the theory $\mathcal{L}_\varphi$ begins by giving the antisymmetric auxiliary field the decomposition

$$\varphi^{a\mu\nu} = \varepsilon^{\mu\nu} E_a, \tag{5.83}$$

where the $\varepsilon^{\mu\nu}$ are the components of the two-dimensional Levi–Civita symbol (1.82), given by $\varepsilon^{01} = -\varepsilon^{10} = 1$ and satisfying the identity

$$\varepsilon^{\mu\nu} = g^{\mu\alpha}g^{\nu\beta}\varepsilon_{\alpha\beta} = -\varepsilon_{\mu\nu} \implies \varepsilon^{\mu\nu}\varepsilon_{\mu\nu} = -2. \tag{5.84}$$

The field E_a is then consistent with the earlier identification $F^{a01} = E_a$ since $\varphi^{a\mu\nu}$ and $F^{a\mu\nu}$ are the same fields. The decomposition (5.83) is possible since the gauge field A_μ^a has only two spacetime degrees of freedom and one is removed by the Weyl gauge condition. Using equations (5.83) and (5.84), the action (5.80) can be written in two forms,

$$\begin{aligned}\mathcal{L}_\varphi &= -\frac{1}{2}E_aE_a - \frac{1}{2}\varepsilon^{\mu\nu}\left(A_\mu^a\partial_\nu E_a - A_\nu^a\partial_\mu E_a + g\varepsilon^{abc}E_aA_\mu^bA_\nu^c\right) + g\dot{A}_0^aA_0^a \\ &= -\frac{1}{2}\left(E_a - \frac{1}{2}\varepsilon^{\mu\nu}F_{\mu\nu}^a\right)\left(E_a - \frac{1}{2}\varepsilon^{\alpha\beta}F_{\alpha\beta}^a\right) - \frac{1}{4}F_{\mu\nu}^aF^{a\mu\nu} + g\dot{A}_0^aA_0^a,\end{aligned} \tag{5.85}$$

where the second version assumes an integration by parts and uses the identity special to two dimensions,

$$\varepsilon^{\mu\nu}F_{\mu\nu}^a\varepsilon^{\alpha\beta}F_{\alpha\beta}^a = -2F_{\mu\nu}^aF^{a\mu\nu}. \tag{5.86}$$

The path integral measure $[\mathrm{d}\varphi]$ will be chosen so that

$$\int[\mathcal{D}\varphi]\exp\left\{-i\int\mathrm{d}^2x\,\frac{1}{2}\left(E_a - \frac{1}{2}\varepsilon^{\mu\nu}F_{\mu\nu}^a\right)\left(E_a - \frac{1}{2}\varepsilon^{\alpha\beta}F_{\alpha\beta}^a\right)\right\} = 1, \tag{5.87}$$

insuring that the two versions of the theory, one with the auxiliary field and the other without, are identical at the *quantum level.* For the case of SU(2), the configuration version of the measure $[\mathrm{d}\varphi]$ is given by the lattice product

$$[\mathcal{D}\varphi] = N\prod_x \mathrm{d}E_1(x)\,\mathrm{d}E_2(x)\,\mathrm{d}E_3(x), \tag{5.88}$$

where N is the usual normalization factor required by the Gaussian integrals discussed in conjunction with the configuration space measure of equation (4.151).

The canonical momenta associated with the action (5.85) are given by

$$\begin{aligned}P_a &= \frac{\partial\mathcal{L}_\varphi}{\partial\dot{E}_a} = A_1^a, \\ \Pi_0^a &= \frac{\partial\mathcal{L}_\varphi}{\partial\dot{A}_0^a} = gA_0^a, \\ \Pi_1^a &= \frac{\partial\mathcal{L}_\varphi}{\partial\dot{A}_1^a} = 0.\end{aligned} \tag{5.89}$$

This gives the Hamiltonian density

$$\mathcal{H} = P_a\dot{E}_a + \Pi_0^a\dot{A}_0^a + \Pi_1^a\dot{A}_1^a - \mathcal{L}_\varphi = \frac{1}{2}E_aE_a + A_0^a\partial_1E_a + g\varepsilon^{abc}E_aA_0^bP_c. \tag{5.90}$$

Using (5.89) shows that the Hamiltonian density (5.90) is consistent with the equations of motion (5.76) and (5.77),

$$\dot{E}_a = \frac{\delta H}{\delta P_a} = gf^{abc}E_bA_0^c = -gf^{abc}A_0^bE_c,$$
$$\dot{P}_a = -\frac{\delta H}{\delta E_a} = -E_a + \partial_1 A_0^a - gf^{abc}A_0^bA_1^c = \dot{A}_1^a. \tag{5.91}$$

However, the Weyl gauge condition, $\Pi_0^a = 0$ requires the implementation of Gauss's law as an additional constraint on E_a,

$$\dot{\Pi}_0^a = -\frac{\delta H}{\delta A_0^a} = -\partial_1 E_a + gf^{abc}E_bA_1^c \equiv G^a = 0. \tag{5.92}$$

The vacuum transition element for the Lagrangian path integral is given by

$$Z[0] = N\int [\mathcal{D}\varphi\,\mathcal{D}A\,\delta(A_0^a)\,\delta(G^a)]\exp\left\{i\int \mathrm{d}^2x\,\mathcal{L}_\varphi\right\}, \tag{5.93}$$

where $\mathcal{L}_\varphi$ is given by the first line of equation (5.85). The integration over A_0^a gives

$$Z[0] = N\int [\mathcal{D}\varphi\,\mathcal{D}A_1\,\delta(G^a)]\exp\left\{i\int \mathrm{d}^2x\left(-\frac{1}{2}E_aE_a + \dot{E}_aA_1^a\right)\right\}, \tag{5.94}$$

where the three Gauss's law constraints $G^a = -\partial_1 E_a + gf^{abc}E_bA_1^c = 0$ are unchanged. These constraints are enforced by changing variables from A_1^c to B^c using a 3×3 linear transformation matrix $\boldsymbol{V}$,

$$A_1 = \boldsymbol{V}^{-1}\cdot B \implies B = \boldsymbol{V}\cdot A_1, \tag{5.95}$$

where the matrix multiplication is on the isospin index, $A_1^a = V_{ab}^{-1}B_b$. Simultaneously, the Gauss's law constraints G are transformed as

$$G = \boldsymbol{V}^{-1}\cdot G' \implies G' = \boldsymbol{V}\cdot G. \tag{5.96}$$

Using the Jacobian of the change of integration $J = \det(\boldsymbol{V}^{-1})$ and the property of the delta function $\delta(G) = \det(\boldsymbol{V})\,\delta(G')$, the measure of the path integral (5.94) becomes

$$[\mathcal{D}\varphi\,\mathcal{D}A_1\,\delta(G)] = [\mathcal{D}\varphi\,\det(\boldsymbol{V}^{-1})\,\mathcal{D}B\,\det(\boldsymbol{V})\,\delta(G')] = [\mathcal{D}\varphi\,\mathcal{D}B\,\delta(G')]. \tag{5.97}$$

The Gauss's law constraints are now given by

$$G^{a'} = -V_{ab}\partial_1 E_b + V_{ab}gf^{bda}E_dV_{ac}^{-1}B_c. \tag{5.98}$$

The unitary transformation $\boldsymbol{V}$ is chosen to diagonalize the matrix $\boldsymbol{M}$ with the elements $M^{ab} = f^{acb}E_c$. This consists of finding the eigenvalues and eigenvectors of the matrix $\boldsymbol{M}$,

$$\boldsymbol{M} = \begin{pmatrix} 0 & -E_3 & E_2 \\ E_3 & 0 & -E_1 \\ -E_2 & E_1 & 0 \end{pmatrix}. \tag{5.99}$$

The three eigenvalues of $\boldsymbol{M}$ are 0 and $\pm iE$, where $E^2 = E_1^2 + E_2^2 + E_3^2$. The orthonormal eigenvectors associated with these three eigenvalues are given by

$$X_0 = \frac{1}{E}\begin{pmatrix} E_1 \\ E_2 \\ E_3 \end{pmatrix}, \quad X_\pm = \frac{1}{\sqrt{2}E\sqrt{E_1^2 + E_3^2}}\begin{pmatrix} E_1E_2 \mp iEE_3 \\ -E_1^2 - E_3^2 \\ E_2E_3 \pm iEE_1 \end{pmatrix}, \tag{5.100}$$

and have the property $X_\pm^* = X_\mp$. The associated matrix V is unitary, so that $V^{-1} = V^\dagger$. It is constructed from the three eigenvectors, which are arbitrarily assigned according to $V = (X_0\ X_+\ X_-)$. Using this assignment in equation (5.98) gives the three transformed constraints $G^{a'} = V_{ab}G_b$,

$$G^{1'} = X_0^a \partial_1 E_a = \frac{1}{E}E_a\partial_1 E_a = \partial_1 E = 0, \tag{5.101}$$

$$G^{2'} = X_+^a \partial_1 E_a + igEB_2 = 0 \implies B_2 = -\frac{1}{igE}X_+^a\partial_1 E_a, \tag{5.102}$$

$$G^{3'} = X_-^a \partial_1 E_a - igEB_3 = 0 \implies B_3 = \frac{1}{igE}X_-^a\partial_1 E_a. \tag{5.103}$$

Using the transformed constraints and $A_a = V_{ab}^\dagger B_b$ allows the fields B_2 and B_3 to be integrated from the path integral (5.94). The action in equation (5.94) becomes

$$\begin{aligned}\mathcal{L} &= -\frac{1}{2}E_aE_a + \dot{E}_a V_{ab}^\dagger B_b \\ &= -\frac{1}{2}E_aE_a + \dot{E}_a X_0^a B_1 + \frac{1}{igE}\dot{E}^a(X_+^a X_-^b - X_-^a X_+^b)\partial_1 E_b.\end{aligned} \tag{5.104}$$

Using the property of the Dirac delta, $\delta(aB) = \delta(B)/|a|$, the measure becomes

$$[\mathcal{D}\varphi\ \mathcal{D}B\ \delta(G')] \to \left[\frac{\mathcal{D}\varphi}{g^2E^2}\ \mathcal{D}B_1\delta(\partial_1 E)\right]. \tag{5.105}$$

The integration over B_1 gives

$$\int [\mathcal{D}B_1] \exp\left\{i\int \mathrm{d}^2x\ \dot{E}_a X_0^a B_1\right\} = [\delta(\dot{E}_a X_0^a)] = [\delta(\partial_0 E)]. \tag{5.106}$$

Using the eigenvectors (5.100) it is tedious but straightforward to show that

$$X_+^a X_-^b - X_-^a X_+^b = \frac{i}{E}\varepsilon^{abc}E_c. \tag{5.107}$$

The final form of the action (5.104) appearing in the path integral is given in terms of E in manifestly covariant form by

$$\mathcal{L}_E = -\frac{1}{2}E^2 + \frac{1}{2gE^2}\varepsilon^{abc}\varepsilon^{\mu\nu}\partial_\mu E_a\, E_b\, \partial_\nu E_c. \tag{5.108}$$

The final form for the path integral in terms of the variable E is given by

$$Z[0] = \int \left[\frac{\mathcal{D}\varphi}{g^2E^2}\,\delta(\partial_0 E)\,\delta(\partial_1 E)\right]\exp\left\{i\int \mathrm{d}^2x\,\mathcal{L}_E\right\}. \tag{5.109}$$

The two vacuum transition elements, the one described by the path integral (5.59) in terms of the gauge field A_μ^a and the one described by the path integral (5.109) in terms of E_a, are equivalent for the case of 1 + 1 dimensions. In the case of equation (5.109) the interaction term is inversely proportional to the coupling constant, demonstrating the manifestly non-perturbative analysis provided by the path integral. The path integral (5.109) can be analyzed exactly by a variant of the WKB method. There are numerous interesting properties of the classical solutions to the equations of motion for E_a. The interested reader is recommended to the literature.

5.6 The Dirac quantization condition

The subject of magnetic monopoles has a long history in theoretical physics. Although there has been no experimental observation of a magnetic monopole to date, they can be accommodated into the gauge field version of electrodynamics. The much studied Dirac quantization condition for electric charge e and magnetic charge g is given by

$$eg = \frac{1}{2}n\hbar, \tag{5.110}$$

where n is an integer. Dirac first derived this result by examining the singularity structure of the classical vector potential $\mathbf{A}$ such that $\mathbf{B} = \nabla \times \mathbf{A} = g\mathbf{r}/r^3$ in the context of the quantum mechanical wave function. In spherical coordinates (r, θ, φ) the vector potential $\mathbf{A}$ can be chosen to be the much studied Dirac string,

$$\mathbf{A}_\pm = \frac{g(\cos\theta \pm 1)}{r\sin\theta}\,\hat{\varphi}, \tag{5.111}$$

which yields $\mathbf{B} = \nabla \times \mathbf{A}_\pm = g\mathbf{r}/r^3$. The two forms of (5.111) are related by the gauge transformation $\mathbf{A}_- = \mathbf{A}_+ - \nabla\Lambda$ with $\Lambda = 2g\varphi$. This approach leads to the Wu and Yang fiber bundle formulation of magnetic monopoles.

However, rather than analyzing the specific details of equation (5.111), the Dirac quantization condition (5.110) will be derived in this section by examining general properties of the WKB approximation for the quantum mechanical configuration transition amplitude in the presence of a magnetic monopole. This takes advantage of the fact there are multiple classical trajectories for an electric point charge between two points in the presence of a magnetic monopole. For the case under consideration, the classical action for the electric charge is given by

$$S = \int_0^T dt\,\mathcal{L} = \int_0^T dt\left(\frac{1}{2}m\dot{\mathbf{r}}\cdot\dot{\mathbf{r}} + e\mathbf{A}\cdot\dot{\mathbf{r}}\right) \tag{5.112}$$

where $\mathbf{A}$ is the vector potential of the monopole. The general result derived in this section will not rely upon a specific form for $\mathbf{A}$. Instead, using the relationship $\mathbf{B} = \nabla \times \mathbf{A} = g\mathbf{r}/r^3$, so that $\nabla \cdot \mathbf{B} = 4\pi g\,\delta^3(r)$, and the time independence of $\mathbf{A}$ will be sufficient to derive the Dirac quantization condition.

The classical motion of a spinless charge in the presence of a magnetic monopole fixed at the origin was first analyzed by Poincarè. The fairly complex properties of trajectories between an initial position $\mathbf{r}_o$ and a final position $\mathbf{r}_f$ in the time T will be reviewed to demonstrate the existence of multiple trajectories. For a magnetic monopole with charge g fixed at the origin, so that the magnetic field is $\mathbf{B} = g\mathbf{r}/r^3$, equation (5.112) gives the equation of motion for a spinless electric point charge e and mass m consistent with the Lorentz force law $\boldsymbol{F} = e\boldsymbol{v} \times \mathbf{B}$,

$$m\ddot{\mathbf{r}} = e\dot{\mathbf{r}} \times \mathbf{B} = \frac{eg}{r^3}\dot{\mathbf{r}} \times \mathbf{r} = -\frac{eg\mathbf{L}}{mr^3}, \tag{5.113}$$

where $\mathbf{L} = m\mathbf{r} \times \dot{\mathbf{r}}$. For simplicity it will be assumed that $eg > 0$, although the opposite choice is possible. The kinetic energy, $E = \frac{1}{2}m\dot{\mathbf{r}}\cdot\dot{\mathbf{r}}$, obeys

$$\frac{\mathrm{d}E}{\mathrm{d}t} = \dot{\mathbf{r}}\cdot m\ddot{\mathbf{r}} = \dot{\mathbf{r}}\cdot\frac{eg}{r^3}(\dot{\mathbf{r}} \times \mathbf{r}) = 0, \tag{5.114}$$

and is therefore conserved. Since equation (5.113) insures that $\mathbf{r}\cdot\ddot{\mathbf{r}} = 0$, it follows that

$$\frac{\mathrm{d}}{\mathrm{d}t}(\mathbf{r}\cdot m\dot{\mathbf{r}}) = m\dot{\mathbf{r}}\cdot\dot{\mathbf{r}} + m\mathbf{r}\cdot\ddot{\mathbf{r}} = 2E \implies m\mathbf{r}\cdot\dot{\mathbf{r}} = R + 2Et, \tag{5.115}$$

where R is the initial value of $\mathbf{r}\cdot m\dot{\mathbf{r}}$. The second relation in equation (5.115) gives

$$\frac{\mathrm{d}}{\mathrm{d}t}(mr^2) = \frac{\mathrm{d}}{\mathrm{d}t}(m\mathbf{r}\cdot\mathbf{r}) = 2\,m\mathbf{r}\cdot\dot{\mathbf{r}} = 2R + 4Et, \tag{5.116}$$

Integrating equation (5.116) yields r as a function of time

$$r(t) = \sqrt{r_o^2 + \frac{2Rt}{m} + \frac{2Et^2}{m}}, \tag{5.117}$$

where $r_o = r(0)$ is the initial radial distance. If v_o is the initial speed of the electric charge then the radial velocity $v_r \to v_o t$ for large t.

Using equation (5.113) and the three-dimensional vector identity $\mathbf{a} \times \mathbf{b} \times \mathbf{c} = (\mathbf{a}\cdot\mathbf{c})\mathbf{b} - (\mathbf{a}\cdot\mathbf{b})\mathbf{c}$, it is straightforward to show that the vector

$$\mathbf{D} = \mathbf{L} - eg\hat{\mathbf{r}} \tag{5.118}$$

is a constant of motion, where $\hat{\mathbf{r}} = \mathbf{r}/r$ is a radial unit vector pointing from the origin to the current position of the electric charge. This follows from the equation of motion (5.113),

$$\frac{\mathrm{d}\mathbf{D}}{\mathrm{d}t} = \frac{\mathrm{d}\mathbf{L}}{\mathrm{d}t} - eg\frac{\mathrm{d}}{\mathrm{d}t}\left(\frac{\mathbf{r}}{r}\right) = m\mathbf{r} \times \ddot{\mathbf{r}} - eg\frac{\dot{\mathbf{r}}}{r} + eg\frac{\dot{r}\mathbf{r}}{r^2} = 0. \tag{5.119}$$

Unlike the central force problem, there are no stable planar orbits since $\mathbf{L}$ is not conserved. However, because $\mathbf{L} \cdot \mathbf{r} = 0$, it follows that $D^2 = L^2 + e^2g^2$ is a constant, which in turn shows that L^2 is constant. The angle α between $\mathbf{L}$ and $\mathbf{D}$ is constant since $\mathbf{L} \cdot \mathbf{D} = LD\cos\alpha = \mathbf{L} \cdot \mathbf{L} = L^2$, giving

$$\cos\alpha = \frac{L}{D} = \frac{L}{\sqrt{L^2 + e^2g^2}}. \tag{5.120}$$

Since $\hat{\mathbf{r}} \cdot \mathbf{D} = -eg$ is a constant, it follows that the angle between $\hat{\mathbf{r}}$ and $\mathbf{D}$ is also constant. Since $\mathbf{r}$ is the position vector of the charge and $\mathbf{D}$ is a fixed vector, the classical trajectory lies on a cone with its tip at the origin and its axis coinciding with $-\mathbf{D}$. The half-angle of the cone, denoted ψ, is given from $-\hat{\mathbf{r}} \cdot \mathbf{D} = -D\cos\psi = -eg$. The angle ψ then determined by L since result (5.120) gives

$$\cos\psi = \frac{eg}{D} = \frac{eg}{\sqrt{L^2 + e^2g^2}}. \tag{5.121}$$

Because $\mathbf{r}$ lies in the surface of the cone and the velocity of the charge is in the tangent space of the cone, it follows that $\mathbf{L} = m\mathbf{r} \times \dot{\mathbf{r}}$ is perpendicular to the surface of the cone at all points along the trajectory. This shows that the angle between $\mathbf{L}$ and $-\mathbf{D}$, defined by equation (5.120), is $\alpha = \pi/2 - \psi$, so that $\cos\alpha = \sin\psi$. When (5.120) and (5.121) are combined with this result, it gives $L = eg\tan\psi$. For the case $eg > 0$ the motion along the cone is therefore right-handed around the $\mathbf{D}$ axis.

There is no loss in generality from choosing the z-axis to coincide with $-\mathbf{D}$, placing the cone of motion in the $+z$ half-space. For such a choice of coordinates the definition of $\mathbf{L}$ gives the azimuthal speed $v_\phi = L/mr$, with $v_\phi = -r\sin\psi\dot{\phi}$, so that the angular velocity of rotation around the z-axis is left handed and given by

$$\dot{\phi} = -\frac{L}{mr^2\sin\psi}. \tag{5.122}$$

Using the form

$$L^2 = \mathbf{L} \cdot \mathbf{L} = m^2(\mathbf{r} \times \dot{\mathbf{r}}) \cdot (\mathbf{r} \times \dot{\mathbf{r}}) = m^2r^2\dot{\mathbf{r}} \cdot \dot{\mathbf{r}} - m^2(\mathbf{r} \cdot \dot{\mathbf{r}})^2 \tag{5.123}$$

shows that

$$L^2 + m^2(\mathbf{r} \cdot \dot{\mathbf{r}})^2 = m^2r^2\dot{\mathbf{r}} \cdot \dot{\mathbf{r}} = 2\,mr^2E \tag{5.124}$$

at all times. Evaluating equation (5.124) at $t = 0$ shows that the constants appearing in result (5.117) are related according to

$$L^2 + R^2 = 2\,mr_o^2 E. \tag{5.125}$$

Combining this with result (5.117) allows equation (5.122) to be integrated to obtain the change in the azimuthal angle of the trajectory,

$$\phi(T) - \varphi(0) = -\frac{1}{\sin\psi}\left[\arctan\left(\frac{R + 2ET}{L}\right) - \arctan\left(\frac{R}{L}\right)\right]. \tag{5.126}$$

The next step in determining the classical trajectory used in the WKB approximation to the propagator is to relate the values of E, R, and ψ to the values of $\mathbf{r}_o$, $\mathbf{r}_f$, and T. The kinetic energy E is found by solving (5.117) at the time T with the identification $r(T) = r_f$,

$$E - \frac{1}{2}m\frac{(r_f^2 - r_o^2)}{T^2} = \frac{R}{T}, \tag{5.127}$$

and combining it with equation (5.125), which gives $R = \sqrt{2mr_o^2 E - L^2}$. The resulting quadratic equation for E has the solutions

$$E = \frac{1}{2}m\frac{(r_f^2 + r_o^2)}{T^2} \pm \frac{mr_f r_o}{T^2}\sqrt{1 - z^2\tan^2\psi}, \tag{5.128}$$

where the relation $L^2 = e^2 g^2 \tan^2\psi$ was used and the dimensionless parameter,

$$z = \frac{egT}{mr_o r_f}, \tag{5.129}$$

has been introduced. The choice of sign in equation (5.128) is dictated by the angle between $\mathbf{r}_o$ and $\mathbf{r}_f$, which will be denoted γ. If $\gamma < \pi/2$ the negative sign is chosen. Although ψ depends upon z, it is clear from the conical nature of the motion that the minimum real value for ψ is $\gamma/2$. As a result, the kinetic energy becomes complex if $z\tan(\gamma/2) > 1$, showing that not all initial and final points are connected by a classical trajectory for arbitrary T.

Result (5.126) may now be simplified by combining equations (5.125) and (5.126) with the trigonometric identity

$$\sin(\arctan X - \arctan Y) = \frac{X - Y}{\sqrt{(1 + X^2)(1 + Y^2)}}. \tag{5.130}$$

The result is that equation (5.126) becomes

$$\varphi(T) - \varphi(0) = -\frac{1}{\sin\psi}\arcsin(z\tan\psi). \tag{5.131}$$

Result (5.131) is consistent with the previously determined emergence of complex values for the case that $z\tan\psi > 1$. For the choice of coordinate system the value of equation (5.131) must coincide with the difference in azimuthal angles for $\mathbf{r}_o$ and $\mathbf{r}_f$ *modulo* 2π. Since ψ is the constant polar angle of both vectors, its value can therefore be found by solving the spherical coordinate formula obtained from $\mathbf{r}_f \cdot \mathbf{r}_o = r_f r_o \cos\gamma$, which gives

$$\begin{aligned}\cos\gamma &= \cos^2\psi + \sin^2\psi\cos(\phi(T) - \phi(0)) \\ &= \cos^2\psi + \sin^2\psi\cos\left(\frac{1}{\sin\psi}\arcsin(z\tan\psi)\right).\end{aligned} \tag{5.132}$$

Solving equation (5.132) analytically for arbitrary z and γ is not possible and numerical methods must be employed. However, for the case that $\mathbf{r}_o$ and $\mathbf{r}_f$ are parallel, i.e. $\gamma \to 0$, it is relatively simple to demonstrate the existence of multiple trajectories. For the case $\psi = \gamma = 0$ there is a radial trajectory consistent with equation (5.128) given by

$$r = \left(\frac{r_f - r_o}{T}\right)t + r_o, \tag{5.133}$$

which corresponds to linear motion and $E = \frac{1}{2}m\dot{r}^2 = \frac{1}{2}m(r_f - r_o)^2/T^2$. In addition, there is a second trajectory if

$$\varphi(T) - \varphi(0) = -\frac{1}{\sin\psi}\arcsin(z\tan\psi) = -2\pi k, \tag{5.134}$$

where k is an arbitrary positive integer, since for such a case the right-hand side of equation (5.132) reduces to unity, corresponding to $\gamma = 0$. If $\psi \to 0$, so that $z\tan\psi \to 0$, then condition (5.134) reduces to

$$\cos\psi = \frac{z}{2\pi k} \Longrightarrow z\tan\psi = z\frac{\sqrt{1 - \cos^2\psi}}{\cos\psi} = \sqrt{4\pi^2k^2 - z^2}. \tag{5.135}$$

Result (5.135) shows that for $z = egT/mr_f r_0 \approx 2\pi k$ there is a cone of motion meeting the criterion $\psi \approx 0$ which corresponds to a trajectory that winds around the infinitesimal cone k times. This result occurs since $\gamma = 0$ corresponds to a cumulation point of the motion. Numerical analysis verifies that there are multiple trajectories for z in a range of values around $2\pi k$, but that in the general case these trajectories are energetically distinct from equation (5.133) since ψ, although small, is nonzero. This is similar to the motion of a free particle on a circle, where for a given time T there is a denumerable infinity of energetically distinct classical trajectories between any two points on the circle characterized by an integer winding number.

However, the existence of multiple trajectories creates a potential anomaly in the WKB approximation for the propagator. This follows from examining the Lagrangian path integral representation of the charged particle's propagator,

$$\langle \mathbf{r}_f, T|\mathbf{r}_o, 0\rangle = \int_{\mathbf{r}_o}^{\mathbf{r}_f}[d^3r]\exp\left\{\frac{i}{\hbar}S\right\}, \tag{5.136}$$

where S is given by equation (5.112). Ignoring the possibility of turning points and the associated Maslov indices, the WKB approximation for the propagator (5.136) is given by the semiclassical form (3.81),

$$\langle \mathbf{r}_f, T|\mathbf{r}_o, 0\rangle = \sum_n \left(\frac{i}{8\pi^3\hbar^3} \left| \det \frac{\partial^2 S_n[\mathbf{r}_f, \mathbf{r}_o, T]}{\partial x_f^j \, \partial x_o^k} \right| \right)^{1/2} \exp\left\{ \frac{i}{\hbar} S_n \right\}, \tag{5.137}$$

where the sum is over all possible classical trajectories connecting $\mathbf{r}_f$ and $\mathbf{r}_o$ in the time T, while $S_n[\mathbf{r}_f, \mathbf{r}_o, T]$ is the value of the action along the nth trajectory. Because the kinetic energy is constant and $\mathbf{A}$ is time-independent, the action (5.112) along a classical trajectory becomes

$$S = \int dt \left(\frac{1}{2} m\dot{\mathbf{r}} \cdot \dot{\mathbf{r}} + e\mathbf{A}(\mathbf{r}) \cdot \dot{\mathbf{r}} \right) = ET + e \int_P \mathbf{A}(r) \cdot d\mathbf{r}, \tag{5.138}$$

where E is the kinetic energy, given by equation (5.128), and the integral is evaluated along the classical trajectory P joining the initial and final points.

It has been specifically established that there may be two or more trajectories or paths, denoted P_1 and P_2, for the case that $\mathbf{r}_o$ and $\mathbf{r}_f$ are parallel. However, the results presented next will not rely upon the specific form for the determinants in the WKB approximation (5.137). Instead, it will be assumed only that there are two trajectories between the initial and final points and that the determinants, denoted D_n for the respective paths, are real valued functions. For such a case equation (5.137) can then be written

$$\begin{aligned}\langle \mathbf{r}_f, T|\mathbf{r}_o, 0\rangle = &\sqrt{\frac{iD_1}{8\pi^3\hbar^3}} \exp\left\{ \frac{i}{\hbar}\left(E_1 T + e\int_{P_1} \mathbf{A}(\mathbf{r}) \cdot d\mathbf{r} \right) \right\} \\ &+ \sqrt{\frac{iD_2}{8\pi^3\hbar^3}} \exp\left\{ \frac{i}{\hbar}\left(E_2 T + e\int_{P_2} \mathbf{A}(\mathbf{r}) \cdot d\mathbf{r} \right) \right\},\end{aligned} \tag{5.139}$$

where E_n is the kinetic energy of the respective path. However, result (5.139) must yield a unique quantum mechanical transition probability density. For equation (5.139) this is given by

$$\begin{aligned}&\left|\langle \mathbf{r}_f, T|\mathbf{r}_o, 0\rangle\right|^2 \\ &= \frac{1}{8\pi^3\hbar^3}\left\{ D_1 + D_2 + 2\sqrt{D_1 D_2} \cos\left(\frac{(E_1 - E_2)T}{\hbar} + \frac{e}{\hbar}\oint_P \mathbf{A} \cdot d\mathbf{r} \right) \right\},\end{aligned} \tag{5.140}$$

where the closed path P is defined by the two paths P_1 and P_2 from $\mathbf{r}_o$ to $\mathbf{r}_f$,

$$\oint_P \mathbf{A} \cdot d\mathbf{r} = \int_{P_1} \mathbf{A} \cdot d\mathbf{r} - \int_{P_2} \mathbf{A} \cdot d\mathbf{r}. \tag{5.141}$$

Using equation (5.141) shows that equation (5.140) possesses the required invariance under the exchange $P_1 \leftrightarrow P_2$. Stokes' theorem and $\mathbf{B} = \nabla \times \mathbf{A}$ reduces equation (5.140) to

$$\left|\langle \mathbf{r}_f,\, T|\mathbf{r}_o,\, 0\rangle\right|^2 = \frac{1}{8\pi^3\hbar^3}\left\{D_1 + D_2 + 2\sqrt{D_1 D_2}\cos\left(\frac{(E_1 - E_2)T}{\hbar} + \frac{e}{\hbar}\int_S \mathbf{B}\cdot d\mathbf{S}\right)\right\}, \tag{5.142}$$

where S is an *arbitrary* surface bounded by P. Result (5.142) is true regardless of the specific form for $\mathbf{A}$.

The right-handed surfaces bounded by the oriented closed path P fall into two equivalence classes in the presence of a magnetic monopole: those with positive magnetic flux and those with negative magnetic flux. Stated another way, two surfaces bounded by P belong to the same equivalence class if the difference in their surface integrals results in a closed surface integral whose volume does not contain the monopole; otherwise they are in separate equivalence classes. In order for equation (5.142) to be unambiguous the choice of equivalence class for the surface integral must be irrelevant. If S_1 belongs to one equivalence class and S_2 to the other, the absence of an anomaly for equation (5.142) requires that

$$\cos\left(\frac{(E_1 - E_2)T}{\hbar} + \frac{e}{\hbar}\int_{S_1}\mathbf{B}\cdot d\mathbf{S}\right) = \cos\left(\frac{(E_1 - E_2)T}{\hbar} + \frac{e}{\hbar}\int_{S_2}\mathbf{B}\cdot d\mathbf{S}\right). \tag{5.143}$$

The identity $\cos\alpha - \cos\beta = -2\sin\frac{1}{2}(\alpha + \beta)\sin\frac{1}{2}(\alpha - \beta)$ shows that equation (5.143) is satisfied if

$$\sin\left(\frac{1}{2}\frac{e}{\hbar}\int_{S_1}\mathbf{B}\cdot d\mathbf{S} - \frac{1}{2}\frac{e}{\hbar}\int_{S_2}\mathbf{B}\cdot d\mathbf{S}\right) = \sin\left(\frac{1}{2}\frac{e}{\hbar}\oint_S \mathbf{B}\cdot d\mathbf{S}\right) = 0, \tag{5.144}$$

where, by definition, the closed surface S contains the monopole. Result (5.144), Gauss's law, and $\nabla\cdot\mathbf{B} = 4\pi g\,\delta^3(\mathbf{r})$ immediately yields the Dirac condition (5.110) since the vanishing of the sine function in equation (5.144) requires

$$\sin\left(\frac{1}{2}\frac{e}{\hbar}\oint_S \mathbf{B}\cdot d\mathbf{S}\right) = \sin\left(\frac{1}{2}\frac{e}{\hbar}\int_{V(S)} d^3r\,\nabla\cdot\mathbf{B}\right) = \sin\left(2\pi\frac{eg}{\hbar}\right) = 0. \tag{5.145}$$

Requirement (5.145) is satisfied if the Dirac quantization condition $eg = \frac{1}{2}n\hbar$ holds, where n is an integer. An identical argument shows that surfaces in the same equivalence class result in equation (5.145) vanishing since $V(S)$ does not contain the monopole. As a result, enforcing the Dirac condition is necessary to remove the anomaly in the WKB approximation for the transition probability.

5.7 The effective potential and spontaneously broken symmetry

In the analysis performed so far on the generating functionals for field theory it was assumed that the path integral represented the transition element between the vacuum or, equivalently, states with zero particle and field content. It was also assumed that the generating functionals were to be evaluated for the case of vanishing external currents. In the case of the scalar field perturbation series and the QED path integral, this led to the absence of vacuum expectation values for the fields. However, this need not be the case. In the derivation of the scalar field path integral (4.121) the in- and out-states were treated as coherent states, and this resulted in an overall exponential factor (4.108) involving the coherent state function at asymptotic times. Such coherent states give rise to an asymptotic value for the associated fields. This allows the path integral to represent a transition between states where the field has *an expectation value*. This expectation value may represent an externally created circumstance, much like the semiclassical treatment of radiation in quantum mechanics. However, it may also represent the presence of a ground state or *true vacuum* with a lower energy *due to interactions*.

If the asymptotic coherent state is to be relativistically invariant, the coherent state must be associated with a scalar field or a relativistic scalar composite of spinor or vector particles. For simplicity, the latter will not be considered. In order that the asymptotic states reflect translational invariance, the expectation value of the asymptotic scalar field $\hat{\varphi}(x)$ must be chosen to be a real constant, denoted $\varphi_0(x) = \nu$. The asymptotic coherent state $|\psi, t_\pm\rangle$ can be expressed using equation (4.87) so that the weak asymptotic condition for the interacting field Φ gives

$$\lim_{t\to t_\pm}\langle\psi, t|\Phi(x, t)|\psi, t\rangle = \nu\sqrt{Z}, \tag{5.146}$$

where the state $|0\rangle$ in equation (4.87) is such that $a_p|0\rangle = 0$. For such a choice of asymptotic coherent state, the phase factor S given by equation (4.108) is unity. The path integral for the scalar field then becomes

$$Z_\nu[J] = \int_\nu^\nu [\mathcal{D}\varphi]\exp\left\{i\int \mathrm{d}^4x\,(\mathcal{L}(\varphi) - J\varphi)\right\}. \tag{5.147}$$

The path integral of equation (5.147) represents the transition amplitude between a state that has a nontrivial vacuum expectation value if $\nu \neq 0$. This path integral can be analyzed using the functional techniques derived so far by writing the Lagrangian as $\mathcal{L}(\varphi) = \frac{1}{2}\partial_\mu\varphi\,\partial^\mu\varphi - \mathcal{H}_I(\varphi)$, where $\mathcal{H}_I$ represents the interaction part of the Hamiltonian and includes any quadratic mass terms that may be present. Simply translating the variables of integration according to $\varphi \to \varphi + \nu$ gives

$$Z_\nu[J] = \int_0^0 [\mathcal{D}\varphi]\exp\left\{i\int \mathrm{d}^4x\left(\frac{1}{2}\partial_\mu\varphi\,\partial^\mu\varphi - \mathcal{H}_I(\varphi+\nu) - J(\varphi+\nu)\right)\right\}. \tag{5.148}$$

The interaction Hamiltonian is then expanded around ν to second order, giving

$$\mathcal{H}_I(\varphi+\nu) = \mathcal{H}_I(\nu) + \frac{\partial\mathcal{H}_I(\nu)}{\partial\nu}\varphi + \frac{1}{2}\frac{\partial^2\mathcal{H}_I(\nu)}{\partial\nu^2}\varphi^2 + \cdots \tag{5.149}$$

For the moment the cubic and higher order terms will be ignored. The path integral then becomes exactly solvable,

$$Z_\nu[J] = \int_0^0 [\mathcal{D}\varphi] \exp\left\{ i \int d^4x \left(\frac{1}{2}\partial_\mu\varphi\, \partial^\mu\varphi - \frac{1}{2}m^2\varphi^2 - (J+\alpha)\varphi \right) \right\}$$
$$\times \exp\left\{ -i \int d^4x\, (\mathcal{H}_I(\nu) - J\nu) \right\}, \tag{5.150}$$

where m^2 and α are functions of ν given by

$$m^2 = \frac{\partial^2 \mathcal{H}_I(\nu)}{\partial \nu^2}, \qquad \alpha = \frac{\partial \mathcal{H}_I(\nu)}{\partial \nu}. \tag{5.151}$$

The path integral of equation (5.150) is identical in form to the quadratic path integral of equation ((4.153) with the identification $J \to J + \alpha$. This path integral was evaluated in equation (4.158), so that

$$Z_\nu[J] = \frac{N}{\sqrt{\det(\square + m^2)}}$$
$$\times \exp\left\{ \frac{1}{2} \int d^4x\, d^4y\, (J(x) + \alpha)\Delta_F(x-y)(J(y)+\alpha) \right\} \tag{5.152}$$
$$\times \exp\left\{ -i \int d^4x\, (\mathcal{H}_I(\nu) + J\nu) \right\},$$

where $\Delta_F(x-y)$ is the Feynman propagator for the scalar field defined in equation (4.178) with its pole occurring at $k^2 = m^2$. Using m^2 as the scale factor in equation (4.158) gives the generating functional

$$W[J] = i \ln Z[J] = \frac{1}{2} i \int d^4x\, d^4y\, (J(x) + \alpha)\Delta_F(x-y)(J(y)+\alpha)$$
$$+ \int d^4x\, (\mathcal{H}_I(\nu) + J\nu) - \frac{1}{2} i \ln \det\left(1 + \frac{\square}{m^2}\right). \tag{5.153}$$

The next step is to determine the vacuum expectation value of φ as a functional of J. This is obtained from the generating functional $W[J]$,

$$\varphi_0[J] = \frac{\delta W[J]}{\delta J(x)} = \nu + i \int d^4y\, \Delta_F(x-y)(J(y)+\alpha). \tag{5.154}$$

The vacuum expectation is $\varphi_0 = \nu$ when $J(y) = 0$ only if $\alpha = \partial\mathcal{H}_I(\nu)/\partial\nu = 0$. In the context of the classical theory, the term α vanishes for a *constant classical solution, $\varphi(x) = \nu$, to the classical equation of motion,* which is given by $\square\varphi + \partial\mathcal{H}_I/\partial\varphi = \partial\mathcal{H}_I(\nu)/\partial\nu = 0$.

This yields an extremely important insight into quantum field theory, which is that the solutions to the *classical* equations of motion yield possible vacuum expectation values for the fields in the quantized theory, since such solutions are consistent with the *vanishing of the external source J*. For the case being considered here, the asymptotic ground state of the interacting system is approximated by the coherent state $|\nu\rangle$, since ν can be chosen so that

$$\langle \nu | \left(\Box\hat{\varphi} + \frac{\partial \mathcal{H}_I(\hat{\varphi})}{\partial \hat{\varphi}} \right) | \nu \rangle \approx \frac{\partial \mathcal{H}_I(\nu)}{\partial \nu} = 0. \tag{5.155}$$

Applying the operator $\Box + m^2$ to equation (5.154) and using $\Box\nu = 0$ and $(\Box + m^2)\Delta_F(x - y) = i\delta^4(x - y)$ yields

$$(\Box + m^2)\varphi_0 = m^2\nu - \alpha - J = m^2\nu - \frac{\partial \mathcal{H}_I(\nu)}{\partial \nu} - J. \tag{5.156}$$

Identifying $\nu = \varphi_0$ reduces equation (5.156) to the original classical equation of motion,

$$\Box\varphi_0 + \frac{\partial \mathcal{H}_I(\varphi_0)}{\partial \varphi_0} = -J. \tag{5.157}$$

If φ_0 is constant then $J + \partial \mathcal{H}_I(\varphi_0)/\partial \varphi_0 = 0$, and so it is consistent to identify the vacuum expectation value as $\varphi_0 \approx \nu$ by ignoring the integral on the right-hand side of equation (5.154).

However, due to the nonlinear interactions, the quantized theory will yield $O(\hbar)$ corrections to the coherent state expectation value $\langle \partial \mathcal{H}_I(\hat{\varphi})/\partial \hat{\varphi} \rangle$ that appears in equation (5.155). Therefore, α does not vanish in the true quantum ground state. However, it will be seen that $\alpha \approx O(\hbar)$, and in that regard the integral involving the Feynman propagator in equation (5.153) will be $O(\hbar)$ and it is consistent to ignore it. The effective action is then given by

$$\Gamma[\varphi_0] = W[J] - \int d^4x\, \nu J(x) = \int d^4x\, \mathcal{H}_I(\varphi_0) - \frac{1}{2} i \ln \det \left(1 + \frac{\Box}{m^2} \right), \tag{5.158}$$

where $m^2 = \partial^2 \mathcal{H}_I(\varphi_0)/\partial \varphi_0^2$. Using the continuum method (4.167) to find the determinant, the second term in equation (5.158) is given by

$$\begin{aligned} -\frac{1}{2} i \ln \det \left(1 + \frac{\Box}{m^2} \right) &= -\frac{1}{2} i \operatorname{Tr} \ln \left(1 + \frac{\Box}{m^2} \right) \\ &= -\frac{1}{2} i \int d^4x \int \frac{d^4k}{(2\pi)^4} \ln \left(1 - \frac{k^2}{m^2} \right). \end{aligned} \tag{5.159}$$

where the trace in the second term is understood as the integral over all spacetime for the continuum matrix $\Box\delta^4(x - y)$ after setting $x = y$. The *effective potential*, denoted $V(\varphi_0)$, is defined by the relation $\Gamma[\varphi_0] = V(\varphi_0) \int d^4x$, and this gives

$$V(\varphi_0) = \mathcal{H}_I(\varphi_0) - \frac{1}{2} i\hbar \int \frac{d^4k}{(2\pi)^4} \ln \left(1 - \frac{k^2}{m^2} \right). \tag{5.160}$$

The factor of $\hbar$ in the Gaussian integrals has been reinserted in order to track the quantum corrections to the effective potential. The functional derivatives of the effective action become partial derivatives for the effective potential. The first term is simply the classical energy density of a constant scalar field, while the second term encapsulates quantum effects due to the nonlinearity of the theory and the possible presence of a nontrivial vacuum structure. It can be shown that the second term is the sum of all one-loop Feynman diagrams which encapsulate the lowest order quantum corrections to the vacuum energy.

The effective potential inherits the properties of the effective action and can be viewed as the effective action evaluated at zero momentum. The condition (5.18) becomes

$$\left.\frac{\partial V(\varphi_0)}{\partial \varphi_0}\right|_{\varphi_0=\varphi_c} = 0, \tag{5.161}$$

so that φ_c is the solution, assumed to be constant, to equation (5.161). In addition, the pole of the propagator is the *physical mass* M^2 of the particle for the case of a single scalar field, and it is given by

$$M^2 = \left.\frac{\partial^2 V(\varphi_0)}{\partial \varphi_0^2}\right|_{\varphi_0=\varphi_c}. \tag{5.162}$$

A stable theory must correspond to $M^2 > 0$, and so this means that the solution to equation (5.161) must correspond to a minimum of the effective potential. The value of the four-point vertex at zero momentum is the quartic coupling constant λ, given by

$$\lambda = \left.\frac{1}{3!}\frac{\partial^4 V(\varphi_0)}{\partial \varphi_0^4}\right|_{\varphi_0=\varphi_c}. \tag{5.163}$$

These relations become apparent for the case that the effective potential is given by the first term in equation (5.160) and the interaction Hamiltonian is given by $\mathcal{H}_I(\varphi_0) = \frac{1}{2}m^2\varphi_0^2 + \frac{1}{4}\lambda\varphi_0^4$. However, the quantum corrections contained in the integral term of equation (5.160) will require further analysis to incorporate.

While the $O(\hbar)$ correction to the vacuum energy density appears to be imaginary, its evaluation using the standard Wick rotation $\omega \overset{W}{\rightarrow} i\,\omega$ gives

$$V(\varphi_0) = \mathcal{H}_I(\varphi_0) + \frac{1}{2}\hbar \int \frac{d^3k\, d\omega}{(2\pi)^4} \ln\left(1 + \frac{\omega^2 + \boldsymbol{k}^2}{m^2}\right). \tag{5.164}$$

The integral, denoted V_1, is badly divergent and must be renormalized. The first step in this process is to control and identify the divergences by placing a *cutoff* Λ on the integrations over ω and $\boldsymbol{k}$. In the evaluation of the integral, any terms that are independent of φ_0 may be discarded since they will have no effect on the value of derivatives of $V(\varphi_0)$, which are the only physically meaningful aspects of the effective potential. The entire dependence on φ_0 is therefore encapsulated in the factor $m^2 = \partial^2\mathcal{H}_I(\varphi_0)/\partial\varphi_0^2$. It is tedious but straightforward to show that the remaining terms are given by

$$\frac{1}{2}\hbar\int_0^\Lambda \frac{k^2\,\mathrm{d}k}{2\pi^2}\int_{-\Lambda}^{\Lambda}\frac{\mathrm{d}\omega}{2\pi}\ln\left(1+\frac{\omega^2+k^2}{m^2}\right)=\frac{\hbar}{32\pi^2}\left[2\Lambda^4\left(1+\frac{m^2}{\Lambda^2}\right)^{3/2}\right.$$
$$\left.-\Lambda^2 m^2\left(1+\frac{m^2}{\Lambda^2}\right)-m^4\ln\left(\frac{\Lambda}{m}+\frac{\Lambda}{m}\sqrt{1+\frac{m^2}{\Lambda^2}}\right)\right]. \tag{5.165}$$

Expanding equation (5.165) in a power series in m^2/Λ^2 and retaining only those terms that do not vanish in the limit $\Lambda \to \infty$ gives the $O(\hbar)$ version of V_1,

$$V_1=\frac{\hbar\Lambda^4}{16\pi^2}+\frac{\hbar\Lambda^2 m^2}{16\pi^2}+\frac{\hbar m^4}{64\pi^2}\ln\left(\frac{m^2}{4\Lambda^2}\right). \tag{5.166}$$

The first term in equation (5.166) is independent of m^2 and is therefore discarded. This gives the *unrenormalized effective potential*,

$$V(\varphi_0)=\mathcal{H}_I(\varphi_0)+\frac{\hbar\Lambda^2 m^2}{16\pi^2}+\frac{\hbar m^4}{64\pi^2}\ln\left(\frac{m^2}{4\Lambda^2}\right). \tag{5.167}$$

The final step is to pick a specific form for $\mathcal{H}_I$ and perform the renormalization procedure to render equation (5.167) finite. A much analyzed form is given by the *Higgs potential*,

$$\mathcal{H}_I=-\frac{1}{2}(\mu^2+\delta\mu^2)\varphi^2+\frac{1}{4}(\lambda+\delta\lambda)\varphi^4, \tag{5.168}$$

where μ^2 and λ are positive quantities. The *counterterms* $\delta\mu^2$ and $\delta\lambda$ are proportional to $\hbar$ and are chosen to complete the renormalization process. For the *classical case* that $\delta\mu^2$ and $\delta\lambda$ are zero, the potential (5.168) has nontrivial minima, $\partial\mathcal{H}_I(\varphi_0)/\partial\varphi_0 = 0$, at $\varphi_0=\pm\sqrt{\mu^2/\lambda}$. It also possesses the discrete reflection symmetry $\varphi_0 \to -\varphi_0$. In what is known as the *minimal subtraction scheme*, the counterterms are chosen to be

$$\delta\mu^2=\frac{\hbar}{8\pi^2}\frac{m^2}{\varphi_0^2}\Lambda^2,$$
$$\delta\lambda=\frac{\hbar}{16\pi^2}\frac{m^4}{\varphi_0^4}\left[\ln\left(\frac{\Lambda^2}{M^2}\right)-\frac{3}{2}\right]. \tag{5.169}$$

A thorough discussion of renormalization using dimensional regularization and the minimal subtraction scheme is available in Peskin and Schroeder. For simplicity, the value of M^2 will be chosen to be $m^2=3\lambda\varphi_0^2-\mu^2$ evaluated at the solution to equation (5.161), so that M^2 coincides with the pole of the Feynman propagator up to $O(\hbar)$ corrections. Substituting the two terms of equation (5.169) into the unrenormalized effective potential (5.167) yields the renormalized effective potential to $O(\hbar)$,

$$V(\varphi_0) = -\frac{1}{2}\mu^2\varphi_0^2 + \frac{1}{4}\lambda\varphi_0^4 + \frac{\hbar}{64\pi^2}(3\lambda\varphi_0^2 - \mu^2)^2\left[\ln\left(\frac{3\lambda\varphi_0^2 - \mu^2}{M^2}\right) - \frac{3}{2}\right]. \quad (5.170)$$

Result (5.170) is real valued as long as $\varphi_0^2 \geqslant \mu^2/3\lambda$.

The vacuum expectation value is the solution, denoted φ_c, of

$$\begin{aligned}\frac{\partial V(\varphi_0)}{\partial\varphi_0} &= -\mu^2\left(1 - \frac{3\hbar\lambda}{16\pi^2}\right)\varphi_0 + \lambda\left(1 - \frac{9\hbar\lambda}{16\pi^2}\right)\varphi_0^3 \\ &+ \frac{3\hbar\lambda}{16\pi^2}\left(3\lambda\varphi_0^3 - \mu^2\varphi_0\right)\ln\left(\frac{3\lambda\varphi_0^2 - \mu^2}{M^2}\right) = 0.\end{aligned} \quad (5.171)$$

Assuming that $\ln[(3\lambda\varphi_c^2 - \mu^2)/M^2] \approx O(\hbar)$ gives the solution

$$\varphi_c^2 = \frac{\mu^2}{\lambda}\left(\frac{1 - 3\hbar\lambda/16\pi^2}{1 - 9\hbar\lambda/16\pi^2}\right) \approx \frac{\mu^2}{\lambda}\left(1 + \frac{3\hbar\lambda}{8\pi^2}\right), \quad (5.172)$$

where the latter approximation holds for small values of the coupling constant λ. This result is in keeping with the assumption made earlier in equation (5.158) that the quantum corrections to the classical solution, $\nu^2 = \mu^2/\lambda$, were $O(\hbar)$.

Because the expansion of the Hamiltonian gave a term that is cubic in the field, the reflection symmetry of the theory, $\varphi \to -\varphi$, has been spontaneously broken by the ground state of the theory. The pole of the Feynman propagator is determined by the value of $\partial^2 V(\varphi_0)/\partial\varphi_0^2$ at the minimum (5.172), which gives

$$M^2 = \left.\frac{\partial^2 V(\varphi_0)}{\partial\varphi_0^2}\right|_{\varphi_0=\varphi_c} = \left(1 - \frac{3\hbar\lambda}{16\pi^2}\right)\left(-\mu^2 + 3\lambda\varphi_c^2\right) \approx \left(1 + \frac{3\hbar\lambda}{8\pi^2}\right)2\mu^2. \quad (5.173)$$

Evaluating the logarithmic term in equation (5.170) at the minimum $\varphi_0 = \varphi_c$ gives

$$\ln\left(\frac{3\lambda\varphi_c^2 - \mu^2}{M^2}\right) = -\ln\left(1 - \frac{3\hbar\lambda}{16\pi^2}\right) \approx \frac{3\hbar\lambda}{16\pi^2}, \quad (5.174)$$

which justifies ignoring the logarithmic term in equation (5.171) when evaluated at φ_c since the logarithmic term becomes $O(\hbar^2)$.

There are many applications of classical solutions in field theory, both as potential ground states and as contributing field configurations. The reason for this was discussed earlier, which is that the quantized field can be treated as a fluctuation around the classical solution, $\varphi_c(x)$, so that $\varphi(x) \to \varphi(x) + \varphi_c(x)$. After an integration by parts the action appearing in the path integral takes the form

$$\int \mathrm{d}^4x\ \mathcal{L}(\varphi + \varphi_c) = \int \mathrm{d}^4x \left(\mathcal{L}(\varphi_c) + \left(\frac{\partial \mathcal{L}(\varphi_c)}{\partial \varphi_c} - \partial_\mu \frac{\partial \mathcal{L}(\varphi_c)}{\partial(\partial_\mu \varphi_c)} \right) \varphi \right.$$
$$\left. + \frac{1}{2} \left(\partial_\mu \varphi\ \partial^\mu \varphi + \frac{\partial^2 \mathcal{L}(\varphi_c)}{\partial \varphi_c^2} \varphi^2 \right) + \ldots \right). \tag{5.175}$$

If the classical configuration satisfies the Euler–Lagrange equation, the term in the expansion linear in φ will vanish. This prevents the coefficient of the linear term from acting as a source that would destabilize the theory by making the one-particle states disappear into the ground state. In many cases, the Wick rotation of the terms quadratic in φ defines an eigenvalue equation for the fluctuation field,

$$\left(\Box_E - \frac{\partial^2 \mathcal{L}(\varphi_c)}{\partial \varphi_c^2} \right) \varphi_n(x) \equiv (\Box_E + V(\varphi_c))\varphi_n(x) = \lambda^{(n)} \varphi_n(x), \tag{5.176}$$

where the eigenfunctions provide a complete orthonormal set,

$$\sum_n \varphi_n(x)\varphi_n(y) = \delta^4(x - y), \qquad \int \mathrm{d}^4x\ \varphi_n(x)\varphi_m(x) = \delta_{nm}. \tag{5.177}$$

Dropping the higher order terms in φ in the action (5.175) allows the path integral to be evaluated using the technique of equation (4.182). The final result is given by equation (4.186) where $\det(\Box_E + V(\varphi_c))$ appears in the prefactor. This determinant determines many of the properties of the transition element, including nonlinear effects encoded in the potential $V(\varphi_c)$ created by the classical solutions. This is the field theory extension of the quantum mechanical result (3.81). This technique has been applied to understand the decay of the false vacuum at $\varphi_c = 0$ in the Higgs potential (5.168), as well as in understanding baryon decay in models of the strong interactions.

Such an approach must deal with the presence of an eigenfunction with a zero eigenvalue for the case that $\partial_\mu \varphi_c(x) \neq 0$. This follows from differentiating the Euler–Lagrange equation to obtain

$$\partial_\mu \left(\Box \varphi_c - \frac{\partial \mathcal{L}(\varphi_c)}{\partial \varphi_c} \right) = \left(\Box - \frac{\partial^2 \mathcal{L}(\varphi_c)}{\partial \varphi_c^2} \right) \partial_\mu \varphi_c(x) = 0. \tag{5.178}$$

The quantity $\partial_\mu \varphi_c$ is known as the *translation mode* and has the potential to render the determinant $\det(\Box_E + V(\varphi_c))$ undefined, which in turn would render the effective potential undefined. This occurs even if the solution is static, in which case the translation modes are proportional to the spatial derivatives $\partial \varphi_c(x)/\partial x^i$.

The origin of the problem is the failure to treat the classical solution quantum mechanically. The original field theory was translationally invariant, but the solution has broken the translational invariance by being placed at a specific location or specific time. The classical solution can be associated with a specific inertial mass given by the value of Hamiltonian found from the static form of the solution. This object can be quantized by treating its location quantum mechanically. The solution to the zero mode problem is therefore to introduce the *collective coordinate* q^i for each translation mode and write the classical solution as $\tilde{\varphi}_c(x) = \varphi_c(x - q) + q^i \partial_i \varphi_c(x)$. The expansion of $\tilde{\varphi}_c$ around a specific value for q^i shows that the translation mode is subtracted from the classical solution. The collective coordinate becomes a quantum mechanical position operator, allowing the object that the classical solution represents to interact with the particle excitations of the field found from equation (5.176). The interested reader is recommended to the monographs by Coleman and Rajaraman.

Further reading

Generating functionals in quantum mechanics and quantum field theory originated with

- J Schwinger 1951 *Phys. Rev.* **82** 664
- K Symanzik 1954 *Z. Natürforsch. A* **9** 10
- G Jona-Lasinio 1964 *Nuovo Cimento* **34** 1790
- J Schwinger 1969 *Particles and Sources* (London: Gordon and Breach)

and the previously referenced monograph by Fried. Almost all modern texts on quantum field theory also include sections on path integrals and functional methods in discussing perturbation theory, renormalization, and nonabelian gauge theories. These include

- S Weinberg 1995 *Quantum Theory of Fields* vol 1 (Cambridge: Cambridge University Press)
- M Peskin and D Schroeder 1995 *An Introduction to Quantum Field Theory* (Reading, MA: Addison-Wesley)
- L H Ryder 1996 *Quantum Field Theory* 2nd edn (Cambridge: Cambridge University Press)
- W Greiner and J Reinhardt 1996 *Field Quantization* (Berlin: Springer)
- M Srednicki 2007 *Quantum Field Theory* (Cambridge: Cambridge University Press)

The relationship of the Gauss's law constraint to gauge fixing is discussed in detail in

- M S Swanson 2014 *Path Integrals and Quantum Processes* (New York: Dover)

The chiral anomaly in field theory is discussed extensively in

- K Fujikawa and H Suzuki 2004 *Path Integrals and Quantum Anomalies* (Oxford: Clarendon)

Non-perturbative aspects of 1 + 1 dimensional Yang–Mills theory are analyzed in

- A Migdal 1975 *Zh. Eksp. Teor. Fiz.* **69** 810
- J Hetrick 1994 *Int. J. Mod. Phys. A* **9** 3153
- M Swanson 2004 *J. Phys. G* **30** 1

The use of instantons to calculate the decay rate of metastable states originated with

- S Coleman 1977 *Phys. Rev. D* **15** 2929

and is presented in the review by

- A Andreassen, D Farhi, W Frost and M Schwartz 2017 arXiv:1604.06090v2 [hep-th]

The quantum mechanical treatment of classical solutions in field theory is presented in

- S Coleman 1988 *Aspects of Symmetry: Selected Erice Lectures* (Cambridge: Cambridge University Press)
- R Rajaraman 1982 *Solitons and Instantons* (Amsterdam: Elsevier)

www.ingramcontent.com/pod-product-compliance
Ingram Content Group UK Ltd.
Pitfield, Milton Keynes, MK11 3LW, UK
UKHW051307130926
8556IPUK00048B/657

9 780750 33545